ADVANCES IN NUCLEAR QUADRUPOLE RESONANCE

Edited by J. A. S. Smith

Volume 1, 1974

Volume 2, 1975

Advances in Nuclear Quadrupole Resonance

Volume 2

Edited by

J. A. S. Smith
**Department of Chemistry,
Queen Elizabeth College,
University of London**

London · New York · Rheine

Heyden & Son Ltd., Spectrum House, Alderton Crescent, London NW4 3XX.
Heyden & Son Inc., 225 Park Avenue, New York, N.Y. 10017, U.S.A.
Heyden & Son GmbH, 4440 Rheine/Westf., Münsterstrasse 22, Germany.

ISBN 0 85501 145 9

Printed in Great Britain by J. W. Arrowsmith Ltd., Bristol BS3 2NT

CONTENTS

Preface   ix

List of Contributors   xi

1. Reorientations and NQR Spectral Parameters
R. Sh. Lotfullin and G. K. Semin

I.   Introduction   1

II. Quadrupole Interaction Hamiltonian   2

III. Effect of Anisotropic Molecular Reorientations on an NQR Frequency   2

IV. Changes in the Asymmetry Parameter due to Anisotropic Hindered Rotations   7

V.   Contribution of Spin–Lattice Relaxation to the Line-width in Crystals with Reorientational Molecular Motions   9

2. Optical Detection of Nuclear Quadrupole Interactions in Excited Triplet States
Charles B. Harris and Michael J. Buckley

I.   Introduction   15
     A. The excited triplet state in organic molecules   16
     B. The historical development of ODMR   18

II. General Considerations   22
     A. Sensitivity considerations in the optical detection of ESR   22
     B. Optically detected ENDOR   30

III. The Zero-Field Spin Hamiltonian   32
     A. $\mathcal{H}_{ss}$—The spin–spin or zero-field splitting Hamiltonian   32

B. $\mathcal{H}_Q$—The nuclear quadrupole Hamiltonian   36
C. $\mathcal{H}_{HF}$—The nuclear–electron hyperfine interaction . .   39
D. The total Hamiltonian, energy levels and transition probabilities   40

IV. Experimental Methods   48
    A. Optically detected magnetic resonance   48
    B. Optically detected ENDOR   50
    C. Variations of the basic experiments   50

V. The ODMR Spectra of 8-Chloroquinoline   52

VI. The ODMR Spectra of *Para* Dichlorobenzene   60

3. Double Resonance Detection of Nuclear Quadrupole Resonance Spectra

    R. Blinc

I. Introduction   71

II. Double Resonance in the Rotating Frame: High-field Quadrupole Double Resonance   73
    A. Experimental procedure and theory   73
    B. Applications   80

III. Double Resonance between the Laboratory and the Rotating Frame: Zero-field Quadrupole Double Resonance . .   82

IV. Double Resonance in the Laboratory Frame: Level Crossing in a Low Magnetic Field   87
    A. Experimental   87
    B. Sensitivity for ^{14}N detection   89
    C. Results   93

V. Double Resonance in the Laboratory Frame via the 'Solid Effect'   95
    A. Introduction   95
    B. Origin of r.f. magnetic field induced coupling between spin systems   96
    C. Experimental   100
    D. Analysis of ^{1}H–^{14}N double resonance spectra   103
      (i) The single coupling process   103
      (ii) The multiple coupling process   105
    E. Results   106

VI. Double Resonance Relaxation Measurements   109
    A. Double resonance relaxation measurements in the rotating frame for $T_{1B} \leqslant T_{1A}$ 109
    B. Double resonance relaxation measurements in the rotating frame for $T_{1B} > T_{1A}$ 111

C. Double resonance relaxation measurements of quadrupolar relaxation times by level crossing in the laboratory frame 112

D. Double resonance relaxation measurements in the laboratory frame via the 'solid effect'. 113

E. Measurement of short quadrupolar relaxation times by ADRF 114

4. Nuclear Quadrupole Resonance of Iodine in Polyiodides
Daiyu Nakamura and Masaji Kubo

I. Introduction 117

II. NQR Frequencies, Assignment, and Crystal Structure . . . 118

III. Quadrupole Coupling Constant and Asymmetry Parameter 126

IV. Charge Distributions in Triiodide and Octaiodide Ions . . . 128

V. Anomalous Temperature Dependence of the Resonance Frequencies 130

5. The *A Priori* Calculation of the Electronic Contribution to the Nuclear Quadrupole Coupling Constant. Theoretical Aspects and Recent Results
R. Moccia and M. Zandomeneghi

Introduction 135

I. Theoretical Background 136

II. First Order Treatment 139
   A. Electronic contribution as an expectation value . . . 139
   B. Variational wave function 144

III. First Order Observables 149
   A. Perturbation approach 149
   B. Variational approach 152

IV. Vibrational and Other Corrections 155

V. Electronic Polarization Effect 157

VI. Summary of the Results of Some Recent Calculations . . . 158

Appendix A. Connection Between q^{00}, q_2^{00} and the Force Constant . 169

Appendix B. Effect of Perturbing Fields and of the Finiteness of the Basis Set upon q in the SCFMO Approximation . . . 171

6. Structural Phase Transitions in RMX$_3$ (Perovskite) and R$_2$MX$_6$ (Antifluorite) Compounds
Robin L. Armstrong and Henry M. Van Driel

I. Introduction 179

II. Geometrical, Symmetry, and Dynamical Considerations . . 180
    A. High temperature structures 180
    B. Nature of the phase transitions 183
    C. Empirical geometrical model 184
    D. Landau and Lifshitz symmetry conditions 186
    E. Dynamical considerations and soft modes 188
    F. Limitations of the Landau theory 190
    G. Beyond the Landau theory 191

III. Static Investigations 191
    A. Nuclear quadrupole resonance line splittings 191
    B. Quadrupole coupling constants for samples in applied static magnetic fields 196
    C. Nuclear quadrupole resonance detection of soft modes 199
    D. Effects of volume changes and π-bonding on nuclear quadrupole resonance frequencies 205
    E. Lattice dynamical model calculations: R$_2$MX$_6$ . . . 211

IV. Dynamic Investigations 223
    A. Resonant versus relaxational soft-mode behaviour . 223
    B. Relaxation theory for nearly harmonic crystalline solids 224
    C. Nuclear quadrupole relaxation detection of soft modes 228
    D. Effect of volume changes and electronic paramagnetism on nuclear quadrupole spin–lattic relaxation times. 236
    E. Critical dynamics and quadrupolar relaxation 238
    F. Quadrupolar relaxation studies of structural phase transitions in perovskites 242
    G. Fluctuations and correlations in SrTiO$_3$ as deduced from electron paramagnetic resonance experiments . 246

V. Concluding Remarks 250

Author Index 255

Subject Index 263

Chemical Compounds Index 271

PREFACE

The second volume in this series, as the first, is devoted to many aspects of nuclear electric quadrupole interactions, both from the chemical and physical points of view. However, unlike the first volume, the pages that follow consist of specially commissioned articles, a pattern that will continue in future issues. We will include reviews that not only survey an established field but also look at new developments and new methods of measuring quadrupole interactions, the intention being to stimulate the advancement of the subject and of all techniques involving radiofrequency radiation.

I am indebted to Mrs. P. Spicer for secretarial assistance and to my wife for her invaluable support.

J. A. S. SMITH

London
June, 1975

LIST OF CONTRIBUTORS

ARMSTRONG, R. L., Department of Physics, University of Toronto, Toronto, Canada M5S 1A7 (p. 179).

BLINC, R., University of Ljubljana, J. Stefan Institute, Ljubljana, Yugoslavia (p. 71).

BUCKLEY, M. J., Air Force Materials Laboratory, Wright–Patterson Air Force Base, Dayton, Ohio 45433, U.S.A. (p. 15).

HARRIS, C. B., Department of Chemistry and Inorganic Materials Research Division of the Lawrence Berkeley Laboratory, University of California, Berkeley, California 94720, U.S.A. (p. 15).

KUBO, M., Department of Chemistry, Nagoya University, Chikusa, Nagoya, Japan (p. 117).

LOTFULLIN, R. Sh., Institute of Organo-Element Compounds, Academy of Sciences of the U.S.S.R., Moscow, U.S.S.R. (p. 1).

MOCCIA, R., Istituto di Chimica Fisica della Università, Via Risorgimento 35, 56100-Pisa, Italy (p. 135).

NAKAMURA, D., Department of Chemistry, Nagoya University, Chikusa, Nagoya, Japan (p. 117).

SEMIN, G. K., Institute of Organo-Element Compounds, Academy of Sciences of the U.S.S.R., Moscow, U.S.S.R. (p. 1).

VAN DRIEL, H. M., Department of Physics, University of Toronto, Toronto, Canada M5S 1A7 (p. 179).

ZANDOMENEGHI, M., Istituto di Chimica Fisica della Università, Via Risorgimento 35, 56100-Pisa, Italy (p. 135).

1. REORIENTATIONS AND NQR SPECTRAL PARAMETERS

R. Sh. Lotfullin and G. K. Semin

Institute of Organo-Element Compounds, Academy of Sciences of the U.S.S.R., Moscow, U.S.S.R.

I. INTRODUCTION

The high sensitivity of a quadrupolar nucleus to its charge environment as well as to the dynamics of molecular motions has made it possible to study the influence on NQR parameters which are governed by molecular vibrations as small as torsional motions.[1] The validity of Bayer's theory[1] of the effects of molecular torsional motions on such quadrupole resonance parameters is now established.[2] However, a number of papers has been published[3-11] in which the changes in spectral parameters were explained entirely in terms of the reorientational motions of molecules in the crystal lattice. Numerous studies have shown that the character of reorientational motions is that of random rotations between the equilibrium positions in the crystal lattice, i.e. only reorientations compatible with the crystal symmetry are allowed. Usually molecular reorientations between two equilibrium positions are thought to be instantaneous, the rotational correlation time τ_c being the time a molecule spends in the equilibrium position. Reorientational molecular motions in the crystal lattice occurring at a frequency τ_c^{-1}, much lower than the NQR frequency ω_Q ($\omega_Q\tau_c \gg 1$—'slow' rotations), do not affect the NQR frequency but cause a broadening of the resonance line by an amount $\delta\omega \sim \tau_c^{-1}$.[4,12] Instantaneous molecular motions do not change the nuclear wave function;[12,13] however, the quadrupole interaction Hamiltonian changes significantly, which causes a very fast energy exchange between the spin system and the crystal lattice.[3,4,8-11] For the case in which the reorientational axis does not coincide with the axis of the maximum electric field gradient and where the conditions $\omega_Q\tau_c \ll 1$ ('fast' reorientations) are fulfilled, the nuclear electric quadrupole moment sees an averaged electric field gradient.[5-7,9,11]

1

II. QUADRUPOLE INTERACTION HAMILTONIAN

The interaction Hamiltonian of a nuclear electric quadrupole moment with an electric field gradient created by a charge cloud around the nucleus in question can be written in the laboratory (x, y, z) coordinate system, which is fixed in space, as[14,15]

$$\mathscr{H} = \tfrac{1}{6} \sum_{n=-2}^{2} (-1)^n Q_{-n} \sum_{m=-2}^{2} \exp(-im\gamma) P_{mn}(\cos\beta) \exp(in\alpha) V'_m \tag{1}$$

where

$$\left.\begin{array}{l} V'_{\pm 2} = \tfrac{1}{2}(V_{x'x'} - V_{y'y'}) \pm i V_{x'y'} \qquad V'_{\pm 1} = \mp V_{x'z'} - i V_{y'z'} \\[2ex] V'_0 = \dfrac{\sqrt{6}}{2} V_{z'z'} \end{array}\right\} \tag{2}$$

and

$$\left.\begin{array}{l} Q_2 = Q^*_{-2} = \tfrac{3}{2} c I^2_+ \qquad Q_1 = -Q^*_{-1} = \tfrac{3}{2} c \{I_z I_+\} \\[2ex] Q_0 = \dfrac{\sqrt{6}}{2} c [3 I^2_z - I(I+1)] \qquad V_{jk} = \dfrac{\partial^2 V}{\partial x_j \, \partial x_k} \\[2ex] (x_j, x_k = x', y', z') \qquad c = \dfrac{eQ}{I(2I-1)} \end{array}\right\} \tag{3}$$

where V is the potential of the electric environment of a nucleus, e the electron charge, Q the nuclear electric quadrupole moment, and I the nuclear spin quantum number. In Eq. (1), V'_m depends on the charge cloud distribution at the nucleus alone, i.e. it is a parameter by which the electron charge distribution of the environment is characterized. α, β, γ are the Eulerian angles determining the orientation of the coordinate system (x', y', z') rigidly attached to the molecule with respect to the laboratory coordinate system (x, y, z). $P_{mn}(\cos\beta)$ are normalized Legendre functions.

III. EFFECT OF ANISOTROPIC MOLECULAR REORIENTATIONS ON AN NQR FREQUENCY

Molecular movements such as torsional oscillations and reorientational motions change the quadrupole interaction Hamiltonian, i.e. the Hamiltonian itself is dependent on the lattice movements. As a result a nucleus sees an averaged electric field gradient. Therefore it is convenient to rewrite the Hamiltonian in the form[1,16]

$$\mathscr{H} = \langle \mathscr{H} \rangle + \mathscr{H}' \tag{4}$$

with

$$\mathcal{H}' = \mathcal{H} - \langle \mathcal{H} \rangle \tag{5}$$

Here $\langle \mathcal{H} \rangle$ is the averaged Hamiltonian in the laboratory coordinate system; this determines the averaged energy levels of the quadrupole interactions, i.e. the NQR spectrum. The perturbation $\mathcal{H}'$ represents the fluctuation of the Hamiltonian from its averaged value, and is therefore a function by means of which the relaxation phenomena are described; it is random with time and its average value is equal to zero. Assuming that β and γ in Eq. (1) determine the rotation axis and α determines the magnitude of the reorientation angle, the matrix elements for transitions $M \to M - n$ between the sublevels of the average value of a quadrupole interaction can be written as

$$\langle \mathcal{H}_{M,M-n} \rangle = \tfrac{1}{6}(Q-n)_{M,M-n} \sum_{m=-2}^{2} \exp{(-im\gamma)} P_{mn}(\cos \beta) V'_m \langle \exp{(-in\alpha)} \rangle \tag{6}$$

The matrix elements of the perturbation Hamiltonian are written analogously

$$\mathcal{H}'_{M,M-n} = \tfrac{1}{6}(Q_{-n})_{M,M-n} \sum_{m=-2}^{2} \exp{(-im\gamma)} P_{mn}(\cos \beta) V'_m [\exp{(-in\alpha)} - \langle \exp{(-in\alpha)} \rangle] \tag{7}$$

One can find an equation for the average energy levels of the nuclear quadrupole Hamiltonian by solving a secular equation based on Eq. (6).

For spins $I = 1, 3/2, 5/2, 7/2$ this problem has already been solved.[15,17,18] Let us consider the solution of the most simple equation for $I = 3/2$. In this case, the expression for the resonance frequency is given by

$$\omega_T = \frac{eQV_{z'z'}}{2\hbar} \left\{ \left[\tfrac{1}{2}(3\cos^2\beta - 1) - \frac{\eta_0}{2}\sin^2\beta\cos 2\gamma \right]^2 \right.$$
$$+ \tfrac{1}{3}[\eta_0^2 \sin^2\beta\sin^2 2\gamma + \tfrac{1}{4}(3 + \eta_0\cos 2\gamma)^2\sin^2 2\beta] \langle \exp{(i\alpha)} \rangle \langle \exp{(-i\alpha)} \rangle$$
$$+ \tfrac{1}{3}[\eta_0^2 \cos^2\beta\sin^2 2\gamma + \tfrac{1}{4}(\eta_0(\cos^2\beta + 1)\cos 2\gamma - 3\sin^2\beta)^2] \cdot \langle \exp{(i2\alpha)} \rangle$$
$$\left. \times \langle \exp{(-i2\alpha)} \rangle \right\}^{1/2} \tag{8}$$

where

$$\eta_0 = \frac{V_{x'x'} - V_{y'y'}}{V_{z'z'}} \tag{9}$$

η_0 being the asymmetry parameter. It is usually accepted that

$$|V_{x'x'}| \leqslant |V_{y'y'}| \leqslant |V_{z'z'}| \tag{10}$$

3

For the case of axial symmetry ($\eta_0 = 0$), we get

$$\omega_T = \frac{\omega_Q}{2}[(3\cos^2\beta - 1)^2 + 3\sin^2 2\beta\langle\exp(i\alpha)\rangle\langle\exp(-i\alpha)\rangle$$

$$+ 3\sin^4\beta\langle\exp(i2\alpha)\rangle\langle\exp(-i2\alpha)\rangle]^{1/2} \quad (11)$$

Equation (11) was also obtained by Ragle,[6] who used this equation to explain the temperature dependence of the NQR frequencies in *trans*-1,2-dichloroethane, assuming that the reorientations occur about the Cl$\cdots$Cl axis through an angle of 180°. Thus the NQR frequency of ^{35}Cl in *trans*-dichloroethane at 77 K is equal to 34.45 MHz and at 240 K to 29.96 MHz. A calculation in which reorientations through an angle of 180° are considered and the contribution of terms containing η_0 is dropped yields a value for the NQR frequency of 28.9 MHz, i.e. a difference of 1 MHz from experimental. This difference drops to 0.1 MHz if $\eta_0 = 0.05$ is used in Eq. (8). The latter value of the asymmetry parameter is reasonable, since the observed spread of η_0 in similar compounds is usually 0.01–0.10. It should also be noted that the agreement between the experimental NQR frequencies and those calculated by means of Eq. (8) with $\eta_0 = 0.05$ is also appreciably improved for *trans*-1,2-dibromoethane.[7] Unfortunately, η_0 values for both compounds have not as yet been measured.

If the model of free anisotropic rotational motions is used, Eq. (8) reduces to:

$$\omega_T = \frac{\omega_Q(3\cos^2\beta - 1)}{2} \quad (12)$$

This expression was also obtained by Bayer[1] and Das and Hahn.[2]

As is seen from Eq. (8), in the case of $I = 3/2$ a closed formula for the NQR frequency is obtained, while the resonance frequency for arbitrary angles and structural parameters is easily calculated. Equation (8) is the most general formula describing the influence of hindered rotations on NQR frequencies, and by its use it is possible to predict the changes in NQR frequencies for several specific cases of hindered molecular rotations. Let us consider the case where $\gamma = 0$ and $\eta_0 \neq 0$. Then we have

$$\omega_T = \frac{eQV_{z'z'}}{2\hbar}\left\{\left[\tfrac{1}{2}(3\cos^2\beta - 1) - \frac{\eta_0}{2}\sin^2\beta\right]^2 + \frac{(3+\eta_0)^2}{12}\sin^2 2\beta\langle\cos\alpha\rangle^2\right.$$

$$\left. + \frac{1}{12}[\eta_0(\cos^2\beta + 1) - 3\sin^2\beta]^2\langle\cos 2\alpha\rangle^2\right\}^{1/2}$$

For certain highly probable values of β, we get

(1) $\beta = 0°$

$$\omega_T = \frac{eQV_{z'z'}}{2\hbar}\sqrt{1 + \frac{\eta_0^2}{3}\langle\cos 2\alpha\rangle^2}$$

4

In this case, hindered rotations around the axis of the maximum principal component of the electric field gradient do not change the NQR frequency if $\langle \cos 2\alpha \rangle = 1$, i.e. if the molecular reorientations occur through an angle of $180°$.[6]

(2) $\beta = 19° \, 16'$

$$\omega_T = \frac{eQV_{z'z'}}{2\hbar}[(0.836 - 0.0544\,\eta_0)^2 + 0.0323(3 + \eta_0)^2 \langle \cos \alpha \rangle^2$$
$$+ (0.545\,\eta_0 - 0.0943)^2 \langle \cos 2\alpha \rangle^2]^{1/2}$$

In this case, the equation agrees well with that obtained by Ragle[6] to explain the temperature dependence of the NQR frequency in *trans*-1,2-dichloroethane for $\eta_0 = 0$ and $\langle \cos 2\alpha \rangle = 1$, namely

$$\omega_T = \frac{\omega_Q}{2}[2.863 + 1.135\langle \cos \alpha \rangle^2]^{1/2}$$

(3) $\beta = 45°$

$$\omega_T = \frac{eQV_{z'z'}}{2\hbar}[0.0625(1 - \eta_0)^2 + 0.083(3 + \eta_0)^2 \langle \cos \alpha \rangle^2$$
$$+ 0.433(\eta_0 - 1)^2 \langle \cos 2\alpha \rangle^2]^{1/2}$$

(4) $\beta = 54° \, 44'$ (the 'magic' angle)

$$\omega_T = \frac{eQV_{z'z'}}{2\hbar}[0.111\,\eta_0^2 + 0.074(3 + \eta_0)^2 \langle \cos \alpha \rangle^2$$
$$+ (0.385\,\eta_0 - 0.577)^2 \langle \cos 2\alpha \rangle^2]^{1/2}$$

This example is especially interesting since according to Ragle's theory a hindered rotation of this type should average the NQR frequency to zero.

(5) $\beta = 71°$

$$\omega_T = \frac{eQV_{z'z'}}{2\hbar}[(0.340 + 0.447\,\eta_0)^2 + 0.0316(3 + \eta_0)^2 \langle \cos \alpha \rangle^2$$
$$+ (0.319\,\eta_0 - 0.774)^2 \langle (\cos 2\alpha \rangle^2]^{1/2}$$

(6) $\beta = 90°$

$$\omega_T = \frac{eQV_{z'z'}}{2\hbar}[\tfrac{1}{4}(1 + \eta_0)^2 + (0.2886\,\eta_0 - 0.866)^2 \langle \cos 2\alpha \rangle^2]^{1/2}$$

Note that if $\eta_0 = 0$ and $\langle \cos 2\alpha \rangle = 0$, then $\omega_T = \omega_Q/2$.

However, for quadrupolar nuclei with $I \neq 3/2$ the energy levels are not determined by closed formulae so that the solution of the secular equations for arbitrary parameters is very difficult. The problem becomes simpler for certain particular cases; for example, for $I = 5/2$, $\beta = \gamma = 0$, and setting $\alpha = \alpha(t)$, the

5

equation for ω_T is:[18]

$$\omega_T(\tfrac{5}{2}\to\tfrac{3}{2})=\frac{3eQV_{z'z'}}{10\hbar}[1-\tfrac{11}{54}\eta_0^2\langle\exp{(-i2\alpha)}\rangle\langle\exp{(i2\alpha)}\rangle+\cdots] \qquad (13a)$$

and

$$\omega_T(\tfrac{3}{2}\to\tfrac{1}{2})=\frac{3eQV_{z'z'}}{20\hbar}[1+\tfrac{59}{54}\eta_0^2\langle\exp{(-i2\alpha)}\rangle\langle\exp{(i2\alpha)}\rangle+\cdots] \qquad (13b)$$

As is seen from these formulae, the temperature dependence is defined by the temperature dependent value of $\langle\exp{(\pm i2\alpha)}\rangle$, in which $\alpha(t)$ angle is a random function of time. In accordance with the theory of random variables,[6]* we have

$$\langle\exp{(\pm i2\alpha)}\rangle=\frac{2}{\pi}\,\text{arc tan}\,(\kappa\omega_Q\tau_c) \qquad (14)$$

where κ is a constant close to unity; note that arc tan $(\omega_Q\tau_c)\to0$ if $\omega_Q\tau_c\ll1$.

It should be noted that if the frequencies of molecular reorientation are lower than the NQR frequencies, they do not cause appreciable changes in the latter. However, more general anisotropic reorientations at frequencies much greater than an NQR frequency lead to nuclear quadrupole precession in the averaged electric field gradient and hence to a sharp decrease in the resonance frequency.[11,19,20] In such cases, it is reasonable to replace the expression defining general reorientations by an equivalent expression describing hindered rotations. In several arsenates, for example, a sharp decrease in the NQR frequencies at a transition is observed (see for example reference 21). Such behaviour can be explained by averaging of the electric field gradient at the nucleus due to modulations caused by proton migration occurring with a frequency higher than the NQR frequency; these proton migrations can be easily represented as an analogue of rotational motion of the electric field gradient. Fast isotropic motion of any kind $(\omega_Q\tau_c\ll1)$ always produces an effective spherically-symmetrical electric field gradient, so that the NQR frequency tends to zero.

It is easy to show that a change in the NQR frequency due to a modulation of gradient caused by motion of adjacent molecules and groups is satisfactorily described by the formula[35]

$$V'_{zz}=\left\{\left[\sum_n(\sigma_n^2+\omega_Q^2)\right]\Big/CT_1\omega_Q^2\sum_n\sigma_n\right\}^{1/2}V_{zz}$$

where $\sigma_n=(1-\cos{n\chi})$, V_{zz} is the EFG at the resonant nucleus of a fixed molecule, V'_{zz} the modulation change of EFG, T_1 the spin–lattice relaxation time ω_Q the NQR frequency of a fixed molecule, τ_c the correlation time, χ the minimum angle of the reorientational motion of a molecule (or its fragment), and is a constant[35] that must be calculated for each specific case.

* The case in which molecular reorientations occur through an angle of 180° and hence $\langle\exp{(\pm i2\alpha)}\rangle=1$ should be excluded.

6

IV. CHANGES IN THE ASYMMETRY PARAMETER DUE TO ANISOTROPIC HINDERED ROTATIONS

Hindered rotations occurring with frequencies higher than the NQR frequencies cause appreciable changes in electric field gradient at the nucleus. In this connection, a study of the influence of molecular hindered rotations on the asymmetry parameter η is of interest. For this purpose, it is necessary to consider the averaging of the V_{jk} ($j, k \equiv x, y, z$) tensor components due to molecular rotation. The values of these components can be found from relationships inverse to Eq. (2), i.e.

$$V_{xx} = \tfrac{1}{2}(V_2 + V_{-2}) - \frac{1}{\sqrt{6}} V_0, \qquad V_{xy} = -\frac{i}{2}(V_2 - V_{-2}),$$

$$V_{xz} = -\tfrac{1}{2}(V_1 - V_{-1}), \qquad V_{yy} = -\tfrac{1}{2}(V_2 + V_{-2}) - \frac{1}{\sqrt{6}} V_0, \qquad (15)$$

$$V_{yz} = \frac{i}{2}(V_1 + V_{-1}), \qquad V_{zz} = \frac{2}{\sqrt{6}} V_0.$$

The transformation rule applied to V_n [14]

$$V_n = \sum_{m=-2}^{2} \exp(-im\gamma) P_{mn}(\cos \beta) \exp(-in\alpha) V'_m \qquad (16)$$

makes it possible to calculate the components of the effective electric field gradient tensor. Molecular motions which change angles determine the averaging of the electric field gradient in the laboratory coordinate system (x, y, z). Thus if the conditions $\langle V_{jk} \rangle = 0$ for $j \neq k$ and $|\langle V_{zz} \rangle| > |\langle V_{yy} \rangle| > |\langle V_{xx} \rangle|$ are fulfilled, the asymmetry parameter in the laboratory coordinate system is defined as in Eq. (9);

$$\langle \eta \rangle = \frac{\langle V_{xx} \rangle - \langle V_{yy} \rangle}{\langle V_{zz} \rangle} \qquad (17)$$

If the frequency of hindered rotation of a molecule exceeds the NQR frequency, the nucleus sees the averaged electric field gradient. If the z-axis is fixed in space, then β, γ are constant, while $\alpha = \alpha(t)$. Under these conditions, the components of the averaged tensor are found by means of expressions (15), (16), and (17), and an equation similar to (14); they are defined by

$$\langle V_{zz} \rangle = -2\langle V_{xx} \rangle = -2\langle V_{yy} \rangle = \frac{V_{z'z'}}{2}[(3 \cos^2 \beta - 1)$$

$$+ 2\eta_0 \sin^2 \beta \cos 2\gamma], \quad \text{and} \quad \langle V_{xy} \rangle = \langle V_{xz} \rangle = \langle V_{yz} \rangle = 0 \qquad (18)$$

Hence one can see that $\langle \eta \rangle = 0$. Thus the study of the effective change of the asymmetry parameter at the investigated nucleus due to anisotropic hindered

7

rotations of a molecule yields information concerning the dynamics of molecular motions in the crystal. For example, the ^{2}H asymmetry parameters in ND_3 groups in triglyclphosphate are 0.247 at 20 °C and 0.021 at 79 °C,[22] a change which could only be explained by rotation of the ND_3 groups about the threefold axis.

The temperature changes in the asymmetry parameter due to reorientations of a molecule through an angle of 180° are rather unusual and worth considering in more detail. Making use of that data in reference 6, according to which reorientations occur through an angle of 180° under conditions in which

$$\omega_Q \tau_c \ll 1, \qquad \langle \exp(\pm i\alpha) \rangle = 0, \qquad \langle \exp(\pm i2\alpha) \rangle = 1$$

and applying Eqs. (15) and (16) with $\gamma = \beta = 0$ to calculate the electric field gradient tensor components in the laboratory coordinate system, one finds[18]

$$\langle V_{xx} \rangle = \tfrac{1}{2}(1 - \eta_0) V_{z'z'},$$
$$\langle V_{yy} \rangle = -\tfrac{1}{2}(1 + \eta_0) V_{z'z'}$$
$$\langle V_{zz} \rangle = V_{z'z'}, \tag{19}$$
$$\langle V_{xy} \rangle = \langle V_{xz} \rangle = \langle V_{yz} \rangle = 0$$

On the other hand, for $\gamma = 0$ and $\beta = 52° \, 30'$, we have

$$\langle V_{xx} \rangle = \tfrac{1}{2}(-1 + \eta_0) V_{z'z'},$$
$$\langle V_{yy} \rangle = (0.4441 - 0.1853\eta_0) V_{z'z'}$$
$$\langle V_{zz} \rangle = (0.059 - 0.3147\eta_0) V_{z'z'}, \tag{20}$$
$$\langle V_{xy} \rangle = \langle V_{xz} \rangle = \langle V_{yz} \rangle = 0$$

Equation (19) corresponds to the case in which the rotation axis and the principal axis of the maximum electric field gradient coincide. In this case, $\langle \eta \rangle = \eta_0$.

Para-dichlorobenzene is also one of the substances where this phenomenon is apparently observed. In reference 23, it is stated that in *p*-dichlorobenzene crystals reorientation of the molecules can occur about a twofold axis passing through the two chlorine atoms. The reorientation barrier U is equal to 2.5 ± 0.5 kcal mol^{-1}[23] and the correlation time τ_c may be calculated according to the equation

$$\tau_c = \tau_0 \exp(U/RT) \tag{21}$$

At 300 K, its value is about 10^{-11}[5]. Here τ_0 is the period of the torsional oscillations corresponding to a frequency of 93 cm^{-1}. At first sight, a hindered rotation of the molecules about the Cl$\cdots$Cl axis in *p*-dichlorobenzene, fast compared to the NQR frequency of ^{35}Cl ($\omega_Q \tau_c = 2 \cdot 10^{-3} \ll 1$), should make V_{xx} and V_{yy} equal ($\langle V_{xx} \rangle = \langle V_{yy} \rangle$), i.e. the asymmetry parameter should vanish

8

$(\langle\eta\rangle = 0)$. However, the experimental values of the asymmetry parameter in p-dichlorobenzene at 77 K, 298 K and 302 K are 0.08 ± 0.02, 0.073 ± 0.003, and 0.075 ± 0.02 respectively;[24] thus not only does the asymmetry parameter not vanish, but it remains almost unchanged. If therefore this motion is occurring, such behaviour is readily explained by Eq. (19).

Equation (20) can be used to explain the temperature dependence of the ²H asymmetry parameter in the case of reorientation of heavy water molecules in crystalline hydrates.[25] As has been shown by NMR studies,[25] the heavy water molecules in several hydrates undergo reorientations through 180° about an axis coinciding with the bisectrix of $\widehat{DOD}$ ($\widehat{DOD}$ is about 105°). At low temperatures, the maximum electric field gradient is close to the O—D bond direction and the asymmetry parameter is small. At high temperatures, when fast reorientations occur ($\omega_0 \tau_c \ll 1$), the maximum electric field gradient is reduced by about a half. It changes its direction and becomes almost perpendicular to the O—D bond; the asymmetry parameter becomes close to unity. As is seen from Eq. (20), $\eta_0 \ll 1$ the condition $\langle|V_{xx}|\rangle > \langle|V_{yy}|\rangle > \langle|V_{zz}|\rangle$ is fulfilled. Since it is generally taken that the z-axis is directed along the maximum and the x-axis along the minimum electric field gradient, it is necessary to change the axis notation in Eq. (20), and in particular to replace x by z^*, y by y^*, and z by x^*. Then $\langle\eta\rangle = 1$. As one can easily see from these calculations, the theory agrees sufficiently well with the experimental data.[25] In conclusion, it follows that the asymmetry parameter as well as the NQR frequencies are very sensitive to the dynamics of molecular motion and in particular cases can give a specific response to a specific kind of motion.

V. CONTRIBUTION OF SPIN–LATTICE RELAXATION TO THE LINE-WIDTH IN CRYSTALS WITH REORIENTATIONAL MOLECULAR MOTIONS

Nuclear quadrupole resonance lines are often appreciably broadened, which decreases their amplitudes and leads to increasing requirements for sensitivity in the apparatus. Therefore, the matter of the line-width is, in essence, the matter of the possibility of observing the phenomenon and it deserves careful consideration. The line-width $\delta\omega$ can be represented in the form[2]

$$\delta\omega \approx \frac{1}{T_2} + \frac{1}{T_1} + S \qquad (22)$$

where T_2 is the spin-phase memory time, T_1 the spin–lattice relaxation time, and S the line-width due to electric broadenings. It has been shown[2,26] that in molecular crystals at low temperatures, torsional oscillations are dominating, so that the main contribution to the line-width comes from T_2 and S and they can be considered independent of temperature.

However, at rather high temperatures when molecular orientations occur, T_1 may make an important contribution to the NQR line-width. It should be noted that T_1 is more sensitive to the dynamics of molecular motions in crystals than are T_2, $\delta\omega$, and S^9 and may vary over a very wide range. Hence the temperature dependence of T_1 can yield the most detailed information concerning the dynamics of molecular motions in crystals, and this dependence as explained by the reorientational movements of molecules or molecular fragments has been most intensely studied during the last decade. In a number of cases, the temperature dependence of T_1 has been treated by means of perturbation theory (see, for example, references 8–12, and in others the density matrix approach has been used (see references 27–31).

Let us now proceed to the calculation by means of perturbation theory of the quadrupole spin–lattice relaxation time, taking into account the effect of anisotropic reorientational motions of molecules in the crystal, the symmetry of the latter, and the asymmetry of the charge environment of the quadrupolar nucleus. The probability of a relaxation transition $M \rightarrow M - n$ is defined by the equation

$$W_{M,M-n} = \hbar^{-2} \int_{-\infty}^{+\infty} K_{M,M-n}(t) \exp(-i\omega_{M,M-n}t) \, dt \tag{23}$$

where $\omega_{M,M-n}$ is the angular resonance frequency. $K_{M,M-n}(t)$ is the time correlation function of the random value $\mathscr{H}'_{M,M-n}$ and according to definition is given by

$$K_{M,M-n}(t) = \frac{1}{2\pi} \int [\mathscr{H}'_{M,M-n}(\alpha_0)][\mathscr{H}'_{M,M-n}(\alpha_t)]^* W(\alpha_t, t|\alpha_0, 0) \, d\alpha_t \, d\alpha_0 \tag{24}$$

Here $W(\alpha_t, t|\alpha_0, 0)$ is the probability density that the orientation of a molecule at the moment t is described by an angle of α_t if at $t = 0$ the orientation of the molecule corresponded to an angle of α_0. The expression for $W(\alpha_t, t|\alpha_0, 0)$ obtained as a result of a solution of the random rotation problem has the form[32]

$$W(\alpha_t, t|\alpha_0, 0) = \frac{1}{2\pi} \sum_l \exp\left[-\frac{|t|}{\tau_c}(1 - \cos l\chi) - il(\alpha - \alpha_t) \right] \tag{25}$$

where χ is the minimum rotation angle allowed by the crystal symmetry. It is obvious that $\chi = 2\pi/k_x$ (k_x is an integer, equal to the order of the one-dimensional rotation group). Using expressions (7), (23)–(25) and those analogous to (14), one can write down the relaxation transition probability as[17,18,33]

$$W_{M,M-n} = \frac{1}{18\hbar^2}|(Q_{-n})_{M,M-n}|^2 \sum_{m=-2}^{2} |P_{mn}(\cos\beta) \exp(-im\gamma) V'_m|^2$$

$$\times \frac{\tau_c^{-1}(1 - \cos n\chi)}{[\tau_c^{-1}(1 - \cos n\chi)]^2 + \omega_{M,M-n}^2} \tag{26}$$

In this equation the asymmetry parameter and the crystal lattice symmetry are taken into account and the expression can be applied to arbitrary orientations of the molecular rotation axis.

It is to be noted that for slow reorientations ($\omega_Q \tau_c \gg 1$), the calculations according to Eq. (25) and those using the density matrix approach[27–31] lead to the same theoretical values of T_1. The fact itself is not surprising since the density matrix method 'is simply a variant of the usual time-dependent perturbation theory'.[34] It is another phenomenon which seems unusual. When slow reorientational motions occur, $T_1 \lesssim \tau_c$; though the condition $(\mathcal{H}'/\hbar)\tau_c < 1$ is fulfilled, the value of $\mathcal{H}'$ may be rather large. However, if one bears in mind that a perturbation occurs suddenly as a result of an instant jump of a molecule from one equilibrium position to another, one can calculate transition probabilities even for those cases where the perturbations are not small.[13]

The fact that $T_1 \lesssim \tau_c$ might apparently be explained by the extra energy of the spin system which can be redistributed over other degrees of freedom during the period of time shorter than τ_c. Such a possibility is plausible for the quadrupole relaxation since a direct interaction of the quadrupole moment with the electric field gradient makes it possible for a spin system to exchange its energy through molecular motions in the same potential well while in NMR the relaxation processes of a magnetic dipole system with reorientational molecular motions are determined by the times a molecule jumps from one equilibrium position to another.[16]

Equation (26) in a slightly changed form can be used to take into account the influence of intermolecular interactions on NQR spectral parameters. For example, the NQR frequency of ^{35}Cl observed above 500 K in KClO$_3$[20] cannot be explained by torsional molecular vibrations[35] while the dependence of T_1 on the inverse temperature is exponential. Such temperature dependence of T_1 can be explained by reorientations of the molecules or their fragments. However, the rotation of the ClO$_3^-$ ion itself must not affect the NQR frequency of ^{35}Cl since the reorientation axis coincides with the maximum EFG axis of the molecule.[2,5,17,18]

We assumed that the behaviour of NQR spectral parameters of ^{35}Cl in KClO$_3$ was caused by the rotation of the neighbouring ClO$_3^-$ groups. For this case Eq. (26) is written in the form:[35]

$$T_1^{-1} = \frac{2}{q} \frac{\omega_Q^2 \tau_c}{1 + (4/q)(\omega_Q \tau_c)^2} \left(\frac{V'_{zz}}{V_{zz}}\right)^2 \tag{27}$$

where V'_{zz} is the modulation change of EFG. The effect of the rotation of molecular fragments was also considered in other studies.[9,36]

The molecules in crystals also undergo torsional oscillations about equilibrium positions (see e.g. reference 2), the effect of which on the NQR spectral parameters has been studied in detail.[2] If the spin–lattice relaxation time T_1 is caused by torsional oscillations, then $T_1 \propto T^{-2}$. Slow reorientational

11

motions of molecules or molecular fragments arising at sufficiently high temperatures (i.e. $\omega_Q \tau_c \gg 1$) result in $T_1 \propto \exp(+U/RT)$. If one therefore considers both torsional oscillations and slow reorientational motions of molecules or molecular fragments to describe T_1, one comes to the equation[9,33]

$$\frac{1}{T_1} = aT^2 + b\exp(-U/RT) \tag{28}$$

where a and b are constants. A mathematical treatment of experimental data according to Eq. (28) makes it possible to obtain more precise values of activation energies (see e.g. references 18, 30, 37–40) than those which consider just molecular reorientations alone (see e.g. reference 27). Moreover, using b, U and the moments of inertia of molecules or their fragments (I_t), one can find torsional oscillation frequencies ν_t and reorientational angles from relationships such as[18]

$$\nu_t = \tfrac{8}{9}b \quad \text{for} \quad I = \tfrac{3}{2} \quad \text{and} \quad \chi = 60\,^\circ C \tag{29}$$

and

$$n = 2\pi\nu_t \sqrt{\frac{2I_t}{u}} \tag{30}$$

where n is the order of the rotation point group.

The spin–lattice relaxation time is so sensitive to the dynamics of molecular motion that it allows us to detect the reorientations of molecules or their fragments at frequencies as low as 1–100 Hz[18,30], which exceeds by several orders of magnitude the experimental possibilities of NMR.[16,41]

REFERENCES

1 H. Bayer, *Z. Physik*, **130,** 227 (1951).
2 T. P. Das and E. L. Hahn, "Nuclear Quadrupole Resonance Spectroscopy" *Solid State Phys.*, Suppl., **1,** 1958.
3 G. D. Watkins and R. V. Pound, *Phys. Rev.*, **85,** 1062 (1952).
4 M. Buyle-Bodin, *Ann. Phys.*, **10,** 533 (1955).
5 H. W. Dodgen and J. L. Ragle, *J. Chem. Phys.*, **25,** 376 (1956).
6 J. L. Ragle, *J. Phys. Chem.*, **63,** 1395 (1959).
7 H. W. Dodgen and R. E. Anderson, *J. Chem. Phys.* **31,** 851 (1959).
8 E. P. Marram and J. L. Ragle, *J. Chem. Phys.*, **37,** 3015 (1962); **41,** 3546 (1964).
9 D. E. Woessner and H. S. Gutowsky, *J. Chem. Phys.*, **39,** 440 (1963).
10 J. L. Ragle and P. Caron, *J. Chem. Phys.*, **40,** 3497 (1964).
11 T. Tokuhiro, *J. Chem. Phys.*, **41,** 438 (1964).
12 Y. Ayant, *Ann. Phys.*, **10,** 487 (1955).
13 L. D. Landau and E. M. Lifshitz, *Kvantovaya Mekhanika*, Moscow, 1963, p. 176.
14 R. Bersohn, *J. Chem. Phys.*, **20,** 1505 (1952).
15 R. Sh. Lotfullin and G. K. Semin, *J. Strukt. Khim.*, **9,** 813 (1968).
16 A. Abragam, *Principles of Nuclear Magnetism*, Oxford University Press, 1961.
17 R. Sh. Lotfullin and G. K. Semin, *Phys. Stat. Sol.*, **35,** 133 (1969).

18 R. Sh. Lotfullin, Dissertation, Inst. Phys. Akad. Nauk Ukr. SSR, Kiev, 1972.
19 A. D. Gordeev, V. S. Grechishkin, I. A. Kyuntsel and Yu. I. Rosenberg, *J. Strukt. Khim.*, **11**, 773 (1970).
20 A. A. Boguslavskii, R. Sh. Lotfullin, Yu. N. Bobkov, and R. V. Magera, *Otbor i peredacha informatsii*, **36**, 72 (1973), Akad. Nauk Ukr. SSR, Kiev.
21 A. P. Zhukov, I. S. Rez, V. I. Pakhomov, and G. K. Semin, *Phys. Stat. Sol.*, **27**, K129 (1968).
22 R. Blinc *et al.*, *J. Chem. Phys.*, **55**, 4843 (1971).
23 A. V. Rakov, *Optika i Spektroskopija*, **10**, 713 (1961); P. A. Bazhulin, A. V. Rakov, and A. A. Rakhimov, *Optika i Spektroskopija*, **16**, 1027 (1964).
24 V. Rehn, *J. Chem. Phys.*, **38**, 749 (1963).
25 S. Ketudat and R. V. Pound, *J. Chem. Phys.*, **26**, 708 (1957); T. Chiba, *J. Chem. Phys.*, **39**, 947 (1963); J. W. McGrath and G. W. Ossman, *J. Chem. Phys.*, **41**, 1352 (1964); **46**, 1824 (1967); T. Chiba and G. Soda, *Magnetic Resonance and Relaxation*, Elsevier, Amsterdam, 1967, p. 722; J. L. Bjorkstam and J. H. Willimorth, *ibid.*, p. 728; V. Sarasvati and R. Vijayaraghava, *ibid.*, p. 767.
26 G. K. Semin and E. I. Fedin, *J. Strukt. Khim.*, **1**, 252, 464 (1960).
27 S. Alexander and A. Tzalmona, *Phys. Rev. Lett.*, **13**, 546 (1964); **138A**, 845 (1965).
28 A. Zussman and S. Alexander, *J. Chem. Phys.*, **49**, 3792, 5179 (1968).
29 A. Tzalmona, *Phys. Lett.*, **20**, 478 (1966); *J. Chem. Phys.*, **50**, 366 (1969); *Phys. Lett.*, **34A**, 289 (1971).
30 N. E. Ainbinder, B. F. Amirkhanov, I. V. Izmestiev, A. N. Osipenko, and G. B. Soifer, *Fiz. Tverd. Tela*, **13**, 424 (1971).
31 N. N. Korst, V. V. Tsyganov, and N. E. Ainbinder, *Fiz. Tverd. Tela*, **14**, 2448 (1972).
32 E. N. Ivanov and K. A. Valiev, *Optika i Spektroskopija*, **19**, 897 (1965).
33 R. Sh. Lotfullin and G. K. Semin, *Kristallographija*, **14**, 809 (1969).
34 C. P. Slichter, *Principles of Magnetic Resonance*, Harper and Row, New York, 1963.
35 A. A. Boguslavskii, R. Sh. Lotfullin and Y. K. Semin, *Phys. Stat. Sol. (B)*, **66**, K95 (1974).
36 J. A. S. Smith, *Advances in NQR*, Vol. 1, pp. 115–132, Heyden, London 1974.
37 I. V. Izmestiev and V. S. Grechishkin, *J. Strukt. Khim*, **II**, 927 (1970).
38 M. Z. Yusupov, V. S. Grechishkin, A. T. Kosulin, V. P. Anferov, and V. S. Versilov, *Org. Magn. Reson.*, **3**, 515 (1971).
39 I. A. Kyuntsel, Yu. I. Rosenberg, and A. D. Gordeev, *Optika i Spectroskopija*, **31**, 67 (1971).
40 I. M. Alymov, V. M. Burbelo, V. A. Egorov, and R. Sh. Lotfullin, *First Specialized 'Colloque Ampère'*, Krakow, Poland, 1973, p. 186.
41 E. R. Andrew, *Nuclear Magnetic Resonance*, Cambridge University Press, 1955.

2. OPTICAL DETECTION OF NUCLEAR QUADRUPOLE INTERACTIONS IN EXCITED TRIPLET STATES

Charles B. Harris

*Department of Chemistry, and Inorganic Materials Research Division
of Lawrence Berkeley Laboratory, University of California,
Berkeley, California, U.S.A. 94720*

and

Michael J. Buckley

*Air Force Materials Laboratory,
Wright-Patterson Air Force Base,
Dayton, Ohio, U.S.A. 45433*

I. INTRODUCTION

The measurement of nuclear quadrupole coupling constants in the excited triplet state of organic molecules has only recently been possible with the development of optically detected magnetic resonance (ODMR) techniques for observing electron spin transitions in zero-field. Since this technique is highly dependent on the nature of the triplet state, a short review of some of the important properties of the triplet state are given. There are several good review articles[1–3] on the triplet state to which the reader is referred for a more complete discussion. The historical development of ODMR and a survey of experimental results are then given, followed by a section that deals with the sensitivity of ODMR and optically detected ENDOR in the framework of intramolecular energy transfer processes. Specifically, the effects of radiation-less, radiative, and spin–lattice relaxation processes on the overall sensitivity of ODMR are considered explicitly. The remainder of this paper deals with the form of the spin Hamiltonian in zero-field followed by an analysis of the excited $\pi\pi^*$ triplet states of 8-chloroquinoline and *para*dichlorobenzene.

15

A. The Excited Triplet State in Organic Molecules

The ground state of most organic molecules consists of a singlet electron configuration in which all the electrons have their spins paired. The molecule may be excited to a higher energy electron configuration by the application of electromagnetic radiation of the appropriate energy. We will primarily be concerned with the excited electron configurations produced when one electron in the highest bonding molecular orbital (ϕ_A) is promoted to the lowest antibonding molecular orbital (ϕ_B). Since electrons have a spin of $1/2$, there are four possible orientations for the two unpaired electrons, which, if we let α equal spin up and β equal spin down, may be represented as

$$
\begin{array}{llll}
\alpha(1)\ \alpha(2) & S_z = 1 & S^2 = 1 & \\
\alpha(1)\ \beta(2) & S_z = 0 & S^2 = 0 & \\
\beta(1)\ \alpha(2) & S_z = 0 & S^2 = 0 & (1) \\
\beta(1)\ \beta(2) & S_z = -1 & S^2 = 1 &
\end{array}
$$

This representation, however, is not satisfactory since the electrons obey Fermi–Dirac statistics and thus the total wave function (orbital times spin) must be antisymmetric with respect to electron exchange. In addition, we would like the spin functions to be eigenstates of S^2 and S_z. The spin functions $\alpha(1)\ \alpha(2)$ and $\beta(1)\ \beta(2)$ are clearly eigenstates of S^2 and S_z since $S^2 = 1$ for both and $S_z = +1$ and -1 respectively. We can generate the $S_z = 0$ component of the triplet spin state by applying the lowering operator to the $\alpha(1)\ \alpha(2)$ state which gives us the desired spin function

$$
{}^3\psi_0 = [1/\sqrt{2}][\alpha(1)\beta(2) + \beta(1)\alpha(2)] \tag{2}
$$

The remaining spin function is a singlet

$$
{}^1\psi = [1/\sqrt{2}][\alpha(1)\beta(2) - \beta(1)\alpha(2)] \tag{3}
$$

and, in contrast to the triplet spin functions, is antisymmetric with respect to electron exchange. The spatial part of the excited state electron wave function may be represented as a symmetric ($+$) and antisymmetric ($-$) linear combination of ϕ_A and ϕ_B as

$$
\psi_\pm = [1/\sqrt{2}][\phi_A(1)\phi_B(2) \pm \phi_A(2)\phi_B(1)] \tag{4}
$$

Since the total wave function must be antisymmetric, there are only four allowed representations of the total wave function; a singlet state with a symmetric spatial function and an antisymmetric spin function

$$
{}^1\Psi = [1/\sqrt{2}][\phi_A(1)\phi_B(2) + \phi_A(2)\phi_B(1)][1/\sqrt{2}][\alpha(1)\alpha(2) - \beta(1)\beta(2)] \tag{5}
$$

and a triplet state with an antisymmetric spatial function and a symmetric spin function

$$^3\Psi = [1/\sqrt{2}][\phi_A(1)\phi_B(2) - \phi_A(2)\phi_B(1)] \begin{Bmatrix} \alpha(1)\alpha(2) \\ [1/\sqrt{2}][\alpha(1)\beta(2) + \beta(1)\alpha(2)] \\ \beta(1)\beta(2) \end{Bmatrix} \quad (6)$$

The repulsive electrostatic interaction between the two unpaired electrons gives rise to a term in the total Hamiltonian equal to e^2/r_{12}, where e is the electron charge and r_{12} is the vector connecting the two electrons. This term removes the degeneracy of the singlet and triplet states and results in the singlet state going to higher energy while the triplet state is shifted to lower energy with an energy separation between the two states of

$$^1E - {}^3E = 2\delta_{12} \quad (7)$$

where δ_{12} is the exchange integral given by

$$\delta_{12} = \langle \phi_A(1)\phi_B(2)|e^2/r_{12}|\phi_A(2)\phi_B(1)\rangle \quad (8)$$

For most organic molecules $2\delta_{12}$ is 1 000 to 10 000 cm^{-1}. As we will see in the discussion of the spin Hamiltonian, the inclusion of the electron dipole–dipole interaction removes the threefold degeneracy of the triplet state. This splitting is usually referred to as the zero-field splitting and is of the order of 0.1 cm^{-1}. An additional contribution to the zero-field splitting arises from the coupling of the spin and orbital electron angular momentum and is of the form $A(\mathbf{L} \cdot \mathbf{S})$ where $\mathbf{L}$ and $\mathbf{S}$ are the spin and orbital angular momenta and A is a constant that depends on the particular molecule being considered. The effect of the spin–orbit Hamiltonian is to mix states of different multiplicity and, therefore, to give singlet character to triplet states and vice versa. The most important consequence of this is to permit the triplet state to undergo weak electric dipole radiation to the ground state (phosphorescence), the intensity and polarization from *each* of the three triplet sublevels being a function of the spin–orbit coupling to both the excited and ground singlet states.

Since the sensitivity of ODMR depends upon the number of molecules in their triplet state, an important consideration is intramolecular energy transfer processes. Following excitation, a molecule may lose energy by radiative or non-radiative pathways. Phosphorescence ($T_1 \rightarrow S_0$) and fluorescence ($S_1 \rightarrow S_0$) comprise the radiative pathways and proceed with rate constants of the order of 10^4 to 10^{-2} s^{-1} and 10^6 to 10^9 s^{-1}, respectively. The longer lifetime for phosphorescence results from the fact that the triplet state is spin-forbidden in first order for electric dipole radiation to the ground state. The molecules may also lose energy through three non-radiative pathways:

(1) *Vibrational Relaxation*—or passage from a non-equilibrium vibrational energy distribution in a given electronic state to the Boltzmann energy

distribution relative to the zero point energy of that same state. This proceeds primarily by a non-radiative mechanism with a rate constant of approximately $10^{12}\,s^{-1}$.

(2) *Internal Conversion*—or radiationless passage between two electronic states of the same spin multiplicity. This pathway also has a fast rate constant of approximately $10^{12}\,s^{-1}$.

(3) *Intersystem Crossing*—or radiationless passage from an electronic state in the singlet manifold to an electronic state in the triplet manifold or vice versa. This pathway is slower than the other two and is of the order of 10^{4} to $10^{12}\,s^{-1}$.

Although the exact mechanisms of intersystem crossing have not been completely elucidated, it is generally found that at liquid helium temperatures (4.2 K) the individual triplet spin sublevels of the lowest triplet states have unequal populations because of unequal intersystem crossing rates into the individual magnetic sublevels via spin–orbit and spin–vibronic coupling and unequal depopulating rates. Consequently, a state of spin alignment exists for the electron spins.[4] The various rate constants for energy transfer, the existence of spin alignment, and the spin–lattice relaxation rate between the triplet spin sublevels are all important factors in determining the sensitivity of ODMR.

B. The Historical Development of ODMR

The development of any field of science is difficult to trace since every advancement is dependent on the work of many previous researchers; however, we will choose for the starting point of this discussion the extensive study of the phosphorescence of organic molecules by Lewis and Kasha[5,6] in 1944. In their series of papers it was proposed that the phosphorescent state of these molecules corresponded to their lowest triplet state. This hypothesis was strongly supported shortly thereafter by magnetic susceptibility measurements[7,8] which showed that small changes in the susceptibility were observed upon irradiation of the samples.

As with any major change in the existing paradigm of science, this hypothesis was not universally accepted.[9] The most distressing aspect of the hypothesis was the failure to observe the predicted electron spin resonance (ESR) of the phosphorescent state. The problem was resolved in 1958 when Hutchison and Mangum[10,11] succeeded in observing the ESR of naphthalene in its phosphorescent state and showed conclusively that the phosphorescent state was a triplet state. The experiment was performed on a single crystal of naphthalene doped in durene using conventional techniques in which the absorption of the microwave energy was monitored while varying the applied magnetic field. Subsequently, the triplet state ESR of many organic compounds was observed; however, most of the work was done on randomly oriented samples. Since only one parameter can usually be measured with randomly-oriented samples, the separation of the three levels of the triplet cannot be determined. In certain cases [12,13] the three levels can be assigned but the assignment is difficult and the method has not been used often. The limited sensitivity of ESR and the difficulty of preparing single crystal samples has restricted the number of molecules investigated. Only a few (about 14) molecules in single crystals have been reported to date using conventional methods and they are all characterized by relatively long-lived $\pi-\pi^*$ triplet states.

The next major change in the existing paradigm occurred in 1965 when Geschwind, Devlin, Cohen and Chinn[14] reported the optical detection of the ESR of the excited metastable $\bar{E}(2E)$ state of Cr^{3+} in Al_2O_3. In this classic experiment they showed that the optical r.f. double resonance techniques first suggested by Brossel and Kastler[15] and widely used in gases[16] could also be applied to solids. The experiment was performed using a high resolution optical spectrometer to monitor the change in intensity of one of the Zeeman components of the phosphorescence $[\bar{E}(^1E \rightarrow {}^4A_2)]$ as $\bar{E}$ was saturated with microwave radiation when the magnetic field was swept through resonance. The resonance signal was observed by modulating the microwave field and detecting the resultant modulation of the optical emission. Since optical rather than microwave photons are detected, the sensitivity can be increased many orders of magnitude over conventional techniques. As an example, at temperatures below the λ-point of helium the resonance could be observed directly on an oscilloscope without the need for phase-sensitive detection. The success in optically detecting the electron spin resonance of a metastable state led several research groups to attempt to apply the same principles to the optical detection of the ESR of organic molecules in their lowest triplet state.

In 1967 the first successful experiment was reported by Sharnoff for the $\Delta M = 2$ transition of naphthalene.[17] In this experiment, a single crystal of biphenyl containing 0.1 mol per cent naphthalene was placed in a microwave cavity where it was immersed in liquid helium maintained at 1.8 K. The crystal was irradiated with the appropriately filtered light from a mercury arc lamp and the phosphorescence isolated with a detector consisting of a linear polarizer and a low-resolution spectrometer. The microwave field was modulated at 40 Hz and the signal detected by feeding the output of the photomultiplier into a phase-sensitive amplifier. In this experiment, it was shown that the radiative matrix elements connecting any triplet sublevel with the ground singlet electronic level are functions of the magnetic quantum numbers of that sublevel.

At this point the development of ODMR of the lowest triplet state of organic molecules entered a new phase. Now that this new method was shown to be applicable to these molecules, the research centred around improving the basic techniques and using this new tool to gain information on a variety of phenomena associated with the triplet state.

Shortly after Sharnoff's paper, Kwiram[18] reported the optical detection of the $\Delta M = 1$ and $\Delta M = 2$ transitions of phenanthrene in its triplet state. In this investigation the experimental methods were the same as those used by Sharnoff except that the microwave field was not modulated while the exciting and emitted light was chopped antisynchronously at 50 Hz. The 50 Hz output of the photomultiplier was converted to d.c. by a phase sensitive detector and fed into a signal averager. The observed change in intensity of the phosphorescence at the three transition frequencies was used to assign the spatial symmetry of the triplet state.

Schmidt, Hesselmann, DeGroot and van der Waals[19] also reported the optical detection of quinoxaline-d_6 in 1967. Their experimental procedure was basically the same as that used by Sharnoff, except that they modulated the magnetic field with and without amplitude modulation of the microwave field. They were able to show (1) that the emission originates from the top spin component (out-of-plane), and (2) from phosphorescence decay studies, that entry into the triplet state by intersystem crossing is also to the top spin component.

In 1968 Schmidt and van der Waals[20] extended the almost zero-field work (3G) of Hutchison's group[21] by optically detecting the zero-field transitions of molecules in their triplet state at zero external magnetic field. Since it is necessary to vary the microwave frequency in order to observe the resonance in zero external magnetic field, a helix was used to couple the microwave power to the sample. The observed signals were extremely sharp and in the case of quinoxaline-d_6, showed fine structure.[22]

Tinti, El-Sayed, Maki and Harris[23] extended the method of optical detection in zero-field by incorporating a high resolution spectrometer and studying the effect of the microwave field on the individual lines of the phosphorescence spectrum of 2,3-dichloroquinoxaline. This method has since been called Phosphorescence Microwave Double Resonance (PMDR) spectroscopy. They showed that the use of a high-resolution spectrometer will give better sensitivity in cases where there is mixed polarization of the phosphorescence, since if the total emission is monitored, the change in intensity due to the microwave field may be partially cancelled. The sensitivity was excellent, and in fact, a very strong signal was observed using c.w. conditions for both the microwave and optical radiations. The observed structure of the zero-field transitions was explained quantitatively in terms of nuclear quadrupole interactions in the excited triplet state in a later paper[24] by Harris *et al.* and by van der Waals and co-workers[25] in which optically detected electron nuclear double resonance (ENDOR) was reported. Several other papers followed on the observation and interpretation of nitrogen ENDOR in zero-field[26,27] and was extended to ^{35}Cl and ^{37}Cl by Buckley and Harris.[28,29] Optical detection of electron–electron double resonance (EEDOR) was reported by Kuan, Tinti and El-Sayed[30] and was demonstrated to be a method of improving the signal strength of weak zero-field transitions if emission is from only one of the triplet sublevels.

Apart from the applications directly associated with magnetic resonance parameters, ODMR techniques have been used to analyse the phosphorescence spectra. Phosphorescence Microwave Double Resonance (PMDR) spectroscopy has already proven itself to be extremely valuable in many applications such as determining the symmetry of the excited state, the pathways of inter- and intramolecular energy transfer for many molecules[31–35] and other phenomena.[36–42] Also, with transient microwave excitation, these techniques have been successfully utilized in studying the coherent interactions of excited triplet states with resonant microwave fields. Theoretical aspects of the problem were first considered by Harris[43] and were followed by a number of experiments including the coherent modulation of the phosphorescence by microwave radiation,[44] the formation of a spin-echo from an excited triplet state,[45] the optical detection of echoes,[46] spin locking,[47] multiple spin-echoes,[48] and adiabatic demagnetization.[49] Other experiments have recently been reported, in which the electron-spin transitions are induced by coherent acoustic waves[50] and heat pulses or incoherent acoustic waves[51] instead of by a microwave field. These techniques hold the promise of providing a much more detailed picture of the spin–phonon interaction in the excited state.

One of the most promising applications of ODMR and PMDR is the study of exciton interactions in molecular crystals. Using PMDR, the coherent nature of energy exchange between pairs of molecules as nearest neighbours in an isotopically dilute system has been observed.[52] In addition, the coherent migration of triplet Frenkel excitons in molecular crystals has been observed[53] and the density-of-states function in the band has been measured.[53]

Recently, even the kinetics and quantum yield for the creation of mobile wavepackets in molecular crystals have been determined using these techniques. The methods have also been extended to ionic solids[54] and there is every reason to believe that the techniques will eventually be extended to surface states, molecules adsorbed on surfaces and semiconductors. In short, the potential uses still remaining to be exploited are many and varied. To date, however, techniques of ODMR and PMDR have developed into several basic areas: (1) the study of the electron distribution of organic molecules in their triplet state by analysis of their zero-field, nuclear quadrupole, and nuclear–electron hyperfine interactions; (2) investigations into the rates and mechanisms of intramolecular processes such as internal conversion and intersystem crossing; (3) as a tool to investigate the energy levels and dynamics properties of exciton bands in molecular crystals; (4) as a method of observing coherent phenomena in the excited states; and (5) as a method of examining spin–phonon interactions. Details of these areas can be found in the work of Sharnoff, Kwiram, van der Waals, Maki, El-Sayed, Harris and others.

The remainder of this article will deal with only the first of the above areas, and in particular, the measurement of the nuclear quadrupole coupling constants by analysis of the optically detected ESR and ENDOR spectra.

II. GENERAL CONSIDERATIONS

A. Sensitivity Considerations in the Optical Detection of ESR

One of the primary advantages of optical detection is the excellent sensitivity of about 10^4 spins as compared to the sensitivity of about 10^{11} spins for conventional ESR. This greatly increased sensitivity permits the detection of the ESR of molecules with very short excited-state lifetimes. In this section, a cursory analysis of the influence of various properties of the excited state on the sensitivity with optical detection are presented. Although many kinetic schemes can be constructed for various experiments, we consider experiments performed under conditions of continuous optical excitation while monitoring the change in intensity of the phosphorescence as a function of the applied microwave field.[33,55] Only the case in which the triplet state is populated by excitation of the sample into the first excited singlet state followed by intersystem crossing into the triplet state will be considered. For molecules with reasonably high symmetry (i.e., D_{2h}, C_{2h}, and C_{2v}) different modes of populating the triplet state produce to varying degrees a different spin alignment. This is illustrated in Fig. 1 where the radiative and non-radiative pathways for

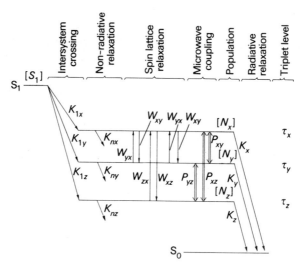

Fig. 1.   Relaxation pathways and rate constants for the triplet state.

energy transfer are depicted. $[S_1]$ is the population of the lowest excited singlet state, $[N_i](i = x, y, z)$ is the steady-state population of the corresponding triplet levels, K_{1x} is the intersystem-crossing rate constant from S_1 to T_1, K_x is the radiative or phosphorescence rate constant for relaxation to S_0, K_{nx} is the non-radiative decay or relaxation rate constant from T_1 to S_0, $W_{x_1x_2}$ $(x_1 \neq x_2)$ is the spin–lattice relaxation rate constant and $P_{x_1x_2}$ $(x_1 \neq x_2)$ is the induced rate constant due to the applied microwave field (H_1). When the microwave field does not connect any two of the zero-field levels of the triplet, the steady-state population is given by setting $P_{x_1x_2} = 0$. The application of the microwave field at a frequency corresponding to the energy separation of two of the levels (i.e., $\nu = (E_x - E_y)/h$) causes a redistribution of the population which in most cases results in a change in the phosphorescence intensity. Since optical, rather than microwave, photons are detected, one expects the sensitivity to be improved in proportion to the ratio of the energies of the photons, which, for a typical molecule, is approximately 3×10^5. The actual change in the phosphorescence intensity, however, is a complex function of the various relaxation channels and rate constants. The actual improvement in sensitivity depends on the molecule under study.

In order to derive a reasonably simple quantitative expression for the change in intensity of the phosphorescence, the three following assumptions will be made:

(1) The splitting of the three triplet zero-field levels by nuclear quadrupole and nuclear hyperfine interactions will be neglected,

(2) Only the two levels connected by the H_1 field will be considered; these are designated τ_x and τ_y, and

(3) Only the steady-state conditions, $dN_x/dt = dN_y/dt = 0$, will be considered for both the cases when $H_1 = 0$ and $H_1 \neq 0$.

The first assumption will predict too great a change in intensity if the individual triplet levels are split by more than the frequency width of the H_1 field, since in this case the H_1 field will allow an additional relaxation pathway for only a fraction of the population of each triplet level at any given frequency. The second assumption will introduce an error in the expression for the percentage change in intensity since the intensity contribution from the level

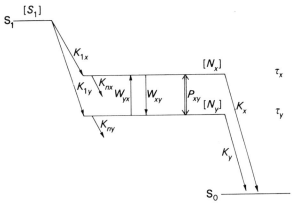

Fig. 2. Relaxation pathways and rate constants considering only two of the three triplet levels (see text).

not connected by the H_1 field (τ_z) is neglected. This assumption also requires that the spin–lattice relaxation rate between τ_z and τ_x and between τ_z and τ_y be neglected. This is usually valid since the experiments are performed at or below 4.2 K. The third assumption requires that the experiment be performed using c.w. microwave conditions or modulating the microwave field with a frequency lower than the total rate constant of the system.

The differential equations describing the population of the levels shown in Fig. 2 are

$$\frac{dN_x}{dt} = S_1 K_{1x} - N_x[K_{nx} + K_x + W_{xy} + P_{xy}] + N_y[W_{yx} + P_{xy}] \tag{9}$$

$$\frac{dN_y}{dt} = S_1 K_{1y} - N_y[K_{ny} + K_y + W_{yx} + P_{xy}] + N_x[W_{xy} + P_{xy}] \tag{10}$$

With the definitions

$$A = K_{nx} + K_x + W_{xy} + P_{xy}$$
$$B = W_{yx} + P_{xy}$$
$$C = K_{ny} + K_y + W_{yx} + P_{xy} \tag{11}$$
$$D = W_{xy} + P_{xy}$$

Equations (9) and (10) may be rewritten

$$\frac{dN_x}{dt} = S_1 K_{1x} - N_x A + N_y B \tag{12}$$

$$\frac{dN_y}{dt} = S_1 K_{1y} - N_y C + N_x D \tag{13}$$

The steady-state assumption allows us to write

$$\frac{dN_x}{dt} = S_1 K_{1x} - N_x A + N_y B = 0 \tag{14}$$

$$\frac{dN_y}{dt} = S_1 K_{1y} - N_y C + N_x D = 0 \tag{15}$$

Upon solving Eqs. (14) and (15) for the population of the triplet levels, we have

$$N_x = \frac{S_1[CK_{1x} + BK_{1y}]}{AC - BD} \tag{16}$$

and

$$N_y = \frac{S_1[AK_{1y} + DK_{1x}]}{AC - BD} \tag{17}$$

24

The intensity of the phosphorescence detected with an optical spectrometer may be written

$$I = \alpha_1 N_x K_x + \alpha_2 N_y K_y \tag{18}$$

where α_1 and α_2 are constants that depend on the polarization of the emission, the orientation of the sample, and the efficiency of the detection system. The assumption will be made that $\alpha_1 = \alpha_2$, which allows the fractional change in the intensity of the phosphorescence upon application of the H_1 field to be written

$$\Delta I = \frac{I - I_0}{I_0} = \frac{I}{I_0} - 1 \tag{19}$$

where I_0 is the intensity of the phosphorescence when $P_{xy} = 0$. With this condition, it is convenient to define the parameters given in Eq. 11 as

$$a = K_{nx} + K_x + W_{xy}$$
$$b = W_{yx}$$
$$c = K_{ny} + K_y + W_{yx} \tag{20}$$
$$d = W_{xy}$$

If both of the triplet levels are monitored, the fractional change in intensity of the emission is given by

$$\Delta I = \left\{ \frac{[K_{1y}(AK_y + BK_x) + K_{1x}(CK_x + DK_y)][ac - bd]}{[K_{1y}(aK_y + bK_x) + K_{1x}(cK_x + dK_y)][AC - BD]} \right\} - 1 \tag{21}$$

In some cases it is possible to monitor only one of the triplet levels connected by the H_1 field, in which case the changes in intensity of emission from the τ_x and τ_y levels are given by

$$\Delta I_x = \left\{ \frac{[CK_{1x} + BK_{1y}][ac - bd]}{[cK_{1x} + bK_{1y}][AC - BD]} \right\} - 1 \tag{22}$$

and

$$\Delta I_y = \left\{ \frac{[AK_{1y} + DK_{1x}][ac - bd]}{[aK_{1y} + dK_{1x}][AC - BD]} \right\} - 1 \tag{23}$$

Three cases will be discussed in order to examine the effect of the magnitude of the various rate constants on the sensitivity of the experiment.

CASE 1. THE EFFECT OF RADIATIVE DECAY

For illustration purposes, we assume that the non-radiative and spin–lattice relaxation rate constants may be neglected. The parameters defined in

Eqs. (11) and (20) become

$$A = K_x + P_{xy} \qquad a = K_x$$
$$B = P_{xy} \qquad b = 0$$
$$C = K_y + P_{xy} \qquad c = K_y \qquad (24)$$
$$D = P_{xy} \qquad d = 0$$

In the absence of the H_1 field, the steady-state populations are given by

$$N_x^0 = S_1(K_{1x}/K_x)$$
$$N_y^0 = S_1(K_{1y}/K_y) \qquad (25)$$

The steady-state population of τ_x is given by Eq. (16) which for this example becomes

$$N_x = \frac{S_1[K_{1x}K_y + P_{xy}(K_{1x} + K_{1y})]}{[K_xK_y + P_{xy}(K_x + K_y)]} \qquad (26)$$

In the limit that P_{xy} is much larger than any of the relaxation rate constants, the populations of τ_x and τ_y are equalized and the transition is saturated. Clearly, the power required to equalize the populations is directly proportional to the relaxation rate of the system and inversely proportional to the lifetime of the excited state. The population of τ_x at saturation is given by

$$N_x^s = \frac{S_1[K_{1x} + K_{1y}]}{[K_x + K_y]} \qquad (27)$$

and the corresponding population of τ_y is given by

$$N_y^s = \frac{S_1[K_{1x} + K_{1y}]}{[K_x + K_y]} \qquad (28)$$

and therefore, $N_x^s = N_y^s$. The change in population of τ_x upon saturation is

$$\Delta N_x = N_x^s - N_x^0 = \frac{S_1[K_xK_{1y} - K_yK_{1x}]}{K_x(K_x + K_y)} \qquad (29)$$

Therefore, if $K_xK_{1y} = K_yK_{1x}$, there is no change in population. If the emissions from τ_x and τ_y are monitored simultaneously, the fractional change in intensity is given by Eq. (21) which, for this example, reduces to

$$\Delta I = \left\{ \frac{[K_{1y}(AK_y + BK_x) + K_{1x}(CK_x + DK_y)]}{[K_{1y}(AC - BD)]} \right\} - 1 \qquad (30)$$

$$= \left\{ \frac{[K_{1x} + K_{1y}][P_{xy}(K_x + K_y) + K_xK_y]}{[K_{1x} + K_{1y}][P_{xy}(K_x + K_y) + K_xK_y]} \right\} - 1 \qquad (31)$$

In such a case, $\Delta I = 0$ and no change in the intensity of emission will be observed. However, if a high-resolution optical spectrometer is used, it is often possible to monitor the emission from just one of the triplet levels via its selective emission to the origin or a vibration of the ground state singlet manifold. Consider for example, emission from τ_x, in which case the change in intensity given by Eq. (22) becomes

$$\Delta I_x = \frac{K_x}{K_{1x}} \left[\frac{P_{xy}(K_{1x} + K_{1y}) + K_{1x}K_y}{P_{xy}(K_x + K_y) + K_xK_y} \right] - 1 \tag{32}$$

In the limiting case where intersystem crossing proceeds primarily to $\tau_x(K_{1x} \gg K_{1y})$, Eq. (32) reduces to

$$\Delta I_x = \left\{ \frac{P_{xy}K_x + K_xK_y}{P_{xy}(K_x + K_y) + K_xK_y} \right\} - 1 \tag{33}$$

At saturation we have

$$\Delta I_x^s = \left(\frac{K_x}{K_x + K_y} \right) - 1 \tag{34}$$

The effect of the ratio of the radiative rate constants (K_x/K_y) on the maximum change in intensity of the emission may be illustrated with the following examples

K_x/K_y	$\Delta I_x^s (\%)$
0.1	91
1	50
10	9

It is apparent that the maximum sensitivity is achieved if the level with the fast intersystem-crossing rate constant has the slower phosphorescence rate constant.

CASE 2. THE EFFECT OF SPIN–LATTICE RELAXATION

The two rate constants for spin–lattice relaxation are not independent and may be related directly to the spin–lattice relaxation time T_1 for any given temperature.

The interaction between the energy and the lattice may be represented schematically as

The conservation of energy requires that for each transition from τ_x to τ_y there be a corresponding lattice transition from X_b to X_a and vice versa. The transition rate for the lattice is written

$$W_{ab} = N_a A$$
$$W_{ba} = N_b A \tag{35}$$

where A is the transition probability. The spin–lattice relaxation rate constants may be written in terms of the population of the lattice as

$$W_{xy} = W_{ba} = N_b A$$
$$W_{yz} = W_{ab} = N_a A \tag{36}$$

Since the lattice is at the temperature of the bath (liquid helium), the normalized population of the lattice is given by

$$N_a = \frac{e^{-\delta/2kT}}{e^{-\delta/2kT} + e^{\delta/2kT}} = f$$

$$N_b = \frac{e^{\delta/2kT}}{e^{-\delta/2kT} + e^{\delta/2kT}} = 1 - f \tag{37}$$

where $\delta = (E_x - E_y)$ and E_x and E_y are the energies of the x and y magnetic sublevels respectively. The spin–lattice relaxation rates may now be written

$$W_{xy} = (1-f)A$$
$$W_{yx} = (f)A \tag{38}$$

The spin–lattice relaxation time is defined by the expression

$$T_1 = \frac{1}{W_{xy} + W_{yx}} = \frac{1}{A} \tag{39}$$

and W_{xy} and W_{yx} may be expressed in terms of T_1 and f as

$$W_{xy} = \frac{1-f}{T_1}$$

$$W_{yx} = \frac{f}{T_1} \tag{40}$$

In the derivation of Eq. (40), it is assumed that only a direct process of energy transfer between the spin system and the lattice exists which is usually the case at the temperatures of the experiments (4.2 to 1.3 K). If Raman or Orbach processes are present, only the explicit temperature dependence of the relaxation must be corrected so that the spin–lattice relaxation may always be defined for a two level system in terms of only T_1 at a given temperature. A short T_1 relaxation time will tend to produce a Boltzmann population distribution between the spin sublevels and will significantly reduce the spin alignment.

This can be seen by considering the simple case where there is only intersystem crossing to τ_x and emission from τ_x and τ_y. Again the non-radiative decay rate constants K_{nx} and K_{ny} are assumed to be negligible. The parameters defining this model are

$$A = K_x + W_{xy} + P_{xy} \qquad a = K_x + W_{xy}$$

$$B = W_{yx} + P_{xy} \qquad b = W_{yx}$$

$$C = K_y + W_{yx} + P_{xy} \qquad c = K_y + W_{yx} \qquad (41)$$

$$D = W_{xy} + P_{xy} \qquad d = W_{xy}$$

and the populations of τ_x and τ_y when $P_{xy} = 0$ are given by

$$N_x^0 = \frac{S_1[(K_y + W_{yx})K_{1x}]}{K_xK_y + K_yW_{xy} + K_xW_{yx}}$$

and $\qquad (42)$

$$N_y^0 = \frac{S_1[(W_{xy})K_{1x}]}{K_xK_y + K_yW_{xy} + K_xW_{yx}}$$

In the limit that $W_{xy} = W_{yx} = 0$, this reduces to

$$N_x^0 = \frac{S_1[K_{1x}]}{K_x}$$

$$N_y^0 = 0 \qquad (43)$$

At high temperatures when $W_{xy} \approx W_{yx} \gg K_x, K_y, K_{1x}$, Eq. (42) becomes

$$N_x^0 = \frac{S_1[K_{1x}]}{K_x + K_y}$$

$$\qquad (44)$$

$$N_y^0 = \frac{S_1[K_{1x}]}{K_x + K_y}$$

Since the change in population is monitored, it is clearly advantageous to perform the experiments at the lowest possible temperature in order to decrease the thermalization of the spin levels and the resulting loss in sensitivity.

CASE 3. THE EFFECT OF NON-RADIATIVE RELAXATION

The final case to be considered is the effect of the non-radiative relaxation rate constants K_{nx} and K_{ny} on the sensitivity of the experiment. It is obvious that since only the radiative emission is detected, a large rate of depopulation by non-radiative relaxation is not desirable, unless it produces enhanced spin alignment. In the case of a sample that relaxes primarily through non-radiative pathways, the sensitivity may be improved by using conventional ESR techniques and monitoring the absorption of microwave power, or in extreme cases by monitoring the change in temperature of the sample. A quantitative measure of the decrease in sensitivity may be calculated by substituting the appropriate rate constants into Eqs. (21), (22) and (23); however, the expressions are rather complex and therefore not particularly useful.

It should be noted that although we have dealt with the rate processes in the discussion of sensitivity, the results can be used to measure the relative rate processes associated with the individual magnetic sublevels. Specifically, the measurement of intensity changes of phosphorescence under the influence of the microwave field can yield the relative intersystem crossing, radiative and radiationless rate constants to and from all three magnetic sublevels. Indeed, this approach has already been widely applied[56] in the limit that spin–lattice relaxation may be neglected and saturation of the transition is achieved. The inclusion of the power factor, however, gives one an additional experimental 'handle' from which to extract information [cf. Eqs. (21), (22) and (23)].

B. Optically Detected ENDOR

The sensitivity of this experiment may be estimated if the assumption is made that there is no nuclear polarization. Since this assumption has yet to be thoroughly investigated, it is reasonable to expect that in some cases it will not be valid. Nuclear polarization may arise through cross relaxation between the electron and nuclear spin systems (the Overhauser effect), or it may be induced by saturation of 'forbidden' transitions (simultaneous electron–nuclear flips). It is also possible that selective intersystem crossing may preferentially populate a particular nuclear spin level if there is strong hyperfine coupling of the electron and nuclear wave functions.

In the absence of nuclear polarization, the sensitivity of the optically detected ENDOR signal may be understood by referring to Fig. 3 in which the

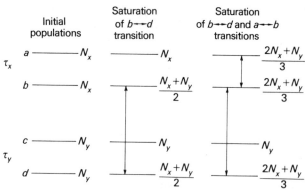

Fig. 3. Population change predicted for ESR ($b \leftrightarrow d$) and ENDOR ($a \leftrightarrow b$) transitions.

τ_x and τ_y triplet levels are now each composed of two levels. This splitting of the triplet levels is due to nuclear quadrupole and hyperfine interactions as will be discussed in the following sections. The results obtained by considering the triplet levels as being split into only two nuclear sublevels are independent of the number of sublevels if the ESR transition connects only one nuclear sublevel in each of the two triplet levels, and the ENDOR transition connects only two nuclear sublevels in one of the triplet levels.

As has already been discussed, the sensitivity of the optical detection technique is dependent on the various relaxation pathways from the triplet state. The same considerations apply in an ENDOR experiment. Since the sensitivity of the ENDOR experiment will be referenced to the sensitivity of the ESR experiment, the explicit dependence of the triplet-state populations on the various rate constants need not be specified. For the system shown in Fig. 3, the phosphorescence intensity may then be written

$$I_0 = 2(N_x K_x + N_y K_y) \tag{45}$$

where $N_x(N_y)$ is now defined as the population of each of the two levels in the $\tau_x(\tau_y)$ manifold.

Upon saturation of the electron–spin transition $(b \leftrightarrow d)$, this becomes

$$I_S = \left(\frac{3N_x + N_y}{2}\right)K_x + \left(\frac{3N_y + N_x}{2}\right)K_y \tag{46}$$

with the change in intensity given by

$$\Delta I = I_S - I_0 = \tfrac{1}{2}(N_x - N_y)(K_y - K_x) \tag{47}$$

If the ENDOR transition $(a \leftrightarrow b)$ is also saturated, the intensity is given by

$$I_E = \tfrac{2}{3}[(2N_x + N_y)K_x + (2N_y + N_x)K_y] \tag{48}$$

Since the ENDOR signal is detected by monitoring the change in intensity of the ESR transition, the signal strength is given by

$$\Delta I_E = I_E - I_S \tag{49}$$

$$= \tfrac{1}{6}[(N_x - N_y)(K_y - K_x)] \tag{50}$$

and the fractional change in intensity of the ESR signal upon saturation of the ENDOR transition is

$$\delta I = \Delta I_E / \Delta I_S = \tfrac{1}{3} \tag{51}$$

If the ENDOR transition $(c \leftrightarrow d)$ is saturated instead of the transition from $(a \leftrightarrow b)$, the same expression is obtained for the change in intensity—Eqs. (50) and (51).

It is interesting to note from Eqs. (47) and (50) that the ESR signal and the ENDOR signal always affect the intensity of the phosphorescence in the same direction. If the forbidden ESR transition from $(b \leftrightarrow c)$ is saturated and if the two ENDOR transitions $(a \leftrightarrow b)$ and $(c \leftrightarrow d)$ occur at the same frequency, the change in phosphorescence intensity is given by

$$\Delta I_E = \tfrac{1}{2}[(N_x - N_y)(K_y - K_x)] \tag{52}$$

and the fractional change in intensity of the ESR signal is unity.

As a final note, if the ESR transitions from $(a \leftrightarrow c)$ and $(b \leftrightarrow d)$ occur at the same frequency, the ENDOR transitions from $(a \leftrightarrow b)$ and $(c \leftrightarrow d)$ must also occur at the same frequency, causing the change in intensity of the ESR signal to be twice as large—Eq. (47)

$$\Delta I = (N_x - N_y)(K_y - K_x) \tag{53}$$

while the ENDOR transitions will not be observed since the populations of the nuclear sublevels are already equal.

III. THE ZERO-FIELD SPIN HAMILTONIAN

The observed magnetic resonance spectra of the excited triplet state of organic molecules in zero external magnetic field may be understood in terms of a Hamiltonian of the form

$$\mathcal{H} = \mathcal{H}_{SS} + \mathcal{H}_Q + \mathcal{H}_{HF}$$

where $\mathcal{H}_{SS}$ is the spin–spin or zero-field interaction between the two unpaired electrons, $\mathcal{H}_Q$ is the nuclear quadrupole interaction, and $\mathcal{H}_{HF}$ is the nuclear–electron hyperfine interaction.

A. $\mathcal{H}_{SS}$—The Spin–Spin or Zero-Field Splitting Hamiltonian

$\mathcal{H}_{SS}$ is primarily due to the magnetic dipole–dipole interaction between the unpaired electrons in the excited triplet state. There can also be a contribution from the spin–orbit coupling between the lowest triplet and other excited states; however, the contribution from the interaction between other *excited* triplet states of the same orbital type shifts the three levels equally,[57] and for our purposes will be neglected.

If the radiative lifetime for fluorescence and phosphorescence is known, the magnitude of the spin–orbit contribution to the zero-field splitting may be estimated by choosing a simple model in which the transition probability for phosphorescence is due only to the spin–orbit coupling of one spin sublevel with only one excited singlet state. In the framework of this model the transition probability for phosphorescence may be expressed as

$$P_P \approx |\langle^3\psi_1|e\mathbf{r}|^1\psi_0\rangle|^2 = \frac{1}{\tau_P} \tag{54}$$

where $e\mathbf{r}$ is the electron dipole moment transition operator, $^3\psi_1$ is the first triplet state, $^1\psi_0$ is the ground singlet state, and τ_P is the phosphorescence radiative lifetime. The wave function for the phosphorescent triplet state is actually a linear combination of the pure triplet state, which is spin forbidden for electric dipole radiation to the ground state, and an admixture of singlet character due to spin–orbit coupling. $^3\psi_1$ may be represented as a linear combination of $^3\psi_1^0$ and $^1\psi_1^0$

$$^3\psi_1 = C_1\,^3\psi_1^0 + C_2\,^1\psi_1^0 \tag{55}$$

where $^3\psi_1^0$ and $^1\psi_1^0$ are the wave functions for the first excited singlet and triplet states respectively in the absence of spin–orbit coupling. In organic molecules the spin–orbit matrix element is generally small so $C_1 \approx 1$ and C_2 is given from perturbation theory as

$$C_2 = \frac{\langle {}^1\psi_1^0|\mathcal{H}_{SO}|{}^3\psi_1^0\rangle}{|{}^1E_1 - {}^3E_1|} = \frac{\delta}{|{}^1E_1 - {}^3E_1|} \tag{56}$$

where $^1E_1$ is the energy of $^1\psi_1$ and $^3E_1$ is the energy of $^3\psi_1$. The phosphorescence transition probability, Eq. (54), is simply

$$\frac{1}{\tau_P} = |\langle C_1\,^3\psi_1^0 + C_2\,^1\psi_1^0|e\mathbf{r}|^1\psi_0\rangle|^2 \tag{57}$$

$$= C_2^2 |\langle {}^1\psi_1^0|e\mathbf{r}|^1\psi_0\rangle|^2$$

while the fluorescence transition probability is given by

$$P_F \approx |\langle {}^1\psi_1|e\mathbf{r}|^1\psi_0\rangle|^2 = \frac{1}{\tau_F} \tag{58}$$

Substituting Eq. (58) into Eq. (57), we have

$$C_2^2 = \frac{\tau_F}{\tau_P} = \frac{\delta^2}{|{}^1E_1 - {}^3E_1|^2} \tag{59}$$

Within the limits of the model, the spin–orbit matrix element is given by

$$\delta = \left(\frac{\tau_F}{\tau_P}\right)^{1/2}({}^1E_1 - {}^3E_1) \tag{60}$$

Also from perturbation theory, the shift in energy of the triplet zero-field level coupled to $^1\psi_1$ may be written

$$\Delta = \frac{\delta^2}{|{}^1E_1 - {}^3E_1|} = \left(\frac{\tau_F}{\tau_P}\right)({}^1E_1 - {}^3E_1) \tag{61}$$

As an example, for benzene,[58] $\tau_P = 30$ s, $\tau_F = 3 \times 10^{-8}$ s, and assuming $|^1E_1 - {}^3E_1| \leq 6\,000$ cm^{-1}, we have

$$\Delta = \frac{3 \times 10^{-8} \text{ s}}{30 \text{ s}} \quad (6\,000 \text{ cm}^{-1})$$

$$= 6 \times 10^{-5} \text{ cm}^{-1}$$

Compared to the measured zero-field splittings of benzene[59] of 0.1644 cm^{-1}, 0.1516 cm^{-1}, and 0.0128 cm^{-1}, the spin–orbit coupling contribution to the zero-field splitting is clearly negligible. An example of the magnitude of the effect is given by *para*dichlorobenzene[60] for which $\tau_P = 16$ ms, $\tau_F = 3 \times 10^{-8}$ s, and $|^1E_1 - {}^3E_1| \leq 7\,800$ cm^{-1}. Substituting these values into Eq. (61), we find that $\Delta = 1.5 \times 10^{-2}$ cm^{-1}. This is still small compared to the observed zero-field splittings[29] of 0.1787 cm^{-1}, 0.1201 cm^{-1}, and 0.0584 cm^{-1}. In addition, since we used the measured lifetime of the phosphorescence which includes both the radiative and non-radiative transition probabilities, the actual contribution of spin–orbit coupling to the zero-field splitting is certainly smaller. For organic molecules in their excited triplet state, the splitting of the zero-field levels due to spin–orbit coupling usually accounts for only a small percentage of the observed zero-field splitting and therefore, we will consider only the magnetic dipole–dipole interaction in explaining the observed spectra. The addition of a heavy atom however will increase the spin–orbit coupling matrix element.

The Hamiltonian for magnetic dipole–dipole interaction between two unpaired electrons may be written[61] as

$$\mathcal{H}_{ss} = g_e^2 \beta_e^2 \left\{ \frac{\mathbf{S}_1 \cdot \mathbf{S}_2}{r^3} - \frac{3(\mathbf{S}_1 \cdot \mathbf{r})(\mathbf{S}_2 \cdot \mathbf{r})}{r^5} \right\} \tag{62}$$

where g_e is the electron g-factor, which has been found to be basically isotropic for aromatic triplet states and equal to the free electron value of 2.002 32, β_e is the Bohr magneton $(e\hbar/2mc)$, and $\mathbf{r}$ is the vector connecting the two electron spins $\mathbf{S}_1$ and $\mathbf{S}_2$. The Hamiltonian is of the same form as any dipole–dipole interaction, and in the case of the interaction between the two triplet-state electrons is expressed as

$$\mathcal{H}_{ss} = \mathbf{S} \cdot \mathbf{D} \cdot \mathbf{S} \tag{63}$$

which may be written in a Cartesian axis system as

$$\mathcal{H}_{ss} = D_{xx}S_x^2 + D_{xy}S_xS_y + D_{xz}S_xS_z + D_{yx}S_yS_x + D_{yy}S_y^2 + D_{yz}S_yS_z + D_{zx}S_zS_x \\ + D_{zy}S_zS_y + D_{zz}S_z^2 \tag{64}$$

The values of the D_{ij} $(i, j = x, y, z)$ are given by averages over the triplet-state electronic wave function[62]

$$D_{xx} = \tfrac{1}{2} g_e^2 \beta_e^2 \left\langle \frac{r^2 - 3x^2}{r^5} \right\rangle$$

$$D_{xy} = \tfrac{1}{2} g_e^2 \beta_e^2 \left\langle \frac{-3xy}{r^5} \right\rangle$$

(65)

and so on. D is a symmetrical tensor $(D_{xy} = D_{yx}$, etc.); therefore, in the principal axis system which diagonalizes the zero-field tensor, the Hamiltonian becomes

$$\mathcal{H}_{ss} = -XS_x^2 - YS_y^2 - ZS_z^2$$

(66)

where

$$X = -D_{xx}, \qquad Y = -D_{yy}, \quad \text{and} \quad Z = -D_{zz}$$

Since the Hamiltonian is traceless, $X + Y + Z = 0$, only two independent parameters are needed to describe the interaction. In conventional ESR, the Hamiltonian in the principal axis system is usually rewritten by defining

$$D = \tfrac{3}{2}(X + Y) - Z \quad \text{and} \quad E = -\tfrac{1}{2}(X - Y)$$

(67)

with the axis convention that $|X| \leq |Y| \leq |Z|$. Therefore, the three components of the Hamiltonian are given by

$$X = D/3 - E$$
$$Y = D/3 + E$$
$$Z = -2/3D$$

(68)

Thus, for the triplet state, the zero-field spin–spin interaction can be written in diagonal form as

$$\mathcal{H}_{ss} = D(S_z^2 - 2/3) + E(S_x^2 - S_y^2)$$

(69)

where the triplet electron representations X, Y, and Z are related to the S_z eigenstates by

$$|X\rangle = 1/\sqrt{2}(|-1\rangle - |1\rangle)$$

$$|Y\rangle = i/\sqrt{2}(|-1\rangle + |1\rangle)$$

$$|Z\rangle = |0\rangle$$

(70)

This form of the Hamiltonian is directly related to the chosen axis system of the molecule and presents a clear picture of the orientational dependence of the energy.

The usual selection rule in ESR of $\Delta S_z = \pm 1$ is not valid in zero magnetic field since the triplet sublevels are not eigenfunctions of S_z. The probabilities of magnetic dipole transitions between the triplet spin sublevels are given by

$$P_{x \to y} = |\langle X|S_z|Y\rangle|^2 = 1$$
$$P_{x \to z} = |\langle X|S_y|Z\rangle|^2 = 1 \qquad (71)$$
$$P_{y \to z} = |\langle Y|S_x|Z\rangle|^2 = 1$$

At this time we should point out an obvious feature about the form of the electron-spin dipolar Hamiltonain. *It is identical in form to the nuclear quadrupole Hamiltonian.* In fact, the zero-field electron-spin Hamiltonian for triplet spins is identical to the ^{14}N nuclear quadrupole Hamiltonian save for the replacement of $\mathbf{I}^2$ operators for $\mathbf{S}^2$ operators. This means that nuclear quadrupole interactions in excited states will appear in zeroth order as satellite transitions split off the zero field electron-spin transition by the appropriate quadrupole interaction energy. As we shall see, however, these satellite transitions are shifted slightly by first-order nuclear electron hyperfine interactions.

B. $\mathscr{H}_Q$—The Nuclear Quadrupole Hamiltonian

As is well known, nuclei with spin quantum numbers $I \geq 1$ have non-spherical charge distribution and therefore an electric quadrupole moment. The quadrupole moment of the nucleus is positive or negative depending on whether the charge distribution is elongated or flattened along the spin axis and each allowed nuclear orientation along the spin axis has associated with it a potential energy due to the surrounding electric field. In the case of a molecule, the electric field is due to non-s electrons which produce a field gradient ($V_{i,j}$) at the nucleus defined by

$$V_{i,j} = \frac{\partial^2 V}{\partial_i \, \partial_j}(i, j = x, y, z) \qquad (72)$$

where V is the electrostatic potential at the nucleus.

In an arbitrary axis system the Hamiltonian[63] is written as

$$\mathscr{H}_Q = B\{V_{zz}(3I_z^2 - I^2) + (V_{zx} + iV_{zy})(I_-I_z + I_zI_-)$$
$$+ (V_{zx} - iV_{zy})(I_+I_z + I_zI_+) + [1/2(V_{xx} - V_{yy}) \qquad (73)$$
$$+ iV_{xy}]I_+^2 + [1/2(V_{xx} - V_{yy}) - iV_{xy}]I_-^2\}$$

where $B = eQ/4I(2I-1)$, e is the electron charge (esu), Q is the quadrupole moment (cm^2), and I is the nuclear spin quantum number. The Hamiltonian being a symmetric tensor, like the electron-spin dipolar Hamiltonian, can be transformed to an axis system such that $V_{i,j} = 0$ for $i \neq j$, where the Hamiltonian is rewritten as

$$\mathcal{H}_Q = B\{V_{zz}(3I_z^2 - I^2) + [1/2(V_{xx} - V_{yy})(I_+^2 + I_-^2)]\} \tag{74}$$

Since the Laplace equation is satisfied

$$V_{xx} + V_{yy} + V_{zz} = 0 \tag{75}$$

only two independent parameters are used to describe the interaction. With the usual definitions

$$eq = V_{zz} \tag{76}$$

$$\eta = \frac{V_{xx} - V_{yy}}{V_{zz}}$$

and

$$|V_{xx}| \le |V_{yy}| \le |V_{zz}| \tag{77}$$

the standard form of the Hamiltonian, Eq. (74), is rewritten as

$$\mathcal{H}_Q = A[(3I_z^2 - I^2) + \eta/2(I_+^2 + I_-^2)] \tag{78}$$

where

$$A = \frac{e^2qQ}{4I(2I-1)}$$

This may also be written in the completely equivalent form

$$\mathcal{H}_Q = A[(3I_z^2 - I^2) + \eta(I_x^2 - I_y^2)] \tag{79}$$

One should note the similarity of Eq. (79) to Eq. (69). The Hamiltonian matrix consists of diagonal terms and off-diagonal terms connecting states differing in I_z by ± 2. The electric potential due to the relative orientation of $\mathcal{H}_Q$ to $\mathcal{H}_{SS}$ can effect the 'apparent' magnitude of $\mathcal{H}_Q$ since $\mathcal{H}_Q$ manifests itself as a perturbation on $\mathcal{H}_{SS}$.

Because we will explicitly deal with ^{35}Cl and ^{14}N quadrupole interactions in excited states, we *review* the explicit form of the Hamiltonian for $I = 1$ and $I = 3/2$. The Hamiltonian for an $I = 1$ nucleus (^{14}N) is expressed in a more convenient form by transforming Eq. (79) to the representation in which the energy is diagonal. In this representation, the Hamiltonian is in the same form as the electron spin–spin Hamiltonian, and is particularly convenient since it may be written in terms of the nuclear angular momentum operators as

$$\mathcal{H}_Q = -xI_x^2 - yI_y^2 - zI_z^2 \tag{80}$$

For a spin of $I = 3/2$ it is easier to use matrix notation, i.e.

$$\mathcal{H}_Q = \frac{e^2qQ}{4} \times$$

	$|3/2\rangle$	$|1/2\rangle$	$|-1/2\rangle$	$|-3/2\rangle$
	1	0	$\eta/\sqrt{3}$	0
	0	-1	0	$\eta/\sqrt{3}$
	$\eta/\sqrt{3}$	0	-1	0
	0	$\eta/\sqrt{3}$	0	1

(81)

The matrix may be rewritten as two separate 2×2 matrices by rearranging the order of the basis states as

$$\mathcal{H}_Q = \frac{e^2qQ}{4} \times$$

	$|3/2\rangle$	$|-1/2\rangle$	$|1/2\rangle$	$|-3/2\rangle$
	1	$\eta/\sqrt{3}$	0	0
	$\eta/\sqrt{3}$	-1	0	0
	0	0	-1	$\eta/\sqrt{3}$
	0	0	$\eta/\sqrt{3}$	1

(82)

The eigenvalues of the Hamiltonian are simply obtained by diagonalizing each of the 2×2 matrices.

$$E_{\pm 3/2} = \frac{e^2qQ}{4}(1 + \eta/3)^{1/2}$$

$$E_{\pm 1/2} = \frac{-e^2qQ}{4}(1 + \eta/3)^{1/2}$$

(83)

The eigenstates are

$$|3/2\rangle' = a|3/2\rangle + b|-1/2\rangle$$
$$|-1/2\rangle' = a|-1/2\rangle - b|3/2\rangle$$
$$|1/2\rangle' = a|1/2\rangle - b|-3/2\rangle$$
$$|-3/2\rangle' = a|-3/2\rangle + b|1/2\rangle$$

(84)

where

$$a = \frac{1+\sqrt{1+x^2}}{[2(1+x^2+\sqrt{1+x^2})]^{1/2}} \tag{85}$$

$$b = x/[2(1+x^2+\sqrt{1+x^2})]^{1/2}$$

and

$$x = \eta/3$$

In contrast to a nucleus with spin $I = 1$, e^2qQ and η cannot be separately determined. It should be noted, however, that the transition frequency is not particularly sensitive to η. The assumption that $\eta = 0$ and the transition energy is equal to $(1/2)e^2qQ$ will produce only a small error for small values of η.

C. $\mathcal{H}_{HF}$—The Nuclear–Electron Hyperfine Interaction

A nucleus with a spin $\geq 1/2$, like an electron, has a magnetic moment and the interaction of this nuclear magnetic moment with the electron magnetic moment leads to both an anisotropic dipole–dipole interaction and a Fermi contact interaction due to a finite electron-spin density at the nucleus. The component of the hyperfine interaction, due to the interaction of the nuclear and electron magnetic moments, is entirely analogous to the zero-field Hamiltonian with the replacement of one of the electron spins with a nuclear spin and the appropriate change of constants. The Hamiltonian[62] may be written as

$$\mathcal{H}_{HF}^{DD} = -g_e\beta_e g_n\beta_n \left[\frac{\mathbf{I}\cdot\mathbf{S}}{r^3} - \frac{3(\mathbf{I}\cdot\mathbf{r})(\mathbf{S}\cdot\mathbf{r})}{r^5} \right] \tag{86}$$

and g_n is the nuclear g-factor and β_n is the nuclear magneton. Since this is identical in form to Eq. (62) for the zero-field Hamiltonian, Eq. (86) is expressed as

$$\mathcal{H}_{HF}^{DD} = \mathbf{S}\cdot\mathbf{A}\cdot\mathbf{I} \tag{87}$$

which can be expanded in the same manner as Eq. (64). The $\mathbf{A}$ matrix is symmetric and therefore, in its principal-axis system, it is written as

$$\mathcal{H}_{HF}^{DD} = A_{xx}S_xI_x + A_{yy}S_yI_y + A_{zz}S_zI_z \tag{88}$$

where the hyperfine elements are given by the average over the spatial distribution of the unpaired spins

$$A_{xx} = -g_e g_n\beta_e\beta_n \left\langle \frac{r^2-3x^2}{r^5} \right\rangle \tag{89}$$

where $\chi = x, y, z$.

The Laplace equation is again satisfied and therefore

$$A_{xx} + A_{yy} + A_{zz} = 0 \tag{90}$$

The unpaired spin density at the nucleus produces an additional contribution to the hyperfine Hamiltonian, the Fermi contact term. This will arise only from spin density in s-orbitals since the other orbitals have a vanishing probability of being at the nucleus. The Fermi contact contribution is usually considered to be isotropic and is written as

$$\mathcal{H}^F_{HF} = C(S_x I_x + S_y I_y + S_z I_z) \tag{91}$$

where

$$C = (8\pi/3) g_e \beta_e g_n \beta_n |\psi_s(0)|^2 \tag{92}$$

$|\psi_s(0)|^2$ is the s-electron-spin density at the nucleus and g_e and g_n are the electron and nuclear g factors, respectively.

The total hyperfine Hamiltonian can be written as

$$\mathcal{H}_{HF} = A'_{xx} S_x I_x + A'_{yy} S_y I_y + A'_{zz} S_z I_z \tag{93}$$

where

$$A'_{xx} = A_{xx} + C, \quad \text{etc.} \tag{94}$$

If the three components of the total hyperfine Hamiltonian are measured, the contribution due to the anisotropic and isotropic components can be separated; however, the absolute signs will not generally be obtained. It should be pointed out that since the nuclei in which we are interested also have quadrupole moments, the Fermi contact term will not be strictly isotropic since the nuclei are distorted, and consequently, the dipole–dipole and contact terms are not completely separable.

D. The Total Hamiltonian, Energy Levels and Transition Probabilities

The total Hamiltonian for two molecules which are examples of the triplet-state electrons interacting with an $I = 1$ and an $I = 3/2$ nuclear spin are considered. In order to simplify the discussion we make the following assumptions for both cases:

(1) The principal axis system of $\mathcal{H}_{SS}$, $\mathcal{H}_Q$ and $\mathcal{H}_{HF}$ are coincident;
(2) Only the out-of-plane component of the hyperfine Hamiltonian need be considered;
(3) The hyperfine interaction due to protons may be neglected.

Assumptions 1 and 2 can be, in many cases, justified on the basis of the single crystal ESR spectra[64] and assumption 3 by the fact that resolved proton hyperfine splitting has not been observed in zero-field ESR.

An example of a molecule which is characterized by the interaction of one $(I = 1)$ nuclear spin with the triplet electrons is the $\pi\pi^*$ state of quinoline (1-azanaphthalene). The spin Hamiltonian for this molecule may be written as

$$\mathcal{H} = \mathcal{H}_{SS} + \mathcal{H}_Q + \mathcal{H}_{HF} \tag{95}$$

where

$$\mathcal{H}_{SS} = -XS_x^2 - YS_y^2 - ZS_z^2$$
$$\mathcal{H}_Q = -xI_x^2 - yI_y^2 - zI_z^2 \tag{96}$$

and

$$\mathcal{H}_{HF} = A_{xx}S_xI_x$$

where x is the out-of-plane axis.

For illustration,[26] we will use for the basis states the product functions $|\mu\nu\rangle = \tau_\mu\chi_\nu$, which form a set of eigenfunctions that diagonalize $\mathcal{H}_{SS}$ and $\mathcal{H}_Q$. τ_μ and χ_ν are the electron and nuclear spin functions while μ and ν correspond to x, y and z. The complete Hamiltonian is, of course, a 9×9 matrix. Since we are only considering the A_{xx} element of the hyperfine interaction,[28] a satisfactory solution is obtained by perturbation theory. As is shown in Fig. 4, the energies of the states $|Zz\rangle$ and $|Zy\rangle$ are shifted by an amount β, where

$$\beta = \frac{A_{xx}^2}{E_y - E_z} \tag{97}$$

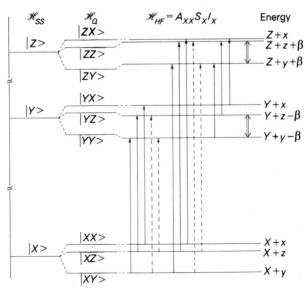

Fig. 4. Energy level diagram for the triplet and one $I = 1$ nuclear spin considering only the A_{xx} hyperfine component.

while the states $|YZ\rangle$ and $|Yy\rangle$ are shifted by an amount $-\beta$. In our axis system the triplet-state energy levels would be ordered $Z > Y > X$ and the nuclear quadrupole energy levels ordered $x > z > y$. The eigenvectors of the states which are coupled by A_{xx} are

$$|Zz\rangle' = (1-\beta)|Zz\rangle - \beta|Yy\rangle$$
$$|Zy\rangle' = (1-\beta)|Zy\rangle - \beta|Yz\rangle$$
$$|Yz\rangle' = (1-\beta)|Yz\rangle + \beta|Zy\rangle$$
$$|Yy\rangle' = (1-\beta)|Yy\rangle + \beta|Zz\rangle$$

(98)

The probability for microwave transitions between the triplet-state magnetic sublevels is given by

$$I \approx |\langle \mu_1 \nu_1 | \mathcal{H}_{RF}(t) | \mu_2 \nu_2 \rangle|^2$$

(99)

where $\mathcal{H}_{RF}(t)$ is the magnetic dipole transition operator defined by

$$\mathcal{H}_{RF}(t) = H_1(t)\hbar(\gamma_n \cdot \mathbf{I} + \gamma_e \cdot \mathbf{S})$$

(100)

γ_n and γ_e are the nuclear and electron gyromagnetic ratios and $H_1(t)$ is the magnitude of the time-dependent magnetic field. The electron-spin magnetic dipole transition operator will connect states with $\mu_1 \neq \mu_2$ and $\nu_1 = \nu_2$, while the nuclear-spin operator will connect states with $\mu_1 = \mu_2$ and $\nu_1 \neq \nu_2$. However, *the mixing of the basis function by A_{xx} allows the observation of 'forbidden' simultaneous electron and nuclear transitions.* This is clearly shown by considering the transition from $|Xz\rangle'$ to $|Yy\rangle'$. The intensity of the transition is given by

$$I \approx |\langle Xz | \gamma_e H_1(t) | [(1-\beta)|Yy\rangle + \beta|Zz\rangle]|^2$$

(101)

$$\approx \beta^2 \gamma_e^2 H_1(t)^2$$

(102)

It should be noted that it is *necessary to have a hyperfine interaction in order to observe the nuclear quadrupole satellites* since the hyperfine term is the only method of coupling the electron- and nuclear-spin Hamiltonians.

In Fig. 5, the spectra expected for the three zero-field transitions are shown in terms of the components of the total Hamiltonian. It is clear that the separation of the quadrupole satellites for both the $\tau_x \to \tau_z$ and $\tau_x \to \tau_y$ transitions is $2(z - y)$ and therefore only one of the three possible nuclear quadrupole transitions equal to $(3/4)e^2qQ(1 - \eta/3)$ is observed. The value of the hyperfine coupling constant A_{xx} is easily obtained from the separation of the two allowed components of each of the three transitions. If we had chosen to use A_{yy} or A_{zz} as the only hyperfine interaction instead of A_{xx}, the spectra would be the same as that shown in Fig. 5 if a cyclic perturbation is applied to our labelling.

Although in this simple example all the parameters in the Hamiltonian can be determined from the three zero-field transitions, in practice this is usually

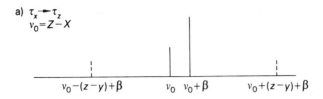

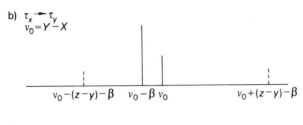

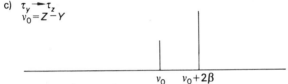

Fig. 5.   ODMR spectra predicted for the energy level diagram shown in Fig. 4.

not the case because of such problems as poor resolution of the spectra or the failure to include enough terms in the Hamiltonian to adequately describe the interactions. Therefore, it is usually advantageous to also perform an electron–nuclear double resonance (ENDOR) experiment to improve the resolution and confirm the assignment of the spectra. The ENDOR transitions are shown in Fig. 4 by the double arrows. Let us consider the intensity of the ENDOR transition. As an example we will treat the transition from $|Yy\rangle'$ to $|Yz\rangle'$:

$$I \approx \|[(1-\beta)\langle Yy| + \beta\langle Zz|]|\mathcal{H}_{RF}(t)|[(1-\beta)|Yz\rangle + \beta|Zy\rangle]\|^2 \tag{103}$$

$$I \approx [(1-\beta)^2\gamma_n H_1 + 2\beta(1-\beta)\gamma_e H_1 + \beta^2\gamma_n H_1]^2 \tag{104}$$

Since H_1 is a constant, we will drop it and may now write

$$I \approx 4\gamma_e^2[\beta^2(1-\beta)] + 4\gamma_e\gamma_n[\beta(1-\beta)^3 + \beta^3(1-\beta)] + \gamma_n^2[(1-\beta)^4 + \beta^4 + 2\beta^2(1-\beta)^2] \tag{105}$$

Since β is usually of the order of 1×10^{-2} for $\pi\pi^*$ triplets, we can reasonably approximate Eq. (105) by

$$I \approx 4\beta^2\gamma_e^2 + 4\beta\gamma_e\gamma_n + \gamma_n^2 \tag{106}$$

43

In contrast, if there were no hyperfine coupling as in the τ_x manifold in our example, the intensity would be given by

$$I \approx \gamma_n^2 \qquad (107)$$

The ratio of the intensity of the ENDOR transitions due to the electron magnetic dipole operator to those due to the nuclear magnetic dipole operator is approximately $4\beta^2\gamma_e^2/\gamma_n^2$ and therefore, unless γ_n^2 is greater than $4\beta^2\gamma_e^2$, the electron dipole moment transition operator will be the major source of the intensity in ENDOR transitions.

As an example, for ^{14}N the ratio of $\gamma_e/\gamma_n = 8.6 \times 10^6$ and therefore, β must be less than 1.57×10^{-3} for the nuclear magnetic dipole transition operator to be comparable to the electron magnetic dipole transition operator in producing intensity in the ENDOR transitions. For a typical separation of $\tau_z - \tau_y$ of 1 000 MHz this would correspond to an extremely small hyperfine element, A_{xx} of only 1.5 MHz, which is much smaller than any out-of-plane hyperfine elements reported for aza aromatics.

As an example of a molecule with one $I = 3/2$ nuclear spin, we will consider the excited $^3\pi\pi^*$ state of chlorobenzene. The spectrum produced in this case is somewhat more complicated to calculate because of the lack of a convenient basis set for both the electron- and nuclear-spin functions. The simplest method with only one hyperfine component is to use the basis set $|\mu\nu\rangle = \tau_\mu\chi_\nu$ where μ corresponds to X, Y and Z and ν to 3/2, 1/2, $-1/2$ and $-3/2$. We will further assume that $\eta = 0$ and therefore both $\mathcal{H}_{ss}$ and $\mathcal{H}_Q$ are again diagonal. In this example, the out-of-plane component of the hyperfine tensor (A_{xx}) couples the basis states in the τ_z manifold with those in the τ_y manifold for which the nuclear spins differ in their I_z quantum number by ± 1. This is easily seen by expanding the hyperfine Hamiltonian as

$$A_{xx}S_xI_x = 1/2[A_{xx}S_x(I_+ + I_-)] \qquad (108)$$

The states in the Hamiltonian that are coupled by A_{xx} may be represented graphically as

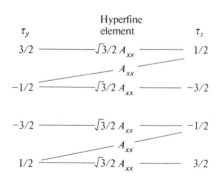

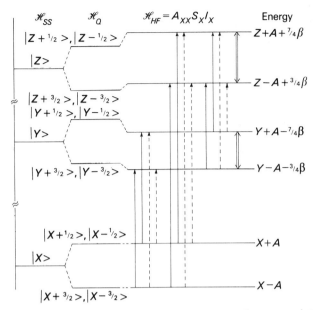

Fig. 6. Energy level diagram for the triplet and one $I = 3/2$ nuclear spin considering only the A_{xx} hyperfine component.

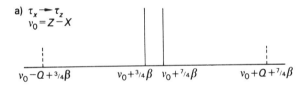

Fig. 7. ODMR spectra predicted for the energy level diagram shown in Fig. 6.

Since the degenerate nuclear levels are not coupled by the same hyperfine element, we may still use non-degenerate perturbation theory to calculate the energy levels and transition moments.

This spin system has a total spin that is a half integer (5/2) ($S = 1$ and $I = 3/2$), it is a Kramers doublet, and therefore all the energy levels are twofold degenerate. The hyperfine coupling will never remove the degeneracy of the $\pm$ nuclear levels in zero-field and consequently we have only six levels to consider.

The energy level diagram resulting from a perturbation treatment of the hyperfine interaction is given in Fig. 6, and the predicted spectra in Fig. 7. The use of the A_{yy} component of the hyperfine tensor instead of the A_{xx} component produces an identical energy level diagram and spectra with the appropriate relabelling. The use of the A_{zz} component of the hyperfine tensor mixes the nuclear sublevels in the τ_x manifold with those in the τ_y manifold having the same I_z quantum number;

τ_x	Hyperfine element	τ_y
3/2 ————	3/2 A_{zz} ————	3/2
1/2 ————	1/2 A_{zz} ————	1/2
−1/2 ————	−1/2 A_{zz} ————	−1/2
−3/2 ————	−3/2 A_{zz} ————	−3/2

therefore, no nuclear quadrupole satellites due to the electron magnetic moment transition operator are observed. The resulting energy level diagram, considering only the A_{zz} component of the hyperfine tensor, is given in Fig. 8 and the predicted spectra in Fig. 9.

The ENDOR transitions permitted by the electron dipole moment transition operator, considering only the A_{xx} hyperfine element, are shown by the double arrows in Fig. 6. The analysis of the ENDOR spectra follows the same method as that for a spin one nucleus, with the same expression for the intensity of the transitions induced by the electron magnetic dipole moment transition operator and the nuclear magnetic dipole moment transition operator. When only the A_{zz} hyperfine element is present, the electron magnetic dipole transition operator is ineffective in producing ENDOR transitions and consequently the intensity of any observed ENDOR signal is due solely to the nuclear magnetic dipole transition operator.

Some generalizations can be made at this point concerning the appearance of 'forbidden' satellites whose separation in the zeroth order is the pure nuclear quadrupole transition frequency of the molecules in an excited triplet

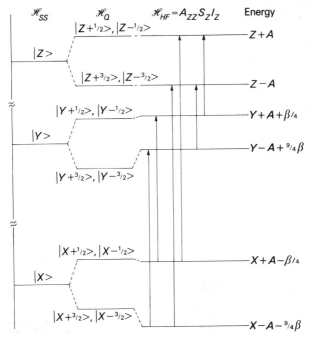

Fig. 8. Energy level diagram for the triplet and one $I = 3/2$ nuclear spin considering only the A_{zz} hyperfine component.

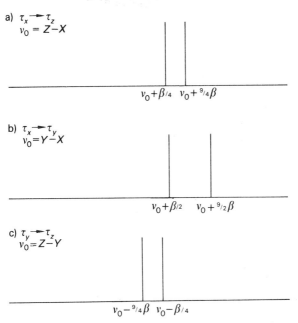

Fig. 9. ODMR spectra predicted for the energy level diagram shown in Fig. 8.

state. (a) For a nuclear spin $I = 1$ (e.g. ^{14}N), a hyperfine element, A_{ii}, associated with a direction i gives intensity into a simultaneous electron–nuclear flip in the plane normal to i. Thus at least two nuclear hyperfine elements must be finite to obtain independently both e^2qQ/h and η. (b) For a nuclear spin $I = 3/2$ (e.g. ^{35}Cl), a nuclear hyperfine element parallel to the principal axis of the field gradient (i.e. A_{zz}) does not introduce mixing between electron–nuclear states that admit intensity into forbidden satellites. (c) For a nuclear spin $I = 3/2$, a nuclear hyperfine element perpendicular to the principal axis of the field gradient introduces intensity into forbidden satellites whose separation in zeroth order is the pure nuclear quadrupole transition frequency; however, e^2qQ/h and η can never be obtained independently in the absence of an external magnetic field. Although we will not discuss the observation in any detail, we should point out that in many cases it is possible to obtain the sign of the nuclear quadrupole moment from an analysis of the zero-field spectra. This feature is discussed fully in reference 78.

Although we have not treated explicitly the case where two nuclei are present on the same molecule, both having nuclear spin $I \geq 1$, the generalizations (a)–(c) hold with one additional feature being manifested, that is the possibility of simultaneous multiple nuclear–electron spin flips. As we will see in the following sections, in 8-chloroquinoline, simultaneous chlorine–nitrogen electron-spin transitions are observed and are easily identified. In addition simultaneous multiple nuclear ENDOR transitions are expected and, indeed, observed.

IV. EXPERIMENTAL METHODS

A. Optically Detected Magnetic Resonance

The basic experimental arrangement is shown in Fig. 10. The sample is mounted inside a helical slow wave structure[65] which is attached to a rigid stainless steel coaxial line suspended in a liquid helium dewar. The exciting light is supplied by a 100-watt mercury short arc lamp. The spectral region of interest is selected by either an interference filter centred at 3 100 Å, or a combination of Corning glass and solution filters.[66] The phosphorescence is collected at a 90° angle to the exciting light and is focused through an appropriate Corning filter (to remove scattered light) onto the entrance slit of a Jarrel–Ash 3/4 m spectrometer. The light at the exit slit is detected with an EMI 6256S photomultiplier cooled to -20 °C whose output is connected to an electrometer through an adjustable load resistor. The output of the electrometer is either monitored directly if c.w. microwave power is used or, if the microwave field is amplitude modulated, connected to the signal channel input of a PAR model HR-8 lock-in amplifier.

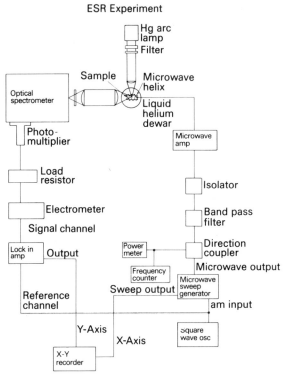

Fig. 10. Experimental arrangement used in performing ODMR experiments in zero-field with amplitude modulation of the microwave field.

The microwave field is generated by a Hewlett–Packard microwave sweep oscillator Model 8690B, amplified with a travelling wave tube and fed consecutively through a directional coupler, band-pass filter, and an isolator to the rigid coaxial line onto which the helix is mounted. The microwave sweep oscillator may be amplitude modulated with a square wave generator which is also connected to the reference channel of the lock-in amplifier. The output of the lock-in amplifier drives the y axis of an x–y-recorder while the ramp voltage from the microwave sweep oscillator drives the x axis.

The temperature of the sample is usually lowered to approximately 1.3 K by pumping on the liquid helium with three Kinney model KTC-21 vacuum pumps operated in parallel.

The experiment is performed by monitoring the change in emission of the sample while varying the frequency of the modulated microwave field. As explained in Section II, the signal strength may either increase or decrease. With a lock-in amplifier, a decrease in emission intensity corresponds to a phase shift of 180 degrees relative to the signal obtained for an increase in emission intensity.

B. Optically Detected ENDOR

The experimental arrangement usually employed is shown in Fig. 11. The optical and microwave equipment is the same as that used in the ODMR experiments with the exception that the microwave field (H_1) is not modulated.

ENDOR experiment

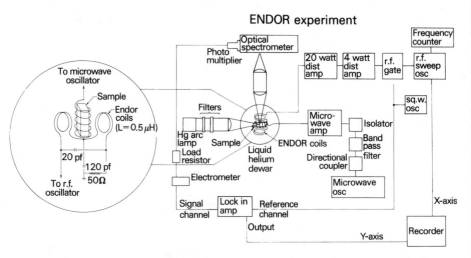

Fig. 11. Experimental arrangement for optically detected ENDOR in zero magnetic field. An enlarged view of the sample and ENDOR coil schematic is shown on the left.

The radiofrequency field (H_2) is supplied by a sweep oscillator that covers the region from 0.1 to 110 MHz. The output is modulated by a linear gate that is driven by a square-wave generator which also drives the reference channel of the lock-in amplifier. The r.f. is then amplified by two broad-band distributed amplifiers, a 4-watt unit and a 20-watt unit, and connected to the ENDOR coils. These amplifiers have the advantage that they operate over the range of 1 to 50 MHz without the need of adjustment. The ENDOR coil consists of a 'bridge T' constant resistance network in a Helmholtz arrangement. This configuration maintains an even r.f. level over a broad-band of frequencies since it looks entirely resistive. The x axis of the recorder is driven by the ramp voltage from the r.f. sweep oscillator and the y axis from the output of the lock-in amplifier.

C. Variations of the Basic Experiments

The optical detection of magnetic resonance permits several additional parameters to be experimentally adjusted. These include the energy and band-width of the phosphorescence that is monitored, as well as the energy, band-width, and intensity of the exciting light. In addition, the power of the microwave field H_1 may be adjusted over a wider range than in experiments in which the absorption of microwave power is monitored. This is due to the fact that saturation gives the maximum signal strength using optical detection techniques, while with absorption experiments the signal strength will decrease as the power is increased above that needed for saturation. The advantage of this is that the signal strength of weak 'forbidden' transitions may be improved by the application of large H_1 fields without a decrease in the signal strength of the allowed transitions.

Some of the most useful variations of the basic experiment are listed in Table 1. If a high-resolution spectrometer is employed to isolate the

Table 1. Techniques of optical detection of ESR

Excitation light	Optical spectrometer	Microwave modulation	Advantages
(1) c.w.	No	No	Measure absolute change in total emission
(2) c.w.	Yes	No	Measure absolute change in emission of particular vibronic bands
(3) Chopped	Optional	No	Improvement in S/N over Methods 1 and 2 by narrow-band phase-sensitive detection of the phosphorescence
(4) c.w.	Optional	a.m.	Detect only the change in emission from either the total emission or a particular vibronic band
(5) c.w.	Optional	f.m.	Detect the derivative of the spectrum, helpful in resolving spectra
(6) c.w.	Sweep	a.m.	Detect only the emission from 2 of the 3 sublevels while sweeping the optical spectrum
(7) Sweep	Yes	a.m.	Useful in studying the pathways of intersystem crossing

phosphorescence emission, the optically detected ESR may be used to simplify the phosphorescence spectrum by amplitude modulation of the H_1 field while saturating an ESR transition. The modulation of the phosphorescence is detected with a phase-sensitive amplifier while sweeping the optical spectrum. Since only two of the three triplet levels are coupled by the H_1 field, only the emission from these two levels will be detected. Therefore, by repeating the experiment while saturating the remaining two ESR transitions, three PMDR spectra are obtained, each including only the emission from two of the three zero-field levels. The information obtained from the analysis of phosphorescence spectra is extremely useful by itself in characterizing the triplet state, and complementary to the information obtained from the analysis of the ESR spectrum.

In ENDOR experiments the radiofrequency field H_2 may also be adjusted. These experiments are usually performed by saturating an ESR transition while varying the frequency of the H_2 field. Either the H_1 or H_2 field may be modulated; however, it is usually preferable to modulate the H_2 field since, in this case, only the change in intensity of the phosphorescence due to the ENDOR resonance is detected with a lock-in amplifier. On the other hand, if the H_1 field is modulated, there is a constant signal due to the ESR transition which changes in intensity when the H_2 field is swept through resonance. A useful modification of this technique is achieved by modulation of the H_2 field while simultaneously saturating an ENDOR transition and sweeping the H_1 field. In this case, only the ESR transitions that connect energy levels simultaneously coupled by the H_1 and the H_2 fields are detected. This method is useful in analysing the structure of the ESR transition since the contribution to the spectrum due to different isotopes and/or nuclei may be isolated.

If both an ESR and an ENDOR transition are saturated while modulating the H_2 field and scanning the phosphorescence spectrum, it is possible to isolate the contribution to the phosphorescence spectrum from molecules containing different nuclear isotopes. As an example, if the phosphorescence from a molecule such as chlorobenzene is monitored and a ^{35}Cl ENDOR transition saturated while modulating the H_2 field, only the contribution to the phosphorescence spectrum from molecules containing the ^{35}Cl isotope will be detected. The same experiment may then be repeated detecting only the contribution from the molecules containing the ^{37}Cl isotope.

V. THE ODMR SPECTRA OF 8-CHLOROQUINOLINE

The zero-field spectra of 8-chloroquinoline is characterized by the interaction of the triplet electrons with both a nitrogen ($I = 1$) and a chlorine ($I = 3/2$) nucleus. The addition of the chlorine atom to quinoline does not appreciably change the lifetime of the phosphorescence (see Table 2). Both quinoline and 8-chloroquinoline show emission primarily from only one of the triplet sublevels and have essentially the same zero-field, nitrogen quadrupole, and nitrogen hyperfine interactions. Although a great deal of information concerning the pathway of intramolecular energy transfer (i.e., intersystem crossing, radiative rate processes, etc.) can be obtained from an analysis of the microwave-induced phosphorescence intensity changes, we will restrict the results and discussion to the salient features of the ODMR spectra in zero-field.

Table 2. Spin Hamiltonian parameters and triplet lifetimes of the $^3\pi\pi^*$ states of 8-chloroquinoline and quinoline

	8-chloroquinoline in durene (1.3 K)	Quinoline in durene (1.35 K)[a]
Y (MHz)	1414.5	1528.5
Z (MHz)	555.5	528.0
X (MHz)	−1970.0	−2056.5
D^b (MHz)	2399.5	2556.75
E^b (MHz)	−429.5	−500.25
A_{xx}^N (MHz)	19.5	22.0[c]
A_{xx}^{Cl} (MHz)	15.0	—
$(e^2qQ/h)\,(^{14}N)^d$ (MHz)	4.27	~4.0[c]
$(e^2qQ/h)\,(^{35}Cl)$ (MHz)	−68.4	—
τ_x (s)	≥1	5.0
τ_y (s)	0.11	0.32
τ_z (s)	≥1	2.7

[a] data from reference 72.
[b] with the definitions $D = -\frac{3}{2}X$ and $E = -\frac{1}{2}(Y-Z)$.
[c] data from reference 64.
[d] with the assumptions $(e^2qQ/h)\,(^{14}N) = \frac{4}{3}(y-z)$.
[e] data from reference 26.

Two of the three electron-spin transitions, those associated with the $\tau_z \to \tau_y$ and $\tau_x \to \tau_y$ manifolds, were observed both with a continuous microwave field while monitoring the intensity of the phosphorescence and with 5 Hz amplitude modulation of the microwave field and phase-sensitive detection of the component of the phosphorescence at the modulation frequency. The $\tau_x \to \tau_z$ transition was only observed in an EEDOR experiment. This was performed by simultaneously saturating the $\tau_x \to \tau_y$ transition with a c.w. microwave field and amplitude modulation of a second microwave field which was swept through the $\tau_x \to \tau_z$ transition. This was necessary since emission originates almost entirely from the τ_y spin manifold. In all cases, the phosphorescence intensity increased when the microwave field coupled the respective electron-spin manifolds. The lifetime of the emission from the τ_y manifold was found to be 0.11 s while the lifetimes of both τ_x and τ_z levels are each more than one second. With the assumption that the radiative lifetimes of the triplet levels are ordered in the same way as the total lifetimes and the observation that the phosphorescence intensity increased while saturating both the $\tau_x \to \tau_y$ and $\tau_z \to \tau_y$ spin manifolds, from Eq. (47) the steady-state population of the τ_y level must be less than the population of either the τ_x or the τ_z level. The spectra obtained with amplitude modulation of the three ESR transitions are shown in Fig. 12. At low microwave powers only the 'allowed' component of each spectrum was observed. As the microwave power was increased, 'forbidden' satellites split off from the major transition were observed. The ^{35}Cl ENDOR resonance observed while saturating the $\tau_z \to \tau_y$

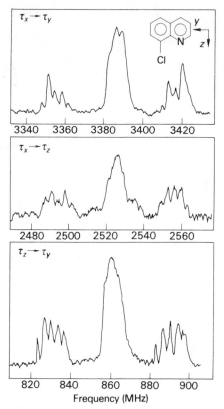

Fig. 12. The $\tau_x \to \tau_y$, $\tau_x \to \tau_z$ and $\tau_z \to \tau_y$ optically detected ESR transitions in 8-chloroquinoline using relatively high microwave power. The $\tau_x \to \tau_z$ transition was obtained by performing an EEDOR experiment.

transition is shown in Fig. 13. This transition was also observed with both a continuous and amplitude-modulated r.f. field.

The phosphorescence of 8-chloroquinoline in durene is due to the two distinct sites,[67] the more intense phosphorescence origin being at 4 795 Å and the weaker origin at 4 792 Å. In order to isolate the emission from the site at 4 795 Å, the ODMR spectra were obtained with the entrance slit of the spectrometer adjusted to 100 μm or less. The ODMR spectra observed may be considered as due to two distinct molecular isotopes, since approximately 75% of the 8-chloroquinoline molecules will have the ^{35}Cl isotope and 25% of the ^{37}Cl isotope. We will initially limit our consideration to only the 8-chloroquinoline molecules that have the ^{35}Cl isotope. The molecular axis system we will use is defined with x, the out-of-plane axis; y, the long in-plane axis; and z, the short in-plane axis. In order to simplify the analysis of the

54

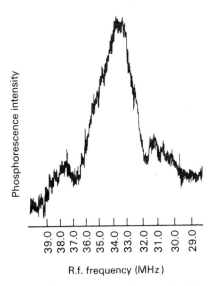

Fig. 13. Optically detected ^{35}Cl ENDOR observed while saturating the $\tau_z \rightarrow \tau_y$ multiplet.

spectra, we will make the following assumptions:

(1) The contribution of the proton hyperfine interactions will be neglected.
(2) The principal axis systems of the spin–spin, nuclear quadrupole, and hyperfine interactions are coincident.
(3) Only the out-of-plane hyperfine element for both nitrogen and chlorine will be considered.
(4) The chlorine asymmetry parameter is assumed to be zero.

The first assumption is justified on the basis of the small contribution to the line-width reported by Hutchison et al.[68] due to the proton hyperfine interaction in zero-field. This effect is smaller than the other terms in the Hamiltonian and would require an extensive computer analysis and excellent resolution of the transitions to justify its consideration.

The second assumption is quite severe, but is reasonable for our purposes since slight non-coincidence of the tensor elements will only produce a small perturbation of the observed spectra in zero-field. In addition, the x-axis is fixed by symmetry to be perpendicular to the plane and in quinoline it has been found that the z-axis of $\mathcal{H}_{ss}$ is within a few degrees of the molecular z-axis.[64,69] It is also reasonable to expect the principal nuclear quadrupole axis for both the nitrogen and chlorine atoms to be along the molecular z-axis.[63]

The third assumption is based on the measured value for the nitrogen hyperfine interaction for the excited triplet state of quinoline for which $A_{xx} \gg A_{yy}, A_{zz},$[64] and on the observation of chlorine hyperfine interactions in organic free radicals in which the principal chlorine hyperfine element has been found to be the out-of-plane element.[70,71] In addition, since in zero-field the hyperfine interaction is an off-diagonal term in the spin Hamiltonian, the magnitude of the effect of the interaction on the observed spectra is in first order inversely proportional to the energy separation of the triplet manifolds that are connected by the respective hyperfine element. In the case of 8-chloroquinoline, even if the hyperfine interaction was isotropic, the effect on the zero-field spectra would still be three times larger for the A_{xx} than the A_{yy} or A_{zz} components. Therefore, only the A_{xx} component of the hyperfine tensor will be included for both the chlorine and nitrogen atoms, since this will account for the major features of the spectra.

The fourth assumption is made on the basis that a finite value of the chlorine asymmetry parameter is a small perturbation that is not easily resolvable and not necessary to explain the main features of the spectra.

With these assumptions, the spin Hamiltonian may be written

$$\mathcal{H} = \mathcal{H}_{SS} + \mathcal{H}_Q^N + \mathcal{H}_{HF}^N + \mathcal{H}_Q^{Cl} + \mathcal{H}_{HF}^{Cl} \tag{109}$$

where

$$
\begin{aligned}
\mathcal{H}_{SS} &= -XS_x^2 - YS_y^2 - ZS_z^2 \\
\mathcal{H}_Q^N &= -xI_x^2 - yI_y^2 - zI_z^2 \\
\mathcal{H}_{HF}^N &= A_{xx}^N (S_x I_x) \\
\mathcal{H}_Q^{Cl} &= \frac{e^2 qQ}{12}\left[3I_z^2 - \frac{15}{4}\right] \\
\mathcal{H}_{HF}^{Cl} &= A_{xx}^{Cl}(S_x I_x)
\end{aligned}
\tag{110}
$$

In the same manner as discussed in Section III-D, the basis states of the spin Hamiltonian are chosen to be the product functions $|u, v, w\rangle = \tau_u, \chi_v, \chi_w$, which diagonalize $\mathcal{H}_{SS}$, $\mathcal{H}_Q^N$ and $\mathcal{H}_Q^{Cl}$. τ_u ($u = x, y, z$) is the electron-spin function, χ_v ($v = x, y, z$) is the nitrogen-spin function and χ_w ($w = \pm 1/2, \pm 3/2$) is the doubly degenerate chlorine-spin function. The total spin of the system is $7/2$ which is therefore a Kramers doublet; consequently there are only 18 energy levels for each of the molecular isotopes.

The similarity of the excited triplet state of 8-chloroquinoline and quinoline leads to the assignment of the order of the triplet energy levels of 8-chloroquinoline as being the same as those of quinoline.[72] With our axis system, the elements of $\mathcal{H}_{SS}$ are ordered $Y > Z > X$. The nitrogen nuclear quadrupole energy levels are also assumed to be in the same order as those reported for the ground state of pyrazine and pyridine[73,74] and thus for $\mathcal{H}_Q^N$ $x > y > z$. Since the chlorine nuclear quadrupole coupling constant $(e^2 qQ/h)$ is negative for all covalently-bonded Cl atoms, the energy of the chlorine spin functions are ordered $\chi_{\pm 1/2} > \chi_{\pm 3/2}$.

In order to treat the out-of-plane hyperfine perturbation due to both the nitrogen and chlorine spins, we will assume that the contribution from each

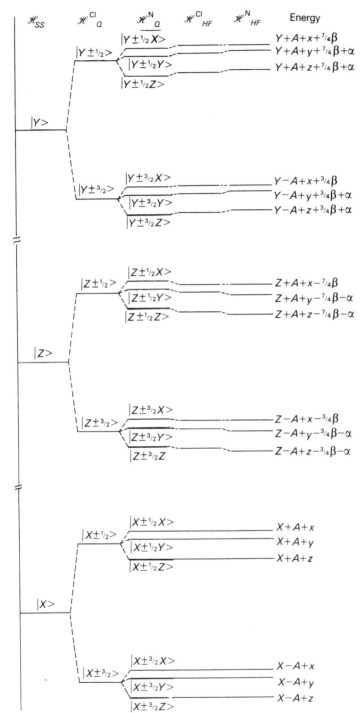

Fig. 14. Energy level diagram for 8-chloroquinoline.

may be considered separately. This is of course not strictly correct, but is satisfactory for the purpose of illustration, and in fact, for the values of A_{xx}^N and A_{xx}^{Cl} used in fitting the spectra, gives values for the energy levels very close to those obtained by diagonalizing the total spin Hamiltonian. An energy level scheme using the perturbation method discussed in Section III-D appropriate for 8-chloroquinoline is given in Fig. 14. There are essentially six types of ESR transitions observed:

(A) electron spin;
(B) electron and ^{14}N spins;
(C) electron and ^{35}Cl spins;
(D) electron and ^{37}Cl spins;
(E) electron, ^{14}N and ^{35}Cl spins;
(F) electron, ^{14}N and ^{37}Cl spins.

Since the chlorine nuclear quadrupole interaction is far larger than the nitrogen nuclear quadrupole interaction, the various types of transitions are easily identified. In Table 3 the measured and calculated frequencies are listed according to their type (A, B, etc). In analysing the spectra, the magnitudes of the components of the spin Hamiltonian were first obtained by perturbation theory and the final results by computer diagonalization of the spin Hamiltonian. The ^{14}N and ^{35}Cl out-of-plane hyperfine elements were found to be approximately 19.5 and 15 MHz respectively. With only one nitrogen hyperfine element, only one nitrogen quadrupole transition is observed corresponding to the in-plane $\chi_z \rightarrow \chi_y$ transition which was found to be 3.2 ± 0.2 MHz. With our assumption that the asymmetry parameter may be neglected the ^{35}Cl nuclear quadrupole coupling constant was found to be -68.4 ± 0.6 MHz (see Table 2). The calculated frequencies listed in Table 3 were obtained by analysis of the components of the observed spectra due to the ^{35}Cl isotope. The transitions associated with molecules possessing the ^{37}Cl isotope were then obtained by using the same values for $\mathcal{H}_{SS}$, $\mathcal{H}_Q^N$, and $\mathcal{H}_{HF}^N$, correcting $\mathcal{H}_Q^{Cl}$ for the difference in the nuclear quadrupole moments and $\mathcal{H}_{HF}^{Cl}$ for the difference in the magnetogyric ratio of the two chlorine isotopes. All calculated frequencies were obtained by collecting all transitions within 0.75 MHz of one another and weighting each by its electron magnetic moment transition probability.

It is difficult to make a comprehensive analysis of the electron distribution in the excited triplet state without a measure of all the components of the hyperfine tensor. The similarity of the nitrogen nuclear quadrupole and hyperfine interactions in 8-chloroquinoline and quinoline and the observation that the chlorine nuclear quadrupole coupling constant is approximately the same as that reported for the ground state of 6-chloroquinoline (69.256 MHz)[75] and 7-chloroquinoline (69.362 MHz)[71] supports the assumption that the excited triplet state of 8-chloroquinoline is essentially the same as that of quinoline.

Table 3. Measured and calculated ESR transitions of the $^3\pi\pi^*$ state of 8-chloroquinoline

	Measured frequency (±0.5 MHz)	Calculated frequency		Classification
(a) $\tau_x \rightarrow \tau_y$				
	3422.8	3422.7		E
	3419.5	3419.5		C
	3416.5	3416.5		E
	3415.3	3415.3		F
	3412.0	3412.1		D
	3409.3	3409.1		F
	3388.9	3388.3		B
	3385.9	3385.1		A
	3382.7	3382.2		B
	3361.2	3361.1		F
	3358.1	3358.0		D
	3354.5	3355.2		F
	3353.9	3353.9		E
	3350.9	3350.8		C
	3347.6	3348.0		E
(b) $\tau_x \rightarrow \tau_z$				
	2562.3	2561.9		E
	2559.3	2558.9		C
	2555.8	2555.7	2554.8	E, F
	2552.3	2551.8		D
	2548.9	2548.6		F
	2525.5	2524.9		A
	2501.7	2500.9		F
	2498.0	2498.1		D
	2494.4	2493.6	2495.0	E, F
	2491.0	2490.8		C
	2487.9	2487.7		E
(c) $\tau_z \rightarrow \tau_y$				
	898.1	897.9		E
	895.6	894.7		C
	891.7	891.6		E
	890.7	890.4		F
	888.2	887.3		D
	884.3	884.2		F
	863.8	864.1		B
	860.1	860.2		A
	857.3	857.4		B
	836.4	836.4		F
	833.5	833.4		D
	830.8	830.3		F
	829.8	829.4		E
	826.5	826.0		C
	822.9	823.2		E

VI. THE ODMR SPECTRA OF *PARA*DICHLOROBENZENE

A detailed analysis of the ODMR spectra of *para*dichlorobenzene (DCB) has been previously reported.[29] In this section those results are summarized with special attention to the determination of the chlorine nuclear quadrupole coupling constant. The DCB ODMR spectrum is a function of the particular trap emission monitored and the host material. We will limit this discussion to the shallow trap emission in neat DCB.

As was the case with 8-chloroquinoline, the observed ODMR spectrum of DCB is due to the interaction of isotopically distinct molecular species. The fractional natural abundances of the ^{35}Cl and ^{37}Cl isotopes are approximately 3/4 and 1/4, respectively. Since there are two chlorine nuclei per molecule, the fractional distribution of the molecular species is:

$$I \quad ^{35}Cl - {}^{35}Cl = 9/16$$
$$II \quad ^{35}Cl - {}^{37}Cl = 6/16$$
$$III \quad ^{37}Cl - {}^{37}Cl = 1/16$$

The spectra are treated as a weighted superposition of the ODMR spectra due to each of the three molecular species.

The $\tau_x \rightarrow \tau_y$ (high frequency) transitions observed using amplitude modulation are shown in Fig. 15. The remaining two electron-spin transitions

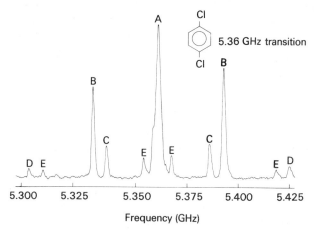

Fig. 15.   ODMR of the $\tau_x \rightarrow \tau_y$ multiplet of *para*dichlorobenzene.

($\tau_x \rightarrow \tau_z$ and $\tau_z \rightarrow \tau_y$) have essentially the same structure as the spectra illustrated for the $\tau_x \rightarrow \tau_y$ transition; however, the signal-to-noise ratio of the $\tau_z \rightarrow \tau_y$ transition was substantially lower. In Table 4, the possible ESR transitions involving the triplet electrons and one or more chlorine nuclei are listed as to type (A, B, C, D, E, or F) and the molecular species (I, II or III) which can undergo each type of transition. The intensity of the transitions involving the

Table 4. ESR transitions and *para*dichlorobenzene

Transition type	Simultaneous transitions	Molecular species
A	Electron spin	I, II, III
B	Electron and ^{35}Cl spins	I, II
C	Electron and ^{37}Cl spins	II, III
D	Electron, ^{35}Cl and ^{35}Cl spins	I
E	Electron, ^{35}Cl and ^{37}Cl spins	II
F	Electron, ^{37}Cl and ^{37}Cl spins	III

electron and one chlorine spin (B and C) and those involving the electron and two chlorine spins (D, E and F) must be considered separately. The ratio of the intensities of the transitions involving a single ^{35}Cl spin (type B) to those involving a single ^{37}Cl spin (type C) should be three to one on the basis of the ratio of ^{35}Cl to ^{37}Cl. The ratio of the intensities of the transitions involving two chlorine spins is likewise $I_D : I_E : I_F = 9 : 6 : 1$.

The structure of the $\tau_x \rightarrow \tau_y$ electron-spin multiplet shown in Fig. 15 is labelled according to the classification given in Table 1. Since the nuclear quadrupole moment of ^{35}Cl is larger than that of ^{37}Cl, the outer pair of the four strong satellites are assigned as type B (^{35}Cl) and the inner pair as type C transitions (^{37}Cl). As can be seen, the ratio of the intensity of the transitions labelled B and C is approximately 3 : 1 as predicted. The outermost satellites in Fig. 15 are assigned to simultaneous double chlorine transitions (labelled D and E on the spectra). The intensity of these transitions is approximately in the predicted ratio of 9 : 6. The transitions corresponding to simultaneous double ^{37}Cl transitions (type F) are not observed, consistent with the small natural abundance of the molecular species responsible for these transitions. The inner pair of satellites (labelled E in Fig. 15) may be considered as simultaneous electron and ^{35}Cl *and* ^{37}Cl transitions. The higher frequency satellite represents a simultaneous electron-spin transition, a ^{35}Cl ($\pm\frac{3}{2} \rightarrow \pm\frac{1}{2}$) and a ^{37}Cl ($\pm\frac{1}{2} \rightarrow \pm\frac{3}{2}$) transition while the lower frequency satellite represents the opposite chlorine transitions. These transitions are separated by the difference between the ^{35}Cl and ^{37}Cl nuclear quadrupole coupling constants. Naturally these occur for only those molecules that have one ^{35}Cl and one ^{37}Cl isotope. Since the matrix elements for these double chlorine transitions are of a different form than those associated with the other double chlorine transitions, the intensity of the inner satellites labelled E in Fig. 15 may not be compared directly with the intensity of the outer satellites labelled D and E. All transitions involving both an electron and a nuclear spin required microwave power greater by several orders of magnitude to obtain intensities comparable to the electron only (type A) transition. Chlorine nuclear transitions were observed via optically detected ENDOR by saturating the ESR transitions associated with the $\tau_x \rightarrow \tau_y$ or $\tau_x \rightarrow \tau_z$ manifolds. Both the ^{35}Cl and ^{37}Cl ENDOR resonances were observed while

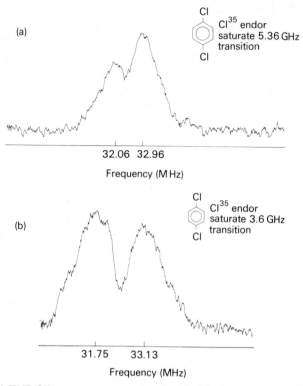

Fig. 16. ^{35}Cl ENDOR resonance associated with (a) the $\tau_x \to \tau_y$ and (b) the $\tau_x \to \tau_z$ electron-spin transitions.

saturating either ESR transition. Figures 16(a) and 16(b) illustrate the ^{35}Cl ENDOR resonances associated with the $\tau_x \to \tau_y$ and $\tau_x \to \tau_z$ transitions, respectively.

As an extension of the ENDOR experiments, a ^{35}Cl ENDOR transition was saturated while sweeping the $\tau_x \to \tau_z$ microwave transition. Since only the

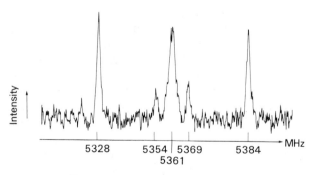

Fig. 17. ^{35}Cl ENDOR pumping of the $\tau_x \to \tau_y$ multiplet.

Table 5. Zero-field splitting parameters (MHz)

	Paradichlorobenzene[a]		Benzene[b]-d_2 in benzene-h_6 (1.95 K)
	X trap (1.3 K)	Y trap (4.2 K)	
X	−2988.75	−2967.7	−3159.8
Z	616.07	654.4	1769.4
Y	2372.68	2313.4	1385.0
D^c	4483.13	4451.6	4739.7
E^c	−878.31	−829.5	192.2
D^a	4733.8	4677.7	4793.2
e^2qQ/h	−64.5	—	—
A_{xx} (^{35}Cl)	22.0	—	—

[a] Phosphorescence origin: X trap (27 868 cm^{-1}) and Y trap (27 807 cm^{-1}).
[b] Data from reference 77 expressed in our axis system.
[c] In order to be consistent with the standard ESR definitions, we have defined

$$D = -3/2X \quad \text{and} \quad E = 1/2(Z - Y).$$

Table 6. Measured and calculated ESR transitions of the $^3\pi\pi^*$ state of para-dichlorobenzene (X trap)

	Measured frequency (MHz)	σ	Calculated frequency	Classification
(a) $\tau_x \rightarrow \tau_y$				
	5426.7	1.0[a]	5426.91	D
	5419.6	1.0[a]	5419.56	E
	5394.56	0.41	5394.62	B
	5387.86	0.41	5387.79	C
	5368.73	0.64	5368.89	E
	5362.20	0.34	5362.14	A
	5355.13	0.25	5355.12	E
	5336.67	0.24	5366.50	C
	5329.74	0.28	5329.75	B
	5303.8	1.0[a]	5304.11	E
	5296.5	1.0[a]	5297.35	D
(b) $\tau_x \rightarrow \tau_z$				
	3636.03	0.07	3636.13	B
	3629.65	0.18	3629.56	C
	3611.18	0.24	3611.04	E
	3604.19	0.25	3604.10	A
	3597.69	0.31	3597.43	E
	3578.90	0.22	3578.89	C
	3571.88	0.34	3571.99	B
(c) $\tau_z \rightarrow \tau_y$				
	1791.1	1.5	1791.13	B
	1758.2	1.0	1758.05	A
	1724.5	1.5	1726.55	B

[a] Estimated value of the standard deviation σ.

Table 7.   Measured and calculated chlorine ENDOR transitions of the $^3\pi\pi^*$ state of *para*dichlorobenzene (X trap)

Measured frequency in MHz (± 0.05)		Calculated frequency in MHz (range)
$\tau_x \rightarrow \tau_y$ manifold		
^{35}Cl 32.06; 32.96		31.56 – 33.03
^{37}Cl 25.12; 26.00		24.94 – 26.09
$\tau_x \rightarrow \tau_z$ manifold		
^{35}Cl 31.75; 33.13		31.53 – 32.94
^{37}Cl 24.94; 26.19		24.79 – 25.90

ENDOR time-dependent magnetic field was amplitude modulated and the change in phosphorescence intensity detected with a lock-in amplifier, only the ESR transitions that involve at least one ^{35}Cl spin transition were detected. The spectrum obtained from this experiment is shown in Fig. 17. As would be expected, satellites assigned as simultaneous electron and ^{37}Cl spin transitions (labelled C in Fig. 15) are not observed. Finally, all measured frequencies associated with the three electron-spin zero-field transitions are given in Table 6, while the ^{35}Cl and ^{37}Cl ENDOR transitions are listed separately in Table 7.

Following the discussion in Section III the observed spectra are explained in terms of a Hamiltonian of the form

$$\mathcal{H} = \mathcal{H}_{SS} + \sum_i \mathcal{H}_Q + \sum_i \mathcal{H}_{HF} \qquad (111)$$

where the summation is over the chlorine nuclei and $\mathcal{H}_{HF} = A_{xx}S_xI_x$. The ODMR spectra were simulated by use of a computer program that diagonalized the spin Hamiltonian and calculated the transition frequencies and intensities. The spin Hamiltonian parameters used in simulating the spectra observed while monitoring the X-trap emission are listed in Table 5 along with the approximate values of $\mathcal{H}_{SS}$ for the Y trap and the values reported for benzene.[77] The best value obtained for the ^{35}Cl nuclear quadrupole coupling constant was -64.50 MHz (^{37}Cl $= -50.84$ MHz) and for the ^{35}Cl hyperfine interaction $A_{xx} = 22$ MHz (^{37}Cl: $A_{xx} = 18.3$ MHz). The experimental and calculated ESR frequencies for the X trap of DCB are listed in Table 6. With the parameters used in the spin Hamiltonian, all of the calculated transition frequencies are within experimental error. However, a small error in the calculated frequencies is introduced since a weighted average of the transitions corresponding to a particular type was made.

The observed and calculated chlorine ENDOR transitions associated with the $\tau_x \rightarrow \tau_y$ and $\tau_x \rightarrow \tau_z$ multiplets are listed in Table 7. Because of the large line-width of the observed ENDOR transitions and because many ENDOR transitions are expected in a small range of frequencies it is difficult to explicitly assign the observed spectra to any particular calculated transition. An additional complication arises when one considers the transition moments associated with the individual transitions. Since they vary, one should weight the calculated spectra according to the square of the ENDOR transition moments and compare these spectra with the observed. However the observed spectra are obtained under saturating conditions, and therefore intensities are practically meaningless. We compared only the range of calculated ENDOR frequencies listed in Table 7 with the experimental results.

From a second-order perturbation analysis of the DCB ODMR spectra,[78] it has been shown that with the assumption that e^2qQ/h is negative, D is positive or $x < z$, y. This result is entirely consistent with previous experimental[11,79] and theoretical[61] studies of aromatic molecules in $\pi\pi^*$ triplet states. Indeed this is what is observed for the lowest $\pi\pi^*$ triplet state of benzene.[77] The ordering of the interaction along the two in-plane molecular axes is however not immediately apparent. From the analysis of the spectra of DCB utilizing phosphorescence microwave double resonance (PMDR) spectroscopy, the component of the electron spin–spin interaction along the molecular y (or short in-plane axis) has been assigned as the larger of the two in-plane components of the electron spin–spin tensor.[39] Since the zero-field splitting parameters D and D^* $(D^* = (D^2 + 3E^2)^{\frac{1}{2}})$ are primarily a function of the size of the π system involved in the excitation,[62] the value of these parameters for both DCB and benzene should be similar if DCB is a $\pi\pi^*$ triplet. As can be seen in Table 5, the values of D and D^* for both traps of DCB differ from the corresponding values for benzene by only a few per cent, which is strong confirmation of the assignment of the excited triplet state of DCB as a $\pi\pi^*$ state. The zero-field splitting parameter E which is a measure of the anisotropy of the triplet electron distribution in the molecular plane is, however, quite different for both molecules. If the benzene molecule possessed D_{6h} symmetry in the excited state, E must be zero by symmetry. The finite value of E for benzene has been explained by de Groot and van der Waals on the basis of a distortion of the benzene ring from the D_{6h} to D_{2h}.[59,77]

A quantitative analysis of the E value of DCB is difficult since accurate wave functions are not available for the chlorine atoms. However, from a simple consideration of the perturbation of the triplet electron distribution in benzene due to the addition of two *para* chlorines, it is expected that the τ_z level will be lowered and the τ_y level raised in energy. Since the E value of DCB is larger than the E value of benzene, this model predicts that in DCB the τ_y level is higher in energy than the τ_z level. This of course gives the opposite sign of E for DCB as compared with benzene, and is consistent with the ordering of the triplet energy levels previously obtained from analysis of the phosphorescence microwave double resonance spectra. It is interesting to note that in 1, 2, 4, 5-tetrachlorobenzene (TCB), the inclusion of chlorine interactions would predict the τ_z level to be higher in energy than the τ_y level; consequently, the E

value would have the same sign as benzene. Other substituted chlorobenzenes should have E values between DCB and TCB. The importance of the zero-field splitting of DCB and TCB is that the presence of the chlorines acting as perturbations on the excited state of benzene raises the possibility that the symmetry of the excited state of DCB and TCB is different.from that of the excited state of benzene. As has been discussed,[39] the sign of E in part answers this interesting question.

The absolute value of the chlorine nuclear quadrupole coupling constant (e^2qQ/h) in the excited state of DCB is significantly reduced compared with the corresponding value for the ground state. With the assumption that the asymmetry parameter (η) may be neglected, the value of e^2qQ/h for the ^{35}Cl nuclei of DCB in its excited triplet state at 1.3 K is -64.5 MHz. The measured pure nuclear quadrupole resonance frequency of DCB in its ground state at 4.2 K is 34.831 MHz[29] which, if η is assumed to equal zero, corresponds to a value of e^2qQ/h of -69.662 MHz. The assumption that η may be neglected is justified on the basis that e^2qQ/h is not changed significantly for small values of η and for the ground state of DCB at room temperature η is only 0.08.[80] Indeed, from the explicit dependence of e^2qQ/h on η, the assumption that e^2qQ/h is simply twice the pure n.q.r. transition frequency causes a positive error in e^2qQ/h of less than 5% for $\eta \leq 0.5$. The increase of 52 kHz in the pure n.q.r. frequency of the ground state of DCB upon lowering the temperature of the sample from 77 K ($\nu = 34.779$ MHz) to 4.2 K ($\nu = 34.831$ MHz) is consistent with Bayer's theory[81] which treats the temperature dependence of the n.q.r. frequency in terms of the molecular torsional motions. More important, however, is the fact that the small change in the pure n.q.r. transition frequency indicates that there is no major physical change in the environment of the chlorine nuclei in DCB upon cooling. Therefore, the difference in e^2qQ/h between the ground and excited states of DCB is clearly due to a change in the electric field gradient (q) at the chlorine nuclei upon excitation. The magnitude of the decrease in the absolute value of e^2qQ/h upon excitation is interesting because: (a) the absolute value of e^2qQ/h in the triplet state of DCB is significantly less than the value reported for the ground state of *any* chlorine bonded to an aromatic molecule;[63] and (b) the decrease in $|e^2qQ/h|$ upon excitation to the lowest $^3\pi\pi^*$ state of 8-chloroquinoline[28] and 1, 2, 4, 5-tetrachlorobenzene[42] is far less than the decrease in $|e^2qQ/h|$ for DCB.

In contrast to the electron spin–spin and hyperfine interactions which are a function of only the triplet electrons, e^2qQ/h is dependent upon the distribution of all electrons. Since electrons in s orbitals have spherical symmetry, they do not contribute to the field gradient. A closed p shell also contributes nothing to the field gradient, and following the analysis of Bersohn[82] the field gradient in DCB can be considered as arising from a hole in the chlorine p_z orbital and a partial hole in the chlorine p_x orbital. The total contribution is due to two axially symmetric tensors whose major axes are perpendicular. In Table 8 the contributions to the field gradient are expressed

Table 8. Contributions to the chlorine nuclear quadrupole coupling constant

Chlorine orbital	No. of holes	Contribution			Total contribution
		V_{xx}	V_{yy}	V_{zz}	
p_x	δ	δq	$-\delta q/2$	$-\delta q/2$	$V_{xx} = (\delta - \sigma/2)q$
p_z	σ	$-\sigma q/2$	$-\sigma q/2$	σq	$V_{yy} = -1/2(\delta + \sigma)q$
					$V_{zz} = (\sigma - \delta/2)q$

in terms of the number of holes in the p_x and p_z chlorine orbitals. The difference in e^2qQ/h for the excited and ground state may be written

$$\Delta(e^2qQ/h) = e^2(q_T - q_G)Q \tag{112}$$

where q_T and q_G refer to the field gradient at the chlorine nuclei in the triplet and ground states of DCB respectively. Equation (112) may be expressed in terms of the number of holes in the p_z and p_x orbitals as

$$\Delta(e^2qQ/h) = e[\sigma_T - \sigma_G] - \tfrac{1}{2}[\delta_T - \delta_G]Q \tag{113}$$

where σ_T and σ_G are the fraction of p_z electron holes in the carbon-chlorine sigma bond in the triplet and ground state, respectively, while δ_T and δ_G are the fraction of p_x electron holes in the π bond for the triplet and ground states, respectively. Since $\Delta(e^2qQ/h)$ is negative, one of the following conditions must be met: (a) $\sigma_G > \sigma_T$, or (b) $\delta_T > \delta_G$. If σ_G is greater than σ_T, the number of holes has decreased along the carbon–chlorine bond, and therefore the chlorine nuclei are more successful in competing for electrons in the excited state. However, since the sigma electrons are not involved in the excitation, this effect should be very small. If δ_T is greater than δ_G, the out-of-plane chlorine p_x orbital has lost electrons. An increase in the number of holes in the p_x orbital would be the most likely explanation of the decrease in e^2qQ/h since the chlorine p_x orbitals are allowed by symmetry to interact with the carbon p_x orbitals. The increase in the number of holes in the chlorine p_x orbital can come about from either an increase in the double bond character of the C—Cl bond or a 'bent' C—Cl bond. Bray, Barnes and Bersohn[80] have shown that although the overlap of the carbon and chlorine p_x orbitals is reduced with a bent C—Cl bond, the chlorine p_x orbitals may overlap with the sigma system, consequently increasing the number of holes in the p_x orbital of chlorine (δ_T) relative to the number of holes in the p_x orbital in the ground state (δ_G). Although it is not possible *a priori* to distinguish between these two possibilities, the interpretation of the change in e^2qQ/h as arising from a bent C—Cl bond is reasonable in view of other experimental results.

The phosphorescence of DCB to the ground state in the O—O band comes from all three triplet levels, which requires that DCB has less than D_{2h} symmetry in its $^3\pi\pi^*$ state.[39] Finally the measured value of the out-of-plane chlorine hyperfine interaction for the $^3\pi\pi^*$ state of 8-chloroquinoline

(15 MHz) is approximately the same as that observed for the $^3\pi\pi^*$ state of DCB (22 MHz). However in 8-chloroquinoline, the chlorine nuclear quadrupole constant is essentially unchanged upon excitation. In view of these observations it seems reasonable to interpret the change in e^2qQ/h as arising from a bent C—Cl bond.

As we can see, ODMR offers many new possibilities for the measurement of the nuclear quadrupole coupling constants in excited triplet states. The excellent sensitivity obtained with optical detection coupled with the accuracy of the measurement in zero-field provides a new technique to obtain a detailed knowledge of the electron distribution and molecular geometry in excited states. We fully expect that these techniques will be applied to a variety of problems associated with organic molecules, inorganic molecules, semiconductors and various colour centres in ionic solids. It has not been our intention to be exhaustive in this paper, but rather to lay down a basic working knowledge of the theory and experimental methods to allow these techniques to be easily adapted to other questions and problems.

REFERENCES

1 S. P. McGlynn, T. Azumi, and M. Kinoshito, *Molecular Spectroscopy of the Triplet State*, Prentice-Hall, Inc., Englewood Cliffs, N.J., 1969.
2 R. S. Becker, *Theory and Interpretation of Fluorescence and Phosphorescence*, Wiley-Interscience, New York, 1969.
3 S. K. Lower and M. A. El-Sayed, *Chem. Rev.*, **66**, 199 (1966).
4 M. S. DeGroot, I. A. M. Hesselmann, and J. H. van der Waals, *Mol. Phys.* **12**, 259 (1967).
5 G. N. Lewis and M. J. Kasha, *J. Am. Chem. Soc.*, **66**, 2100 (1944).
6 G. N. Lewis and M. J. Kasha, *J. Am. Chem. Soc.*, **67**, 994 (1945).
7 G. N. Lewis and M. Calvin, *J. Am. Chem. Soc.*, **67**, 1232 (1945).
8 G. N. Lewis, M. J. Kasha, and M. Calvin, *J. Chem. Phys.*, **17**, 804 (1949).
9 H. F. Hameka, Thesis, Leiden (1956).
10 C. A. Hutchison, Jr. and B. W. Mangum, *J. Chem. Phys.*, **29**, 952 (1958).
11 C. A. Hutchison, Jr. and B. W. Mangum, *J. Chem. Phys.*, **34**, 908 (1961).
12 Ph. Kottis and R. Lefebvre, *J. Chem. Phys.*, **39**, 393 (1963).
13 Ph. Kottis and R. Lefebvre, *J. Chem. Phys.*, **41**, 379 (1964).
14 S. Geschwind, G. E. Devlin, R. L. Cohen, and S. R. Chinn, *Phys. Rev.*, **137**, (1965).
15 J. Brossel and A. Kastler, *Compt. Rend.*, **229**, 1213 (1949).
16 J. Brossel and F. Bitter, *Phys. Rev.*, **86**, 308 (1952).
17 M. Sharnoff, *J. Chem. Phys.*, **46**, 3263 (1967).
18 A. L. Kwiram, *Chem. Phys. Lett.*, **1**, 272 (1967).
19 J. Schmidt, I. A. M. Hesselmann, M. S. DeGroot, and J. H. van der Waals, *Chem. Phys. Lett.*, **1**, 434 (1967).
20 J. Schmidt and J. H. van der Waals, *Chem. Phys. Lett.*, **2**, 640 (1968).
21 R. W. Brandon, R. E. Gerkin, and C. A. Hutchison, Jr., *J. Chem. Phys.*, **41**, 3717 (1968).
22 J. Schmidt and J. H. van der Waals, *Chem. Phys. Lett.*, **3**, 546 –1969).

23 D. S. Tinti, M. A. El-Sayed, A. H. Maki, and C. B. Harris, *Chem. Phys. Lett.*, **3,** 343 (1969).

24 C. B. Harris, D. S. Tinti, M. A. El-Sayed, and A. H. Maki, *Chem. Phys. Lett.*, **4,** 409 (1969).

25 I. Y. Chan, J. Schmidt, and J. H. van der Waals, *Chem. Phys. Lett.*, **4,** 269 (1969).

26 M. J. Buckley, C. B. Harris, and A. H. Maki, *Chem. Phys. Lett.*, **4,** 591 (1970).

27 I. Y. Chan and J. H. van der Waals, *Chem. Phys. Lett.*, **20,** 157 (1973).

28 M. J. Buckley and C. B. Harris, *Chem. Phys. Lett.*, **5,** 205 (1970).

29 M. J. Buckley and C. B. Harris, *J. Chem. Phys.*, **56,** 137 (1972).

30 T. S. Kuan, D. S. Tinti, and M. A. El-Sayed, *Chem. Phys. Lett.*, **4,** 507 (1970).

31 J. Schmidt, W. S. Veeman, and J. H. van der Waals, *Chem. Phys. Lett.*, **4,** 341 (1969).

32 D. S. Tinti and M. A. El-Sayed, *J. Chem. Phys.*, **54,** 2529 (1971).

33 C. B. Harris and R. J. Hoover, *J. Chem. Phys.*, **56,** 2199 (1972).

34 C. J. Winscom and A. H. Maki, *Chem. Phys. Lett.*, **12,** 264 (1971).

35 D. A. Antheunis, J. Schmidt, and J. H. van der Waals, *Chem. Phys. Lett.*, **6,** 255 (1970).

36 A. H. Francis, C. B. Harris, and A. M. Nishimura, *Chem. Phys. Lett.*, **14,** 425 (1972).

37 M. A. El-Sayed, A. A. Gwaiz and C. T. Lin, *Chem. Phys. Lett.*, **16,** 281 (1972).

38 G. Kothandaraman and D. S. Tinti, *Chem. Phys. Lett.*, **19,** 225 (1973).

39 M. J. Buckley, C. B. Harris, and R. M. Panos, *J. Am. Chem. Soc.*, **94,** 3692 (1972).

40 A. I. Attia, B. H. Loo, and A. H. Francis, *Chem. Phys. Lett.*, **22,** 537 (1973).

41 A. A. Gwaiz and M. A. El-Sayed, *Chem. Phys. Lett.*, **19,** 11 (1973).

42 A. H. Francis and C. B. Harris, *J. Chem. Phys.*, **57,** 1050 (1972).

43 C. B. Harris, *J. Chem. Phys.*, **54,** 972 (1971).

44 J. Schmidt, W. G. van Dorp, and J. H. van der Waals, *Chem. Phys. Lett.*, **8,** 345 (1971).

45 J. Schmidt, *Chem. Phys. Lett.*, **14,** 411 (1972).

46 W. G. Breiland, C. B. Harris, and A. Pines, *Phys. Rev. Lett.*, **30,** 158 (1973).

47 C. B. Harris, R. L. Schlupp, and H. Schuch, *Phys. Rev. Lett.*, **30,** 1019 (1973).

48 C. A. van't Hoff, J. Schmidt, P. J. F. Verbeck, and J. H. van der Waals, *Chem. Phys. Lett.*, **21,** 437 (1973).

49 H. C. Brenner, J. C. Brock, and C. B. Harris, *J. Chem. Phys.*, **60,** 4448 (1974).

50 M. J. Buckley and A. H. Francis, *Chem. Phys. Lett.*, **23,** 582 (1973).

51 S. J. Hunter, H. Parker, and A. H. Francis, *J. Chem. Phys.*, **61,** 1390 (1974).

52 A. H. Zewail and C. B. Harris, *Chem. Phys. Lett.*, **28,** 8 (1974).

53 A. H. Francis and C. B. Harris, *Chem. Phys. Lett.*, **9,** 181 & 188 (1971).

54 C. B. Harris, M. Glasbeek, and E. B. Hensley, *Phys. Rev. Lett.*, **33,** 537 (1974).

55 M. A. El-Sayed, *J. Chem. Phys.*, **54,** 680 (1971).

56 M. A. El-Sayed, D. S. Tinti, and E. M. Yee, *J. Chem. Phys.*, **51,** 5721 (1969); M. A. El-Sayed, D. S. Tinti, and O. V. Owens, *Chem. Phys. Lett.*, **3,** 339 (1969); M. A. El-Sayed, *J. Chem. Phys.*, **52,** 6438 (1970); M. A. El-Sayed and O. F. Kalman, *J. Chem. Phys.*, **52,** 4903 (1970).

57 H. F. Hameka, in: *The Triplet State* (proc. Intl. Symposium on the Triplet State, 1967), Cambridge Univ. Press, 1967, p. 25.

58 E. Gilmore, G. Gibson, and D. McClure, *J. Chem. Phys.*, **20,** 829 (1952); **23,** 399 (1955).

59 M. S. de Groot and J. H. van der Waals, *Mol. Phys.*, **6,** 545 (1963).

60 M. J. Buckley and C. B. Harris, *unpublished results.*

61 M. Tinkham and M. W. P. Strandberg, *Phys. Rev.*, **97,** 937 (1955); M. Gouterman and W. Moffitt, *J. Chem. Phys.*, **30,** 1107 (1959).

62 A. Carrington and A. D. McLachlan, *Introduction to Magnetic Resonance*, Harper and Row. London, 1967, p. 117.

63 T. P. Das and E. L. Hahn, *Nuclear Quadrupole Resonance Spectroscopy*, Academic Press, New York, 1958; M. H. Cohen and F. Reif, *Solid State Physics*, Vol. V, Academic Press, New York, 1957; E. A. C. Lucken, *Nuclear Quadrupole Coupling Constants*, Academic Press, New York, 1969.

64 J. S. Vincent and A. H. Maki, *J. Chem. Phys.*, **42,** 865 (1965).

65 R. H. Webb, *Rev. Sci. Instr.*, **33,** 732 (1962).

66 M. Kasha, *J. Opt. Soc. Am.*, **38,** 929 (1948).

67 D. Owens, M. A. El-Sayed, and S. Ziegler, *J. Chem. Phys.*, **52,** 4315 (1970).

68 C. A. Hutchison, Jr., J. V. Nicholas, and G. W. Scott, *J. Chem. Phys.*, **53,** 1906 (1970).

69 Y. Gondo and A. H. Maki, *J. Chem. Phys.*, **50,** 3270 (1969).

70 D. Pooley and D. H. Whiffen, *Spectrochimica Acta*, **18,** 291 (1962).

71 H. R. Falle, G. R. Luckhurst, A. Horsfield, and M. Ballester, *J. Chem. Phys.*, **50,** 258 (1969).

72 J. Schmidt, W. S. Veeman, and J. H. van der Waals, *Chem. Phys. Lett.*, **4,** 341 (1969).

73 E. A. C. Lucken, *Trans. Faraday Soc.*, **57,** 729 (1961).

74 E. Schempp and P. J. Bray, *J. Chem. Phys.*, **46,** 1186 (1967).

75 S. L. Segel, R. G. Barnes, and P. J. Bray, *J. Chem. Phys.*, **25,** 1286 (1956).

76 M. Dewar and E. A. C. Lucken, *J. Chem. Soc.* (*London*), 2653 (1958).

77 M. S. De Groot, I. A. M. Hesselmann, and J. H. van der Waals, *Mol. Phys.*, **16,** 45 (1969).

78 D. S. Tinti, G. Kothandaraman, and C. B. Harris, *J. Chem. Phys.*, **59,** 190 (1973).

79 N. Hirota, C. A. Hutchison, Jr., and P. Palmer, *J. Chem. Phys.*, **40,** 3717 (1964).

80 P. J. Bray, R. G. Barnes, and R. Bersohn, *J. Chem. Phys.*, **25,** 813 (1956).

81 H. Bayer, *Z. Physik*, **130,** 227 (1951).

82 R. Bersohn, *J. Chem. Phys.*, **22,** 2078 (1954).

3. DOUBLE RESONANCE DETECTION OF NUCLEAR QUAD-RUPOLE RESONANCE SPECTRA

R. Blinc

University of Ljubljana, 'J. Stefan' Institute, Ljubljana, Yugoslavia

I. INTRODUCTION

Pulsed nuclear double resonance—as introduced by Hartmann and Hahn,[1] and later by Lurie and Slichter[2]—is a trigger detection method which uses the 'strong' NMR signal of the I spin system to detect the 'weak' nuclear resonance of the spins S, which might have either a low natural abundance or a low resonance frequency. The sensitivity of this method may be several orders of magnitude greater than the sensitivity of ordinary magnetic or quadrupole resonance techniques. Double resonance thus allows an extension of nuclear quadrupole resonance (NQR) spectroscopy into a region which is hardly accessible by classical NQR techniques.

The method is based on the fact that, if certain conditions are met, the effect of a radiofrequency perturbation of the S system can be via I–S dipole–dipole interactions transferred to the I system, integrated, and detected as a change in the NMR signal of the I spins. The transfer of the r.f. perturbation will be maximized if the separation between the energy levels of the S spins (in any given representation in which the S spins are quantized) exactly matches the separation between the energy levels of the I spins in the representation in which these are quantized.

Depending on the representation in which the resonance transfer of the r.f. perturbation is performed, we may distinguish:

i.   Double resonance in the rotating frame;
ii. Double resonance in the laboratory frame;
iii. Double resonance between the laboratory and the rotating frames.

In case (i) both spin systems are quantized in their respective rotating frames, i.e. frames which rotate around the direction of the external magnetic field $\mathbf{B}_0$ with the Larmor frequency $\omega_I = \gamma_I B_0$ and $\omega_S = \gamma_S B_0$ respectively. We thus deal with NQR in a high magnetic field.

In case (ii) we have to get frequency matching in the laboratory frame and therefore at least one of the spin species involved must exhibit a non-zero quadrupole splitting in addition to the Zeeman splitting. We have now a case of low-field or zero-field NQR.

In case (iii) one of the spin species is quantized in the laboratory and the other in the rotating frame. This is a case of zero-field NQR.

Each double resonance experiment involves three major steps:

a. First, the I spins are ordered;
b. Second, the S spin system is transformed into a state of maximum possible disorder and is coupled to the I system;
c. Third, the remaining order in the I spin system is detected.

These steps can be realized in various different ways which will be discussed separately for the cases (i), (ii) and (iii).

A simple thermodynamic analogy is as follows:

The I and the S systems represent two thermal reservoirs, a large one and a small one, which are connected with a heat conducting link. The S spin system can be heated externally by irradiation with a resonant r.f. field, but its heat capacity is too small to allow a direct measurement of the resulting change in its temperature. If, however, the S system is continuously heated for a long enough time, and the thermal conductivity of the link to the I system is sufficiently high, the temperature of the I system will make a detectable change. The effect is the larger the smaller the initial temperature of the I reservoir. The increase in the I spin temperature means that resonant r.f. absorption has taken place in the S system: the S resonance is thus detected via its effect on the I system. The increase in sensitivity is due to the fact that the I system integrates over many S spin resonance absorption cycles.

In the above experiment the increase in the temperature of the I system depends on the heat capacity of the I system and the thermal conductivity of the link connecting the two reservoirs, but not on the heat capacity of the S system. In another version of the double resonance experiment the thermal link between the two systems is repeatedly broken, the S system heated to a known high temperature, and then reconnected to the I system. After a sufficiently long time, the two systems and the link will reach a common temperature, which depends only on their heat capacities and initial temperatures.

There are several quantities which can be measured in a double resonance experiment:

a. The fact that the temperature of the I system has changed shows that resonant r.f. absorption in the S spin system has occurred. In this way the S spin NQR spectrum can be measured.

b. In addition, one can measure the ratio of the heat capacities of the two systems, the thermal conductivity of the link, and the time the two thermal reservoirs remain isolated from the surrounding lattice reservoir, i.e. the I and S spin–lattice relaxation times.

II. DOUBLE RESONANCE IN THE ROTATING FRAME: HIGH-FIELD QUAD-RUPOLE DOUBLE RESONANCE

A. Experimental procedure and theory

In the present case both spin systems are quantized in their respective rotating frames, i.e. frames which rotate around the direction of the external magnetic field with the corresponding Larmor frequencies (Fig. 1). Since the experiment is performed in a large external magnetic field single crystalline samples are required.

The basic physics of a double resonance experiment in the rotating frame may be perhaps understood in the following way:

The Hamiltonian of a spin system, for instance I, in the rotating frame is time independent. At exact resonance it is equivalent to the Hamiltonian of a spin system in a static magnetic field with the r.f. field $H_1 \parallel x$ replacing the static field $H_0 \parallel z$. The direction (x) of H_1 thus determines the axis of quantization. A field which oscillates perpendicular to the direction of quantization (x) with a frequency which, multiplied by Planck's constant, matches the energy level separation in the rotating frame will induce resonance transitions in the I system. Such a field is created by the magnetization of the S spins if irradiated with another r.f. field of appropriate frequency and amplitude, which satisfies the double resonance condition.

We shall in the following limit ourselves to the special case where the spin–lattice relaxation times of both species are long compared to the spin–spin relaxation times involved.

The experimental procedure is illustrated in Figs. 1 and 3. The two spin species, I (or A) and S (or B) are first polarized in a strong external magnetic field $H_0 \parallel z$. Then we apply along the x direction a radiofrequency field pulse $H_1(t)$ which is resonant, $\omega_I = \gamma_I H_0$, with respect to the I spins and has an amplitude H_{1I} and a pulse length t

$$\gamma H_{1I} t = \pi/2 \tag{II.1}$$

so that it turns the I magnetization from the z into the y direction. Next we change the phase of the above r.f. field by $90°$ so that the r.f. field and the I magnetization are parallel to each other. For an uncoupled system we would now have a stationary situation, i.e. the I magnetization and the r.f. field are locked and would remain parallel to each other for ever. In view of the coupling

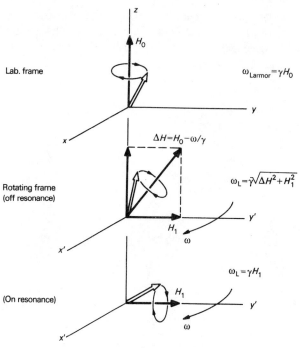

Fig. 1.   Motion of a spin as viewed from the laboratory and rotating frames of reference.

of the *I* spins to the lattice, however, this state decays with a time constant $T_{1\rho}$, which we call the spin–lattice relaxation time in the rotating frame. For times short compared to $T_{1\rho}$ a spin temperature Θ can be defined in the rotating

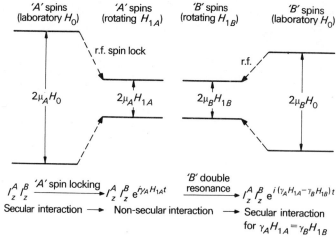

Fig. 2.   Double resonance in the rotating frame ($A \equiv I$ and $B \equiv S$).

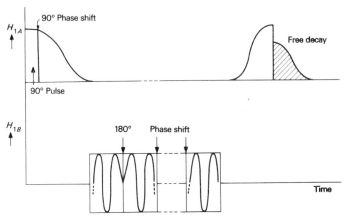

Fig. 3.   Pulse sequence for double resonance in the dipolar frame.

frame which is related to the spin magnetization in this frame by

$$M = C \cdot H_{1I}/\Theta \tag{II.2a}$$

C being a Curie constant

$$C = \gamma_i^2 \hbar^2 I (I+1) N / 3k \tag{II.2b}$$

It should be noted that the spin magnetization in the rotating frame is the same as it was in the laboratory frame

$$M = C \cdot H_0/\Theta_L \tag{II.3a}$$

before the $\pi/2$ pulse. Hence the spin temperature in the rotating frame is related to the spin temperature in the laboratory frame Θ_L by

$$\Theta = \Theta_L H_{1I}/H_0 \tag{II.3b}$$

and $\Theta \ll \Theta_L$, i.e. the I spin system is cooled as a result of transfer from the laboratory to the rotating frame. The splitting between the energy levels of this system is now γH_{1I} if we can for the moment neglect the local dipolar fields.

The above procedure has had so far no effect on the S spins provided that the I and S Larmor frequencies are far apart. The S magnetization is still parallel to H_0.

Let us now turn on a second r.f. field, perpendicular to H_0, which is resonant with respect to the S spins, $\hbar \omega_S = \Delta E_S$. Here ΔE_S corresponds to a given quadrupole perturbed Zeeman splitting of the energy levels of the S system. The amplitude of the r.f. field should be H_{1S}. The initial S spin magnetization along H_{1S} is zero. According to Eq. (II. 2) this means that the spin temperature in the rotating frame of the S spins is infinitely high. If resonant energy transfer is possible, i.e. if the energy level splittings in the two rotating frames $\Delta E'_I$ and $\Delta E'_S$ are matched (see Fig. 2), then

$$\Delta E'_I = \Delta E'_S$$

or

$$\gamma_I H_{1I} = \gamma_S H_{1S} \tag{II.4a}$$

with

$$\hbar \omega_S = \Delta E_S \tag{II.4b}$$

the spin temperatures of the I and S system will soon become equal because of dipolar interactions, and a finite S magnetization along H_{1S} is created. In order that the heat flow from the S to the I system does not stop, we invert the phase of H_{1S} for 180°. After that, the S magnetization and the field H_{1S} have opposite directions so that the S spin system has a negative spin temperature [see Eq. (II.2a)]. Since negative temperatures are, as a matter of fact, 'hotter' than positive temperatures, the heat flow from the S to the I system continues. The procedure is repeated for a time of the order of $T_{1\rho}$.

The requirement of having two simultaneous strong r.f. fields during the double resonance experiment can be avoided by performing an adiabatic demagnetization in the rotating frame of the I spins (Fig. 3). The I spins are in this case spin locked in their dipolar frame and the double resonance condition (II.4a) has to be replaced by

$$\gamma_I H_{\text{loc}} \approx \gamma_S H_{1S} \tag{II.4c}$$

where H_{loc} is the local dipolar field acting on the I spins. The double resonance conditions (II.4a) and (II.4c) are strictly valid only for nuclei with no quadrupole interactions. Rigorously one should of course match the rotating-frame energy level splitting of the I system to the rotating-frame energy level splitting of the S system, which is a quadrupolar system in a high magnetic field. The difference is, however, small [see Eq. (III.2)]. Since the heat flow from S to I does not change the total energy of the combined system, it can be easily shown that after n 180° phase changes (Fig. 3) the common spin temperature Θ_n equals

$$\left(\frac{1}{\Theta_n}\right) = \left(\frac{1}{\Theta_0}\right)/(1+\varepsilon)^n \approx \frac{1}{\Theta_0} \exp\left(-n\varepsilon\right) \tag{II.5}$$

where $\varepsilon \ll 1$ is the ratio of the magnetic heat capacities of the two systems. After that, the I spins are adiabatically remagnetized and the free precession signal of the I spins is measured as:

$$M \approx M_0 \exp\left(-n\varepsilon\right) \tag{II.6}$$

if spin–lattice relaxation effects can be neglected. In practice ΔE_S is not known and the frequency ω_S as well as the amplitude H_{1S} have to be adjusted to satisfy both the 'resonance' (II.4b) and the 'double resonance' (II.4c) conditions.

Normally H_{1S} is fixed and ω_S is varied. The resonant value of ω_S is detected as the value where the S and, through it, the I spin systems are heated resulting in a decrease in the free precession signal of the I spins.

A more quantitative description of the above procedure is as follows. Let the spin Hamiltonian of the total system be a sum of Zeeman, radiofrequency, dipole–dipole and nuclear electric quadrupole terms

$$\mathcal{H} = \mathcal{H}_{ZI} + \mathcal{H}_{ZS} + \mathcal{H}_{r.f.I} + \mathcal{H}_{r.f.S} + \mathcal{H}_{dII} + \mathcal{H}_{dSS} + \mathcal{H}_{dIS} + \mathcal{H}_{QS} \tag{II.7}$$

Here

$$\mathcal{H}_{ZI} = -\hbar\gamma_I H_0 \sum_k I_{zk} \tag{II.8a}$$

$$\mathcal{H}_{ZS} = -\hbar\gamma_S H_0 \sum_l S_{zl} \tag{II.8b}$$

$$\mathcal{H}_{r.f.I} = -2\hbar\gamma_I H_{1I} \sum_k I_{xk} \cos\omega_I t \tag{II.8c}$$

$$\mathcal{H}_{r.f.S} = -2\hbar\gamma_S H_{1S} \sum_l S_{xl} \cos\omega_S t \tag{II.8d}$$

etc.

The transformation from the laboratory to the doubly 'rotating' frame is performed by the operators T_I and T_S, where

$$T_I = \exp\left(+i\mathcal{H}_{ZI}t\right) \tag{II.9a}$$

and

$$T_S = \exp\left(+i\mathcal{H}_S t\right) \tag{II.9b}$$

for the I and S systems respectively. Here $\mathcal{H}_S = \mathcal{H}_{ZS} + \mathcal{H}_{QS}$. In view of the quadrupole term H_{QS}, T_S does not perform a transformation into a 'pure' rotating frame, but into a more general interaction representation. For the sake of simplicity, however, we shall continue to call this frame a rotating one.

The time dependent Schrödinger equation in the laboratory frame is

$$i\hbar|\dot{\psi}\rangle = \mathcal{H}|\psi\rangle \tag{II.10}$$

The wave function of the total system in the rotating frame $|\phi\rangle$, which is obtained from $|\psi\rangle$ as

$$|\phi\rangle = T|\psi\rangle \tag{II.11a}$$

with

$$T = T_S T_I \tag{II.11b}$$

obeys the following equation

$$i\hbar|\dot{\phi}\rangle = \mathcal{H}'|\phi\rangle \tag{II.12}$$

where

$$\mathcal{H}' = T^{-1}\mathcal{H}T - T^{-1}\dot{T} \tag{II.13}$$

and explicitly

$$\mathcal{H}' = -\hbar\gamma_I H_{1I} \sum_k I_{xk} - \hbar\gamma_S H_{1S} \sum_l S_{xl} + \mathcal{H}^0_{dII} + \mathcal{H}^0_{dSS} + \mathcal{H}^0_{dIS} \tag{II.14a}$$

plus small time-dependent parts, which can be neglected. The Hamiltonian in the doubly rotating frame, where the I spins rotate around the z-axis with a frequency ω_I and the S spins with a frequency ω_S, is thus given by

$$\mathcal{H}' = \mathcal{H}'_I + \mathcal{H}'_S + \mathcal{H}'_{IS} \tag{II.14b}$$

Here $\mathcal{H}^0_{dII}$, $\mathcal{H}^0_{dSS}$ and $\mathcal{H}^0_{dIS}$ represent the secular part of the corresponding dipole–dipole operators, i.e. the part which commutes with the Zeeman operators $\mathcal{H}_{ZI} + \mathcal{H}_{ZS}$. The neglected terms oscillate at frequencies ω_I, ω_S, $2\omega_I$, $2\omega_S$, $\omega_I - \omega_S$, $\omega_I + \omega_S$, which are all large as compared to the resonance frequencies in the rotating frame ($\omega_{1I} = \gamma_I H_{1I}$, $\omega_{1S} = \gamma_S H_{1S}$) and can hence be safely neglected. The transformed Hamiltonian $\mathcal{H}'$ is, in contrast to $\mathcal{H}$, time independent. Double resonance will occur when the differences between the eigenvalues of $\mathcal{H}'_I$ and $\mathcal{H}'_S$ become equal.

Another important fact is that in view of the time independence of $\mathcal{H}'$, the system can be described by a spin temperature in the rotating frame Θ for times which are short compared to $T_{1\rho}$. The density matrix of the total system

$$\sigma = \exp\left(-\mathcal{H}'/k\Theta\right)/\text{Tr}\,\exp\left(-\mathcal{H}'/k\Theta\right) \tag{II.15}$$

is practically always equal to

$$\sigma = (1 - \mathcal{H}'/k\Theta)/\text{Tr}\,1 \tag{II.16}$$

as $\langle\mathcal{H}'\rangle \ll k\Theta$. The total energy of the system is

$$E' = \text{Tr}\,(\sigma\mathcal{H}') = E'_I + E'_S + E'_{IS} \approx E'_I + E'_S \tag{II.17}$$

where, for instance,

$$E'_I = -\text{Tr}\,(\mathcal{H}'^2_{\text{r.f.}I} + \mathcal{H}^{02}_{dII})/k\Theta\,\text{Tr}\,1 = -C_I(H^2_{1I} + H^2_L)/\Theta \tag{II.18a}$$

with the local field H_L being defined as

$$C_I H^2_L/\Theta = \text{Tr}\,(\sigma\mathcal{H}^0_{dII}) \tag{II.18b}$$

The effective field for the I spins is thus

$$H_{eI} = (H^2_{1I} + H^2_L)^{1/2}$$

and the condition for resonant energy transfer

$$\Delta E'_I = \Delta E'_S \tag{II.19}$$

becomes

$$\hbar \gamma_I H_{el} = \Delta E'_s = E'_{S,p} - E'_{S,q} \qquad (II.20)$$

It should be pointed out that for a double resonance experiment to be successful, spin diffusion within the I spin system must take place. The S spins interact significantly only with their nearest neighbours, and a uniform spin temperature within the I spin system can only result from fast spin diffusion.

So far we have neglected the fact that the I spin magnetization in the rotating frame is not really constant, but decays with a time constant $T_{1\rho}$. If we wish to estimate the signal-to-noise ratio with which the S spectra can be measured by the double resonance method, this decay has to be taken into account.

After the end of the double resonance cycle the I magnetization precessing around H_0 is not really given by Eq. (II.6) but by

$$M = M_0 \exp - (n\varepsilon + t/T_{1\rho}) \qquad (II.21a)$$

or, in another notation, by

$$M = M_0 \exp - (Wt + t/T_{1\rho}) \qquad (II.21b)$$

The difference ΔM between the free precession signal of the I spins without $(W = 0)$ and with $(W \neq 0)$ resonant energy transfer from the S system is maximum when

$$t = \ln (1 + WT_{1\rho})/W \qquad (II.22)$$

Then

$$\frac{\Delta M}{M_0} = \frac{WT_{1\rho}}{[1 + WT_1]^{(1+1/WT_{1\rho})}} \qquad (II.23)$$

Double resonance can be observed if the decrease of the I magnetization ΔM due to the heat flow from the S system is larger than the noise ΔM_0 of the I signal M_0. The signal-to-noise ratio $(S/N)_S$ with which the S spectra can be measured is thus proportional to the signal-to-noise ratio $(S/N)_I$ of the I system. From Eq. (II.23), we easily find that

$$(S/N)_S = (S/N)_I \cdot \frac{WT_{1\rho}}{(1 + WT_{1\rho})} \cdot f(x) \qquad (II.24)$$

where

$$f(x) \to 1/2.7 \quad \text{for } x = WT_{1\rho} \ll 1 \quad \text{and}$$

$$f(x) \to 1 \qquad \text{for } x \gg 1$$

Thus

$$(S/N)_S \approx (S/N)_I \qquad (II.25)$$

if $WT_{1\rho} \to \infty$ and

$$(S/N)_s \to 0 \qquad \qquad (II.26)$$

if $WT_{1\rho} \to 0$. A necessary condition for the usefulness of double resonance techniques in the rotating frame is thus the presence of a spin species with not only a large signal but also a long spin–lattice relaxation time in the rotating frame.

B. Applications

By measuring the I–S double resonance spectra as a function of the orientation of the single crystal in the magnetic field H_0, both the magnitude of the S spin quadrupole coupling tensor and its orientation with respect to the crystal axes can be determined[3,4] with a sensitivity which exceeds that of quadrupole perturbed NMR.

A typical example is provided by the ^{14}N-proton double resonance study[4] of the ferroelectric phase transition in triglycine sulfate (TGS). The angular dependence of the ^{14}N transition frequencies in paraelectric and ferroelectric TGS is shown in Fig. 4 for $c \perp H_0$. The components of the ^{14}N quadrupole coupling tensors were determined from crystal rotation data about three mutually perpendicular axes in both phases. In the paraelectric phase, there are three physically non-equivalent nitrogen sites (glycine I, glycine II, and glycine III), two of which are also chemically non-equivalent. As it was found that the direction of the largest principal axis of the ^{14}N electric field gradient tensor agrees with the C—N bond direction, it was possible to assign the observed ^{14}N electric field gradient tensors to the various glycine groups in a unique way. The chemical equivalence of the ^{14}N sites in G(II) and G(III), which are connected

Table 1. Components of the ^{14}N quadrupole coupling tensors V_{ij} (in kilohertz) in TGS expressed in the crystal-fixed coordinate system $a \sin \beta, b, c$

	$T<T_c$			$T>T_c$		
	$a \sin \beta$	b	c	$a \sin \beta$	b	c
G(I)	−747	0	280	−740	0	208
	0	−107	±350	0	−140	0
	280	±350	853	208	0	880
G(II)	960	0	−243	960	0	−208
	0	−493	±73	0	−480	±52
	−243	±73	−467	−208	±52	−480
G(III)	880	0	−243	960	0	−208
	0	−467	±73	0	−480	±52
	−243	±73	−453	−208	±52	−480

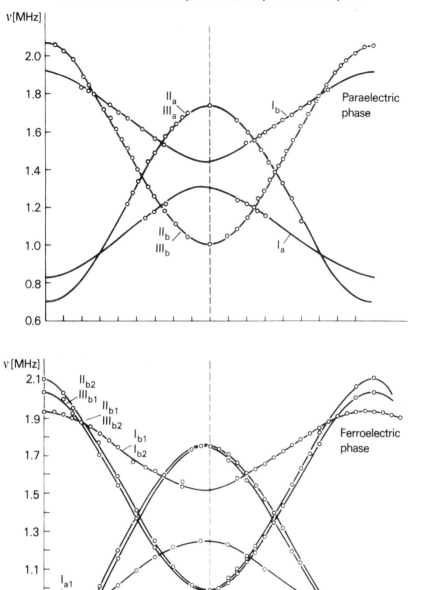

Fig. 4.   Double NMR ^{1}H–^{14}N in a TGS single crystal. Angular dependence of the ^{14}N transition frequencies in para- and ferroelectric TGS for $c \perp H_0$ ($H_0 = 4\ 228$ gauss).

with a short O—H ··· O hydrogen bond, demonstrates the existence of fast proton exchange, O—H ··· O⇌O ··· H—O, between these two glycine groups. This proton exchange stops in the ferroelectric phase and all three glycines become chemically non-equivalent. The protons thus move in a symmetric double minimum potential above T_c which becomes asymmetric below T_c. There are six physically non-equivalent ^{14}N sites in the unit cell below T_c (Table 1). The paraelectric ^{14}N electric field gradient tensors are very nearly the average of the ferroelectric ones, thus suggesting that the transition is of the order–disorder type. The temperature dependence of the ^{14}N transition frequencies on cooling through the Curie point is shown in Fig. 5.

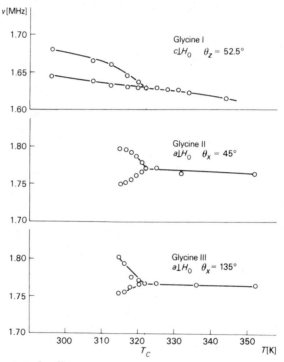

Fig. 5. Double NMR ^{1}H–^{14}N in a TGS single crystal. Temperature dependence of the ^{14}N transition frequencies on cooling through the Curie point.

III. DOUBLE RESONANCE BETWEEN THE LABORATORY AND THE ROTATING FRAMES: ZERO-FIELD QUADRUPOLE DOUBLE RESONANCE

The main disadvantage of double resonance in the rotating frame is that single crystals are required. The application of an external magnetic field to a powder sample results namely in a smearing out of the quadrupole energy levels of the S spins, making the S spectrum unobservable.

As single crystals are in many cases hard to obtain, it is often desirable to use zero-field quadrupole double resonance spectroscopy[5,6] where one can work with polycrystalline samples. This method combines the simplicity of pure NQR with the sensitivity of nuclear double resonance. The information extracted is however not as complete as in high-field nuclear quadrupole double resonance, since now the orientation of the principal axes of the electric field gradient tensors with respect to the crystal axes cannot be obtained.

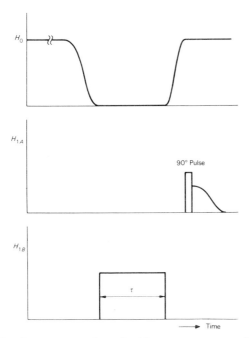

Fig. 6.   Field and pulse sequence in a double resonance experiment between the laboratory and the rotating frame.

The experimental procedure is as follows (Fig. 6). The I (or A) spins are first ordered in a high external magnetic field and then cooled by adiabatic demagnetization to zero external field. This is usually performed by moving the sample out of the magnet. In the demagnetized state, the S (or B) spins are exposed to a search field H_{1S}, which induces nuclear transitions if its frequency ω_S matches the quadrupole energy-level splitting of the S spins:

$$\hbar\omega_S = (\Delta E)_S \qquad\qquad (\text{III.1})$$

H_{1S} thus heats up the S spin system and cross-relaxation causes a loss of order within the I system. By inverting the phase of H_{1S} for 180° after a time of the order of the I–S cross-relaxation time, the heat exchange can be prolonged for a time of the order of the dipolar spin–lattice relaxation time $(T_{1D})_I$. Finally, the sample is moved back into the magnet and the remaining I magnetization is determined by measuring the I free induction decay signal following a 90° pulse. The S spin system can pass on the absorbed energy to the I spin system only if the energy level splitting of the I spins in the local dipolar field H_L is matched by the quadrupole energy level splitting of the S spins in the rotating frame

$$(\gamma_I H_{\text{loc}})_I = \sqrt{S(S+1) - \tfrac{1}{4}(4m_q^2 - 1)}\gamma_S H_{1S} \tag{III.2}$$

This is of course just the 'double resonance' condition for a quadrupolar system, which is not too different from the simple expression (II.4c). It should be noted that expression (III.2) is experimentally hard to fulfil if $\gamma_I \gg \gamma_S$, which is usually the case. In such a case, it is advisable to increase the effective field in the rotating frame of the S spins by working slightly off resonance:

$$(\gamma H_{\text{eff}})_S = \sqrt{(\gamma H_1)_S^2 + (\Delta\omega)^2}, \qquad \text{where } \Delta\omega = \omega_{\text{res}} - \omega \tag{III.3}$$

An example of the use of this method is the determination of the ^{17}O NQR spectra in KH$_2$PO$_4$ via proton $-^{17}$O$(S = 5/2)$ double resonance.[7]

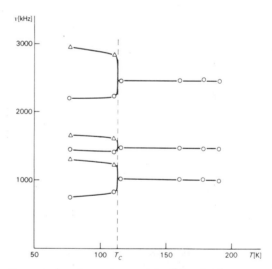

Fig. 7. Temperature dependence of the ^{17}O NQR spectra in KH$_2$PO$_4$.

The results are shown in Fig. 7. The ^{17}O NQR spectra consist in the paraelectric phase of three lines, the positions of which do not depend on temperature. At the ferroelectric transition temperature T_c, there is an abrupt change in the ^{17}O spectrum. In the ferroelectric phase, we have six ^{17}O lines the positions of which are again nearly temperature independent. The principal values of the ^{17}O quadrupole coupling tensor are determined from the

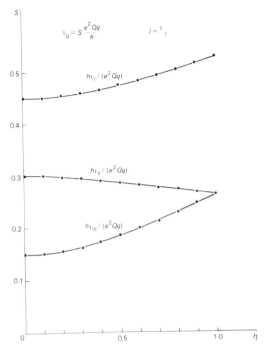

Fig. 8.   The ratio between the NQR transition frequencies and the quadrupole coupling constant for $S = 5/2$ as a function of the asymmetry parameter, η.

observed transition frequencies with the help of Fig. 8, which gives the ratio of the transition frequencies to the quadrupole coupling constant as a function of the asymmetry parameter. Above T_c, all ^{17}O sites in the unit cell are chemically equivalent. The ^{17}O quadrupole coupling constant is equal to $e^2qQ/h = 5.16\,MHz$ and the asymmetry parameter $\eta = 0.55$. Below T_c, we have two chemically non-equivalent ^{17}O sites: $(e^2qQ/h)_1 = 5.96\,MHz$, $\eta_1 = 0.72$ and $(e^2qQ/h)_2 = 4.85\,MHz$, $\eta_2 = 0.18$.

The temperature dependence of the ^{17}O spectra can be described by the following model. The electric field gradient tensor at a ^{17}O site in KH_2PO_4 is determined by the position of the proton in the hydrogen bond linking the ^{17}O site to neighbouring PO_4 tetrahedra. The hydrogen bond potential has two equilibrium sites, $^{17}O-H\cdots O$ and $^{17}O\cdots\cdots H-O$, so that the proton can be found either in a 'far' or in a 'close' position. Above T_c, the proton moves between these two sites. With the help of the normalized $O-H\cdots O$ bond dipole moment $p_i(t)$, we can express the time dependence of the ^{17}O electric field gradient tensor $T_i(t)$ as:

$$T_i(t) = \tfrac{1}{2}[1 + p_i(t)]T(1) + \tfrac{1}{2}[1 - p_i(t)]T(2) \tag{III.4}$$

where $T(1)$ is the tensor for the ^{17}O site to which the proton is directly attached ($-^{17}O-H\cdots O-$), whereas $T(2)$ is the tensor for the ^{17}O site to which the proton is only hydrogen bonded ($^{17}O\cdots H-O$). For $p_i = 1$, $T_i = T(1)$, whereas $T_1 = T(2)$ for $p_i = -1$.

Since the frequency of the proton motion is much larger than the difference in the quadrupole splitting between sites 1 and 2, the ^{17}O nucleus sees only an effective, time-averaged, electric field gradient. Since for $T > T_c$ the proton spends an equal amount of time in each of these two equilibrium sites, $\langle p(t) \rangle = 0$ in the paraelectric phase and $\langle T_i(t) \rangle$ is temperature independent:

$$T > T_c : \langle T \rangle = T_0 = \tfrac{1}{2}[T(1) + T(2)] \tag{III.5}$$

Below T_c, on the other hand, $\langle p(t) \rangle$ is different from zero and equals the normalized spontaneous polarization p. We find

$$T < T_c : \langle T \rangle = T_0 + A \cdot p \tag{III.6}$$

where

$$A = \tfrac{1}{2}[T(1) - T(2)] \tag{III.7}$$

The temperature dependence of the ^{17}O electric field gradient tensor on going through T_c should be thus equal to that of the spontaneous polarization, as is indeed observed. It should be noted that if $p = 1$, $\langle T \rangle = T(1)$, whereas $\langle T \rangle = T(2)$ for those ^{17}O sites where $p = -1$.

The above results thus confirm the existence of a double-minimum type $O-H\cdots O$ potential in KH_2PO_4. They further show that the ^{17}O electric field gradient tensor is nearly completely determined by the motion of its closest proton and does not depend on the arrangement of the four protons surrounding a PO_4 group.

The above technique has one important limitation. The magnetic dipole–dipole coupling of integer spin nuclei in a non-axially symmetric electric field gradient to neighbouring nuclei of half-integer spin is quenched[8] in zero magnetic field: $\langle a | \mathcal{H}_{dis} | b \rangle = 0$. A double resonance experiment in the above form thus cannot be performed as the I and the S spins are decoupled. This prevents, for example, the use of the above technique for proton–^2H or proton–^{14}N double resonance. This difficulty can be avoided in two different ways:

(a) by level crossing in a low magnetic field

and

(b) by the 'solid effect' using strong radiofrequency fields at $\omega = \omega_I \pm \omega_S$.

In the following we shall discuss these two techniques separately. It should be noted that they both allow the use of polycrystalline samples.

IV. DOUBLE RESONANCE IN THE LABORATORY FRAME: LEVEL CROSSING IN A LOW MAGNETIC FIELD

A. Experimental

The method of level crossing in the laboratory frame[9-12] can be used as a high-sensitivity double resonance detection method in cases where the nuclei of one, say S, of the two spin species present in the sample (I and S) has a non-zero electric quadrupole moment in addition to magnetic dipole moments. The I system must be chosen to have a larger gyromagnetic ratio, γ_I, than the S system. In the absence of an external magnetic field, the Zeeman splitting of the spin levels is zero, while the quadrupole splitting in the local electric field gradient is non-zero. In a high magnetic field, on the other hand, the Zeeman splitting may be larger than the quadrupole one. Somewhere between these two extremes, the Zeeman splittings of the I system and the Zeeman perturbed electric quadrupole splittings of the S system are the same. At this value of the magnetic field, resonant energy transfer between the two spin species is possible, and if the spin temperatures of the two systems differ, they soon equalize. The effect of a radiofrequency perturbation of the S system can now be transferred via I–S dipole–dipole coupling to the I system, integrated, and detected as a change in the NMR signal of the I spins.

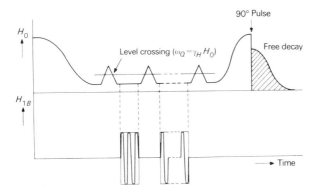

Fig. 9. Pulse sequence and variation of the Zeeman magnetic field with time for level-crossing double resonance in the laboratory frame.

The procedure is illustrated in Fig. 9 and the block diagram of a typical spectrometer is given in Fig. 10. The proton system (I spins) is first polarized in a high magnetic field H_0 and then adiabatically demagnetized to zero field by moving the sample out of the magnet. In this process, the spin temperature of the proton system decreases from T_L to a rather low value $T_I = T_L(H_L/H_0)$ where H_L is the local field at the protonic sites. The spin system (S or B) is now irradiated with a strong radiofrequency field. If the frequency ν_S of this field

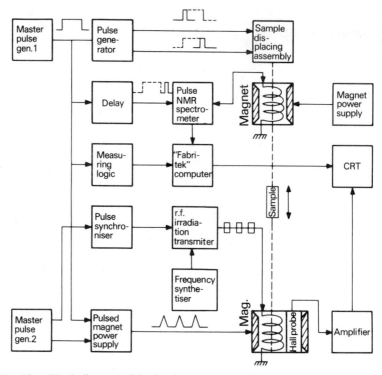

Fig. 10. Block diagram of the level-crossing double resonance spectrometer.

corresponds to a given splitting of the quadrupole energy levels, $\Delta E_a = h\nu_S$, the populations of the two levels in question equalize, resulting in an infinite spin temperature of the S system. After that, the proton system is partially remagnetized by applying a magnetic field H_0' so that the level splittings of the S and I systems match: $(\omega_Q)_S = \gamma_H H_0'$. The spin populations will now exchange by mutual 'spin-flips' induced by dipolar interactions until the ratios of the populations of the upper and lower levels of the two spin systems will be the same. This means that the spin temperature of the proton system is raised and some of its magnetization is lost. The whole process can be repeated many times by applying a low-frequency magnetic field to the sample, so that multiple level crossing occurs. The S system is irradiated when the magnetic field is close to zero. Since the effect of destroying the proton magnetization by bringing it into contact with the 'hot' S spin reservoir is cumulative, the total effect is larger than after a single level crossing. Finally, the sample is moved back into the magnet—where the field equals H_0—and the remaining proton magnetization is determined by applying a 90° pulse and storing the free induction decay signal into the memory of a Fabritek computer. Now the whole procedure is repeated without irradiating the S system and the difference between the proton signals with and without S irradiation is recorded. After that, the whole cycle is started again with a different S irradiation frequency. The S NQR spectrum is obtained by plotting the difference in the proton signal with and without S irradiation versus the S irradiation frequency ν_S. One complication comes from the fact that level crossing has to occur in a finite magnetic field H_0', which broadens out the energy levels of the quadrupolar nuclei in a powdered sample where each crystallite makes a different angle with the direction of the external magnetic field.

Let us now try to estimate this effect for a spin $S = 1$ nucleus in a highly asymmetric field gradient. Level crossing occurs when

$$(\gamma H_0)_I \approx \tfrac{3}{4}(e^2 qQ/h)_S$$

The magnetic field has a negligible effect on the energy levels of the S system as long as the quadrupole coupling constant is much larger than the Zeeman energy: $(e^2 qQ/4h)_S \geq 5(\gamma H_0)_S$. These two conditions can be simultaneously fulfilled if the gyromagnetic ratio of the I spins is much larger than the gyromagnetic ratio of the S spins

$$\gamma_I \geq 15\gamma_S$$

The smearing out of the quadrupole lines of polycrystalline samples in a level-crossing experiment is thus negligible for ^{14}N, but may not be quite negligible for ^{2}H as first pointed out by Edmonds *et al.* (11).

B. Sensitivity for ^{14}N detection

^{14}N is a naturally abundant spin-1 nucleus, the NQR spectrum of which exhibits in zero magnetic field three allowed transitions:

$$\nu_3 = \frac{3}{4} \cdot \frac{e^2 qQ}{h}\left(1 + \frac{1}{3}\eta\right) \tag{IV.1a}$$

$$\nu_2 = \frac{3}{4} \cdot \frac{e^2 qQ}{h}\left(1 - \frac{1}{3}\eta\right) \tag{IV.1b}$$

$$\nu_1 = \nu_3 - \nu_2 = \frac{1}{2}\frac{\eta e^2 qQ}{h} \tag{IV.1c}$$

Here η is the asymmetry parameter and $e^2 qQ/h$ the ^{14}N quadrupole coupling constant.

In view of the interest in measuring the ^{14}N NQR spectra of complex organic molecules, it seemed worthwhile to estimate the signal-to-noise ratio with which the three ^{14}N transitions can be observed by single or multiple proton–nitrogen level crossing in the laboratory frame. Let us designate the initial populations of the upper and lower protonic Zeeman levels before level crossing by

$$n_2 = n(1 - \alpha) \tag{IV.2a}$$

and

$$n_1 = n(1 + \alpha) \tag{IV.2b}$$

respectively, where $2n$ is number of protons per unit volume. Similarly we denote the initial populations of the three ^{14}N energy levels by

$$N_3 = N(1 - \alpha) \tag{IV.3a}$$

$$N_2 = N(1 - \beta) \tag{IV.3b}$$

$$N_1 = N(1 + \alpha + \beta) \tag{IV.3c}$$

where $3N$ is the number of chemically equivalent nitrogens per unit volume and $E_3 > E_2 > E_1$, with $h\nu_1 = E_3 - E_2$, $h\nu_2 = E_2 - E_1$, and $h\nu_3 = E_3 - E_1$.

Before level crossing with, for instance, the lowest quadrupole transition, we have

$$N_3/N_2 \neq n_2/n_1 \tag{IV.4}$$

whereas after level crossing $(\Delta E)_H = (h\nu_1)_{14_N}$ the following conditions have to be met:

$$N_3'/N_2' = n_2'/n_1' \tag{IV.5a}$$

$$N_3' + N_2' = N_3 + N_2 \tag{IV.5b}$$

$$n_2' + n_1' = n_2 + n_1 \tag{IV.5c}$$

$$N_3 + n_2 = N_3' + n_2' \tag{IV.5d}$$

Equations (IV.5b) and (IV.5c) express the conservation of the number of particles, whereas Eq. (IV.5d) represents the energy conservation law. Expression (IV.5a) could be interpreted as saying that the spin temperature of the two '2-level' spin systems become equal. From these expressions, the new equilibrium populations after level crossing are found.

The simplest 'single' level-crossing cycle contains in fact six level crossings: when the system is demagnetized, the separation between the protonic energy levels crosses successively $h\nu_3$, $h\nu_2$, and $h\nu_1$, whereas on remagnetization $h\nu_1$, $h\nu_2$, and $h\nu_3$ are crossed. If a given ^{14}N quadrupole transition is saturated by irradiation with a strong radiofrequency field during the time the system is in zero magnetic field, the population of the proton levels after level crossing is different from the one obtained without nitrogen irradiation, thus allowing a precise determination of the ^{14}N NQR spectrum. Though the pure level-crossing method without nitrogen irradiation works as well, we shall limit ourselves to the more precise method with irradiation. The signal-to-noise ratio $(S/N)_{14_N}$ with which the ^{14}N NQR spectra can be measured is now given as

$$(S/N)_{14_N} = (S/N)_H (\Delta\alpha/\alpha_0) \tag{IV.6}$$

where $(S/N)_H$ is the signal-to-noise ratio with which the proton signal can be measured and $\Delta\alpha = (\alpha_f) - (\alpha_f)_{irr}$ is the difference in the remaining proton magnetization at the end of the level-crossing cycle without and with radiofrequency saturation of a given ^{14}N transition. α_0 is the equilibrium proton magnetization in the magnetic field H_0.

In the following, we shall assume that $N/n \ll 1$ and evaluate $\Delta\alpha/\alpha_0$ for the cases of single and multiple level crossing. The sensitivity depends critically on the values of the proton spin–lattice relaxation time in zero-field T_{1D} and on the nitrogen spin–lattice relaxation time T_{1Q}, a dependence which limits the useful length of the level-crossing experiment.

(i) SINGLE LEVEL CROSSING CYCLE

$$1.\quad T_{1D} \gg \tau, \qquad T_{1Q} \gg \tau$$

This is the case where both the proton and the nitrogen spin–lattice relaxation times are long compared to the level-crossing cycle time τ; under these conditions, the double resonance method works at its best.

For the three ^{14}N quadrupole transitions ν_1, ν_2, and ν_3, we find

$$(\Delta\alpha/\alpha_0)_{\nu_1} = (12/8)(N/n) \tag{IV.7a}$$

$$(\Delta\alpha/\alpha_0)_{\nu_2} = (2/8)(N/n) \tag{IV.7b}$$

$$(\Delta\alpha/\alpha_0)_{\nu_3} = (27/8)(N/n) \tag{IV.7c}$$

The relative intensities of the three ^{14}N NQR lines are

$$I_1 : I_2 : I_3 \approx 6 : 1 : 13 \tag{IV.7d}$$

The line with the highest frequency is the most intense. The low-frequency line is about half as strong, whereas the medium frequency line is much weaker and usually hard to detect.

$$2.\quad T_{1D} \ll \tau, \qquad T_{1Q} \gg \tau$$

The nitrogen spin–lattice relaxation time is long, whereas the proton relaxation time is short. This is the case which is often found in amino acids and other biological molecules in the region between room temperature and 77 K. Here the protons are remagnetized by the final level crossing with the nitrogens.

We get

$$(\Delta\alpha/\alpha_0)_{\nu_1} = (96/64)(N/n) \tag{IV.8a}$$

$$(\Delta\alpha/\alpha_0)_{\nu_2} = (25/64)(N/n) \tag{IV.8b}$$

$$(\Delta\sigma/\alpha_0)_{\nu_3} = (221/64)(N/n) \tag{IV.8c}$$

The surprising result is that in this case the sensitivity is not lower than in the first case though the proton relaxation time is short. This situation is different from the one usually found in nuclear double resonance, where the signal of the strong spin species is destroyed by thermal contact with the weak spin species the spectrum of which we wish to measure.

The relative intensities of the three NQR lines are

$$I_1 : I_2 : I_3 \approx 4 : 1 : 9 \tag{IV.8d}$$

3. $T_{1D} \gg \tau, \qquad T_{1Q} \ll \tau$

Here the situation is just the opposite to the one before. The proton T_{1D} is long and the ^{14}N T_{1Q} short, so that the nitrogens are at the lattice temperature at the end of the cycle. Now,

$$(\Delta\alpha/\alpha_0)_{\nu_1} = (12/16)(N/n)(\nu_1/\nu_0) \tag{IV.9a}$$

$$(\Delta\alpha/\alpha_0)_{\nu_2} = (5/16)(N/n)(\nu_2/\nu_0) \tag{IV.9b}$$

$$(\Delta\alpha/\alpha_0)_{\nu_3} = (17/16)(N/n)(\nu_3/\nu_0) \tag{IV.9c}$$

The sensitivity is here reduced—in comparison with the former cases—by a factor $\nu_i/\nu_0 (i = 1, 2, 3)$, i.e. by the ratio of the quadrupole transition frequency to the proton Larmor frequency in H_0. Exactly the same result is obtained if $T_{1D} \ll \tau, T_{1Q} \ll \tau$.

(ii) MULTIPLE LEVEL CROSSING

The sensitivity of the method can be greatly enhanced if the proton spin–lattice relaxation time is long enough that a multiple proton–nitrogen level-crossing technique can be used.

1. $T_{1D} \gg \tau, \qquad T_{1Q} \gg \tau$

Here both relaxation times are long, and instead of a single cycle containing six level crossings, p such cycles are applied before the proton magnetization is measured. The optimum number of level-crossing cycles and the total multiple level-crossing time, $\tau' = \tau p$, is larger, the longer is T_{1D} and T_{1Q}.

For the relative change in the proton magnetization after p cycles coupled with irradiation of a given ^{14}N transition, we find

$$(\Delta\alpha/\alpha_0)_{\nu_1} = (45/32)p(N/n) \tag{IV.10a}$$

$$(\Delta\alpha/\alpha_0)_{\nu_2} = (7/32)p(N/n) \tag{IV.10b}$$

$$(\Delta\alpha/\alpha_0)_{\nu_3} = (102/32)p(N/n) \tag{IV.10c}$$

If T_{1D} and T_{1Q} are really long, so that p can be made large, the ^{14}N NQR spectrum can be measured with a signal-to-noise ratio, Eq. (IV.6), which is not much smaller than the one associated with the proton signals. It is this case which seems to offer the best chances to study the NQR spectra of proteins and other biological macromolecules.

The ratio of sensitivities of the single and multiple level-crossing techniques is given by the number of the crossing cycles p

$$\frac{(\Delta\alpha)\text{ multiple level crossing}}{(\Delta\alpha)\text{ single level crossing}} \approx p \qquad\qquad (IV.11)$$

The relative intensities of the three NQR lines are now

$$I_1 : I_2 : I_3 \approx 6 : 1 : 12 \qquad\qquad (IV.12)$$

i.e. practically the same as in the single level-crossing case [Eq. (IV.7d)].

2. $T_{1D} \gg \tau'$, $\qquad \tau < T_{1Q} \ll \tau'$

Here the nitrogen spin–lattice relaxation time is long as compared to a single level-crossing cycle time τ but short in comparison with the multiple level-crossing time, $\tau' = p\tau$. The proton spin–lattice relaxation time in zero-field, on the other hand, is long.

Assuming that T_{1Q} is also short as compared to the time interval between two successive level-crossing cycles so that the nitrogen system is in thermal equilibrium with the lattice before each level crossing, we get

$$(\Delta\alpha/\alpha_0)_{\nu_1} = (48/64)p(N/n)(\nu_1/\nu_0) \qquad\qquad (IV.13a)$$

$$(\Delta\alpha/\alpha_0)_{\nu_2} = (17/64)p(N/n)(\nu_2/\nu_0) \qquad\qquad (IV.13b)$$

$$(\Delta\alpha/\alpha_0)_{\nu_3} = (65/64)p(N/n)(\nu_3/\nu_0) \qquad\qquad (IV.13c)$$

The intensities of the various transitions are again lowered by a factor $\nu_i/\nu_0 (i = 1, 2, 3)$ in comparison with the preceding case. The relative intensities of the NQR lines are, in the case of a small asymmetry parameter η,

$$I_1 : I_2 : I_3 = \tfrac{3}{2}\eta : 1 : 3 \qquad\qquad (IV.14a)$$

whereas we get for a large η

$$I_1 : I_2 : I_3 = 3 : 1 : 8 \qquad\qquad (IV.14b)$$

In the case $T_{1D} \ll \tau'$, $T_{1Q} \lesssim \tau'$, multiple level crossing makes no sense and we have to use the single level-crossing technique.

C. Results

The proton–nitrogen double resonance NQR spectra of thymine, cytosine, uracil, and guanine are shown in Fig. 11(a)–(d). The ^{14}N spectra of these compounds are of particular interest in view of the base pairing in DNA

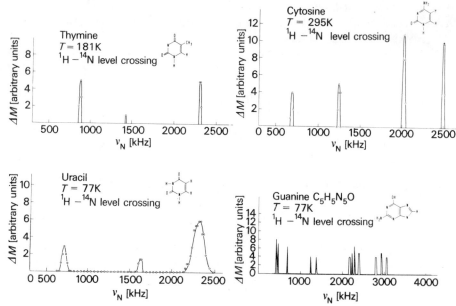

Fig. 11. ^{14}N NQR spectra of (a) thymine, (b) cytosine, (c) uracil and (d) guanine.

and RNA which occurs through N—H⋯O hydrogen bonds. The observed transition frequencies, quadrupole coupling constants, e^2qQ/h, and asymmetry parameters η are collected in Table 2. As all three ^{14}N NQR lines ν_1, ν_2, and ν_3 have not been observed in all cases, some of the assignments are only tentative.

Table 2. Observed ^{14}N NQR transition frequencies (in kilohertz), quadrupole coupling constants e^2qQ/h (in kilohertz), and electric field gradient asymmetry parameters η for some nucleic bases. Guanine and uracil have been measured at 77 K, thymine at 181 K, and cytosine at 295 K. In the case of guanine with its five nitrogen sites, the assignment of the lines is only tentative

Substance	ν_3	ν_2	ν_1	e^2qQ/h	η
Guanine (1)	3040	2380	660	3610	0.37
(2)	2940	2280	660	3480	0.38
(3)	2790	2160	630	3300	0.38
(4)	1360	1235	—	1730	0.15
(5)	740	700	—	960	0.08
Thymine (1, 2)	2300	1420	880	2490	0.71
Uracil (1, 2)	2330	1620	720	2630	0.55
Cytosine (3)	2500	—	650	2895	0.44
(1, 2)	2020	1240	—	2170	0.72

The level-crossing method has been also successfully used for the detection of the pure NQR spectra of ^{2}H ($I = 1$) in organic molecules by Edmonds and co-workers.[11] These authors have also studied the ^{14}N spectra of a large number of amino acids.[10]

V. DOUBLE RESONANCE IN THE LABORATORY FRAME VIA THE 'SOLID EFFECT'

A. Introduction

Nuclear double resonance is—as mentioned before—based on the existence of nuclear dipole–dipole coupling between the two spin systems in a frame of reference in which the energy levels of the I and S spins are equally spaced, which makes entropy transport possible. Dipolar coupling between integer nuclear spins (S spin system) in an asymmetric electric field gradient and any other non-resonant spins (I spin system) is highly reduced in zero magnetic field.[8] This effect is usually called spin quenching. The level-crossing technique discussed in the preceding section,[6, 10, 12] where the I and S spin systems couple in a non-zero magnetic field, is therefore usually used to detect pure NQR spectra of integer spin nuclei by nuclear double resonance.

In this section, we wish to show[13] that the two spin systems couple even though spin quenching occurs when a strong r.f. magnetic field is applied to the sample with a frequency which is either (a) near to one of the quadrupole transition frequencies of the S nuclei when the sample is in zero magnetic field

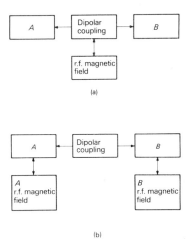

Fig. 12. Dipolar coupling between the A and B spin systems in the case of the solid effect (a) and in the case of rotating frame nuclear double resonance (b).

95

and the I spins exhibit no quadrupole coupling, or (b) equal to $\omega_I \pm \omega_S$ when the Zeeman or quadrupole coupling of the I and S nuclei is non-zero. Here ω_I is one of the transition frequencies between the energy levels of the S nuclei, and ω_S is one of the transition frequencies between energy levels of the S nuclei. This effect is one of the forms of the well-known solid effect.[14] It has already been observed by Abragam and Proctor[15] in LiF, by Landesman[16] in *para*-dichlorobenzene, and by Koo[6] and Edmonds *et al.*[10] in zero magnetic field nuclear double resonance detection of ^{14}N nuclei.

A schematic representation of the r.f. magnetic field induced coupling between the I (or A) and S (or B) spin systems as well as a schematic representation of the coupling between the I and S spin systems in the case of rotating-frame nuclear double resonance is shown in Fig. 12. This technique is of particular importance in the detection of pure NQR spectra of integer spin nuclei, since it can be used in many cases when the level-crossing technique fails.

B. Origin of the r.f. Magnetic Field Induced Coupling between Spin Systems

In this section we are going to describe the physical origin of the r.f. magnetic field induced coupling between spin systems. In the following description, we shall limit ourselves to the case of well-resolved spectra of the I and S nuclei, i.e. to the case when the energy differences between Zeeman, quadrupole or mixed energy levels of both nuclei are much larger than the resonance linewidths. The case of coupling between purely magnetic I spins

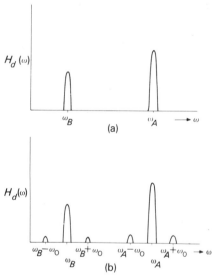

Fig. 13.   Frequency spectra of the dipolar magnetic fields produced by the A and B spins without (a) and with (b) an r.f. magnetic field applied.

and quadrupole S spins in zero magnetic field will be treated at the end of this section.

The I spins precess in a static magnetic field or in local electric field gradients at the I spin sites or in both with their resonance frequencies ω_I. The frequency spectra of the dipolar magnetic fields produced by the I spins consist therefore of discrete peaks at the frequencies ω_I. Similarly the frequency spectra of the dipolar magnetic fields produced by the S spins consist of discrete peaks at the S spin resonance frequencies ω_S. The frequency spectra of the dipolar magnetic fields produced by the I and S spins are shown in Fig. 13(a). Only one resonance frequency of the I spins and one resonance frequency of the S spins are shown.

A strong r.f. magnetic field $H_1 \cos \omega_0 t$ modulates the precession of the I and S spins. In the frequency spectra of the dipolar magnetic fields new r.f. magnetic field induced peaks appear at the frequencies $\omega_I \pm \omega_0$ and $\omega_S \pm \omega_0$. This situation is shown in Fig. 13(b). These new peaks are weak compared with the original ones, since

$$\gamma_I H_1, \quad \gamma_S H_1 \ll \omega_0 \tag{V.1}$$

which is usually the case during a double resonance experiment.

The position of the r.f. magnetic field induced peaks can be changed by changing the r.f. field frequency ω_0. At certain frequencies ω_0, the r.f. magnetic field induced peaks in the frequency spectra of the dipolar magnetic fields produced by the I spins match the resonance frequencies of the S spins and vice versa. This happens when

$$\omega_0 = \omega_I \pm \omega_S \tag{V.2}$$

When such a situation occurs, the I spins, driven by the r.f. magnetic field, induce transitions between energy levels of the S spins and at the same time the S spins, driven by the r.f. magnetic field, induce transitions between energy levels of the I spins. Since a quantum of energy $\hbar\omega_0$ gained by the I spin system from the r.f. magnetic field is different from a quantum of energy $\hbar\omega_S$ needed to produce a transition in the S spin system, the energy difference $\pm\hbar\omega_I$ should be absorbed or emitted from the I spin system during such a process, depending on whether $\omega_0 = \omega_I + \omega_S$ or $\omega_0 = \omega_I - \omega_S$. Every transition in the S spin system due to this process is therefore accompanied by a transition in the I spin system and similarly every transition in the S spin system produced by the S spins is accompanied by a transition in the I spin system. These simultaneous transitions in both spin systems are schematically represented in Fig. 14 for the case of $\omega_0 = \omega_I + \omega_S$ and for the case of $\omega_0 = \omega_I - \omega_S$.

This effect may be interpreted as follows. In the oscillating frame of reference of the I spins, the energy levels of the S spins seem to be $\hbar\omega_I$ apart and thus resonant with the energy levels of the I spins. Similarly in the oscillating frame of reference of the S spins, the energy levels of the I spins seem to be $\hbar\omega_S$ apart. Under the influence of the r.f. magnetic field, the two spin

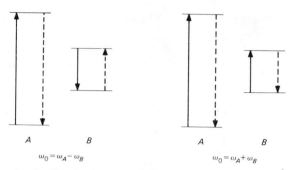

Fig. 14. Mutual spin flips in the A and B spin systems for $\omega_0 = \omega_A + \omega_B$ and $\omega_0 = \omega_A - \omega_B$.

systems relax to a common equilibrium state with a common spin temperature in a frame of reference in which the energy levels of the I and S spins are equally spaced. Here the spin temperature is defined only over the energy levels touched by the coupling process.

When only two energy levels of the I spin system and two energy levels of the S spin system are touched by the coupling process, the equilibrium is reached after

$$n_{1I}n_{2S} = n_{2I}n_{1S} \qquad \text{for} \qquad \omega_0 = \omega_I - \omega_S$$

and
$$(\text{V.3})$$

$$n_{1I}n_{1S} = n_{2I}n_{2S} \qquad \text{for} \qquad \omega_0 = \omega_I + \omega_S$$

Here n_{1I} is the number of the I nuclei on the lower energy level and n_{2I} is the number of I nuclei on the upper energy level. Similarly n_{1S} and n_{2S} are the numbers of the S nuclei on the lower and upper energy levels respectively.

For two purely magnetic spin systems with arbitrary spins in a strong magnetic field, the equilibrium is reached when

$$T_I/T_S = -\omega_I/\omega_S \qquad \text{for} \qquad \omega_0 = \omega_I + \omega_S$$

and
$$(\text{V.4})$$

$$T_I/T_S = \omega_I/\omega_S \qquad \text{for} \qquad \omega_0 = \omega_I - \omega_S$$

Here T_I and T_S are the spin temperatures, and ω_I and ω_S are the Larmor frequencies of the I and S spin systems respectively.

When a system of purely magnetic I spins is put in zero magnetic field, the only motion the spins perform is due to magnetic dipolar coupling between the I spins themselves and between the I spins and any other spins which are not decoupled from the I spins by spin quenching. The frequency spectra of the dipolar magnetic fields produced by the I spins consist of a single peak at zero frequency with a width equal to the I spin zero-field line-width $(\Delta\omega)_I$. The S spins, for which we assume non-zero quadrupole coupling, precess at their quadrupole resonance frequencies ω_{QS} in the local electric field gradients. The frequency spectra of the dipolar magnetic fields produced by the S spins consist of discrete peaks at the frequencies ω_{QS}.

The r.f. magnetic field $H_1 \cos \omega_0 t$ induces some extra peaks at the frequencies ω_0 and $\omega_{QS} \pm \omega_0$ in the frequency spectra of the dipolar magnetic fields. The peak at the frequency ω_0 has a width equal to the I spin zero-field linewidth $(\Delta\omega)_I$, whereas the S spin resonance linewidths are usually small compared with the $(\Delta\omega)_I$. The two spin systems couple whenever a quadrupole resonance line lies within the peak at the frequency ω_0. This happens for all frequencies ω_0 in the range

$$|\omega_0 - \omega_{QS}| \lesssim (\Delta\omega)_I \qquad (V.5)$$

For a given frequency ω_0 within this range, every transition in the S spin system with frequency ω_{QS} is accompanied by a transition in the I spin system with a frequency $\omega_0 - \omega_{QS}$. In this case, equilibrium in the common IS spin system is reached when

$$n_1/n_2 = 1 - \hbar(\omega_{QS} - \omega_0)/kT_I \qquad (V.6)$$

Here n_1 and n_2 are the populations of the two corresponding quadrupole energy levels of the S nuclei with energies $E_1 - E_2 = \hbar\omega_{QS}$, and T_I is the final spin temperature of the I spin system.

The two spin systems couple:

(1) when $\omega_I \mp \omega_S = \omega$. In this case, an r.f. magnetic field induced peak and an original peak in the frequency spectra of the dipolar magnetic field match;

(2) when $\omega_I \mp \omega_S = 2\omega$. In this case, two r.f. magnetic field induced peaks match.

The cross relaxation rate W_{IS} is, in case (1), approximately equal to

$$W_{IS} \approx (\Delta\omega)_{IS}[(\omega_{1I}/\omega)^2 + (\omega_{1S}/\omega)^2] \qquad (V.7)$$

whereas in case (2) it is approximately equal to

$$W_{IS} \approx (\Delta\omega)_{IS}(\omega_{1I}^2 \omega_{1S}^2/\omega^4) \qquad (V.8)$$

Here $(\Delta\omega)_{IS}$ is the broadening of the I spin resonance line due to the S spins.

For a typical case of ^{14}N and ^1H spin systems in a static magnetic field of 4 kilogauss with a typical value $(\Delta\omega)_{NH} = 6 \text{ kHz}$ and for an r.f. magnetic field amplitude 100 gauss, the cross-relaxation rate W_{NH} is in case (1) approximately equal to 0.3 s, whereas in case (2) it is of the order of 10^5 s and is thus negligibly long. For that reason, we consider only case (1) in this article.

The cross-relaxation times in the present case are approximately a factor

$$1/(\omega_{1I}^2/\omega^2 + \omega_{1S}^2/\omega^2)$$

longer than in the case of rotating-frame nuclear double resonance. This is the main disadvantage of the measuring technique using r.f. induced coupling between spin systems. However, in the case of the nuclear double resonance detection of integer spin nuclei in zero magnetic field, the presently discussed coupling mechanism is the strongest one because of spin quenching. The measuring technique using this coupling is in this case the only possible double resonance detection scheme. In other cases, the main advantage of the new measuring technique is its simplicity and convenience.

C. Experimental

(i) THE CASE OF WELL-RESOLVED SPECTRA OF THE I AND S SPINS

The technique described in this section can be used in all cases when the Zeeman, quadrupole, or mixed resonance frequencies of the I and S nuclei are significantly larger than the resonance linewidths.

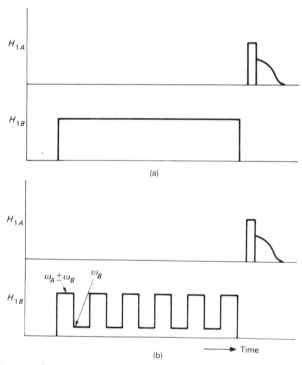

Fig. 15. Schematic representation of the nuclear double resonance process in the case of well-resolved spectra of the A and B nuclei. (a) Single coupling process; (b) multiple coupling process.

The experimental procedure is shown in Fig. 15(a). The I and S spins are left until thermal equilibrium of both spin systems with the lattice is reached. At this moment, a strong r.f. magnetic field pulse at a frequency ω is applied to the sample. After the end of this pulse, a 90° pulse is applied to the I spin system at some I spin resonance frequency ω_A and the I spin free induction decay amplitude is measured. Now the whole procedure is repeated without the strong r.f. magnetic field pulse applied and the I spin free induction decay amplitude at a frequency ω_I is again measured. The difference between the two free induction decays, which is called the double resonance signal, is recorded as a function of frequency ω.

When $\omega = \omega_I \pm \omega_S$, ω_S being a resonance frequency of the S spins, a new state in the I spin system is reached during the strong r.f. magnetic field pulse due to the r.f. magnetic field induced coupling between spin systems. This new state results in a changed I spin free induction decay amplitude and therefore in a non-zero double resonance signal. In order to get an optimum double resonance signal, the strong r.f. field pulse should last several cross-relaxation times.

The sensitivity and the resolution of the technique can be increased by applying a train of strong r.f. coupling pulses at the frequency $\omega_I \pm \omega_S$ with weak r.f. field pulses at the frequency ω_S in between (Fig. 15(b)) in order to saturate the particular transition between energy levels of the S spins. The weak r.f. field pulses at the frequency ω_S can, for example, be obtained by simultaneous attenuation and amplitude- or frequency- or phase-modulation of the carrier signal at the frequency $\omega_I \pm \omega_S$ with the known frequency ω_I. No r.f. at the frequency ω_I should get to the irradiation coil in order to prevent direct saturation of the I spin energy levels. The same r.f. field sequence can be obtained with two signal generators working at two different frequencies which are always ω_I apart.

During the experiment, the modulation frequency or the frequency difference is set to the fixed value ω_I and the frequency of the carrier signal is swept. The double resonance spectra consist in this case of broad single-coupling lines with strong narrow lines added at the frequencies $\omega_I \pm \omega_S$. The widths of these additional lines are equal to the linewidths of the corresponding S nuclei resonance lines.

In the multiple coupling process, thermal equilibrium between the S spins and the lattice does not need to be reached. When weak enough saturating r.f. field pulses are used, the resonance line-shapes of the S nuclei can be measured directly.

(ii) Double Resonance between Purely Magnetic I Spins and Quadrupole S Spins in Zero Magnetic Field

Here the basic cycle of the new measuring technique is essentially the single level-crossing cycle[5] with a strong r.f. magnetic field applied to the sample in zero static magnetic field.

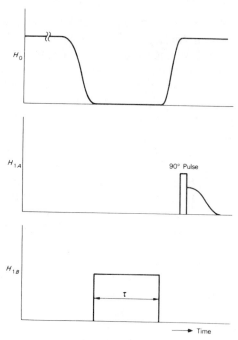

Fig. 16. Field sequence in a zero magnetic field nuclear double resonance. Single coupling process.

The procedure is illustrated in Fig. 16. The I (or A) spin system is first polarized in a high magnetic field H_0 and then adiabatically demagnetized by moving the sample out of the magnet. During this process the spin temperature of the I spin system decreases from the lattice temperature T_L to a rather low value $T_I = T_L(H_L/H_0)$. Here H_L is the local magnetic field at the I spin sites. The sample is now irradiated with a strong r.f. magnetic field at a frequency ω for a time τ. After the irradiation the sample is moved back into the magnet and the remaining I spin magnetization is determined by measuring the I spin free induction decay amplitude following a 90° pulse. Then the whole procedure is repeated without the strong r.f. magnetic field applied. The difference between the I spin free induction decay amplitudes with and without the strong r.f. magnetic field applied is recorded as a function of the r.f. field frequency ω. It is different from zero only when the r.f. magnetic field couples the two spin systems, i.e. when $\omega \approx \omega_{OS}$. Here ω_{OS} is any of the quadrupole transition frequencies of the S (or B) nuclei. If the S spin quadrupole resonance frequencies are lower than the I spin Larmor frequency in high magnetic field H_0, the two spin systems also couple when their energy levels cross, which has some additional influence on the double resonance spectra.

The sensitivity and the resolution of the technique is increased when one uses instead of the strong r.f. magnetic field coupling pulse, a train of coupling

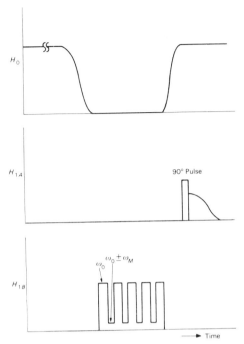

Fig. 17.   Field sequence in a zero magnetic field nuclear double resonance. Multiple coupling process.

pulses at a frequency ω near to a quadrupole resonance frequency ω_{OS} with the weak r.f. field pulses at the frequency ω_{OS} in between (Fig. 17). The purpose of the weak r.f. field pulses is to saturate the particular quadrupole transition. Experimentally a weak r.f. signal at the frequency ω_{OS} can be obtained from the carrier signal at the frequency ω by its simultaneous attenuation and amplitude- or phase- or frequency-modulation by the carrier signal at the frequency $\omega_M = \omega - \omega_{OS}$. A sideband of the modulated signal is used to saturate the particular quadrupole transition. During the experiment, the modulation frequency ω_M is set to a fixed value and the frequency of the carrier signal is swept through the resonance frequencies ω_{OS}. The double resonance spectra appear as strong sharp lines at the frequencies $\omega_{OS} \pm \omega_M$ on the broad single-coupling lines. Also in this case the resonance lineshapes of the S nuclei can be obtained directly.

D. Analysis of ^{1}H–^{14}N Double Resonance Spectra

(i) THE SINGLE COUPLING PROCESS

In this section we present the results of a calculation of the double resonance spectra for a ^{1}H–^{14}N system in a single level-crossing cycle for the case of a strong applied r.f. magnetic field. The analogous treatment for the weak r.f. magnetic field case has been given in Section IV. For the sake of simplicity, spin–lattice relaxation is omitted from the calculation.

The following conditions, usually met in a double resonance experiment, have been considered to be valid throughout the calculation.

(i) The number $3N$ of chemically non-equivalent nitrogen nuclei per unit volume is much smaller than the number $2n$ of protons per unit volume.
(ii) The nitrogen quadrupole transition frequencies are much lower than the Larmor frequency of protons in the strong static magnetic field H_0 in which the protons are polarized.
(iii) An equilibrium in the common NH spin system is reached during every level crossing.

In addition, the cross-relaxation rate W_{NH}, i.e. the rate at which the two spin systems are relaxing towards the common equilibrium state under the influence of a strong r.f. magnetic field, was assumed to have the Gaussian form

$$W_{NH} = \frac{1}{T_0} \sum_i \exp\left[-(\omega - \omega_{QN_i})^2/\omega_L^2\right] \tag{V.9}$$

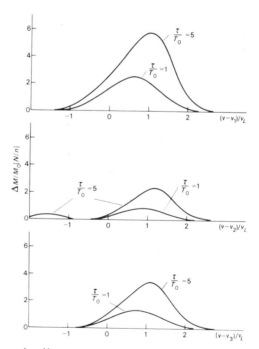

Fig. 18. Calculated ^{1}H–^{14}N double resonance lineshapes obtained by zero magnetic field nuclear double resonance in a single coupling process.

Here ω is the r.f. field frequency, T_0 is a constant proportional to $(\gamma_H H_1/\omega)^2$, ω_{QN_i} are the nitrogen quadrupole resonance frequencies and the proton local frequency $\omega_L = \gamma_H H_L$ measures the proton zero-field linewidth.

The changes in proton magnetization ΔM at the end of the level-crossing cycle due to the double resonance process are for the three nitrogen quadrupole transitions to first order in N/n equal to

$$(\Delta M)_1 = M_0 \frac{N}{n}\left(\frac{169}{64}+\frac{13}{4}x_1+x_1^2\right)\left[1-\exp\left(-\frac{\tau}{T_0}e^{-x_1^2}\right)\right]$$

$$(\Delta M)_2 = M_0 \frac{N}{n}\left(\frac{25}{64}+\frac{5}{4}x_2+x_2^2\right)\left[1-\exp\left(-\frac{\tau}{T_0}e^{-x_2^2}\right)\right] \qquad \text{(V.10)}$$

$$(\Delta M)_3 = M_0 \frac{N}{n}(1+2x_3+x_3^2)\left[1-\exp\left(-\frac{\tau}{T_0}e^{-x_3^2}\right)\right]$$

M_0 being the equilibrium proton magnetization in the static magnetic field H_0, $x_i = (\omega - \omega_{QN_i})/\omega_L$ and $\omega_{QN_1} > \omega_{QN_2} > \omega_{QN_3}$. The calculated double resonance lines are plotted in Fig. 18 for two different ratios τ/T_0.

The double resonance signals are at $x_i = 0$ for a high enough ratio τ/T_0, equal to the corresponding level-crossing signals.

(ii) THE MULTIPLE COUPLING PROCESS

In the calculation of the double resonance sensitivity, we shall in this case completely neglect the level-crossing process, but we can no longer neglect the proton spin–lattice relaxation in zero magnetic field. Spin–lattice relaxation of nitrogen is still assumed to be slow as compared with the double resonance process.

With the assumption that during a single coupling pulse, an instantaneous equilibrium in the common NH spin system is reached, the proton spin temperature at the end of the coupling pulse Θ_f is related to the proton spin temperature at the beginning of the coupling pulse Θ_i through the equation

$$\Theta_f = \Theta_i(1+\varepsilon) \qquad \text{(V.11)}$$

Here the heat-capacities ratio ε is equal to

$$\varepsilon = \frac{N}{n}(\omega_M/\omega_L)^2 \qquad \text{(V.12)}$$

where ω_M is the modulation frequency, i.e. the frequency difference between the corresponding nitrogen quadrupole resonance frequency and the r.f. field frequency. After the nth coupling, the final spin temperature of the proton system $\Theta_f(n)$ is related to the initial proton spin temperature Θ_i through the equation

$$\Theta_f(n) = \Theta_i(1+\varepsilon)^n \approx \Theta_i \exp(n\varepsilon) \qquad \text{(V.13)}$$

Let τ_1 be the length of a strong r.f. field pulse, τ_2 the length of a weak r.f. field pulse and τ the duration of the whole pulse train. In order to reach an equilibrium in the common NH spin system during each coupling pulse, the time τ_1 should be equal to several cross-relaxation times W_{NH}^{-1}. In the following, we shall assume $\tau_1 + \tau_2$ to be equal to $3W_{NH}^{-1}$. The number of couplings is thus equal to

$$n = \tau/(\tau_1 + \tau_2) = \tau W_{NH}/3 \qquad (V.14)$$

In addition to the double resonance process, the proton spin temperature also rises because of spin–lattice relaxation. Both processes are independent. The inverse proton spin temperature, which is proportional to the proton magnetization in the high magnetic field, is at the end of the irradiation equal to

$$\Theta_f^{-1} = \Theta_i^{-1} \exp\left(-\varepsilon W_{NH}/3 - \tau/T_{1H}\right) \qquad (V.15)$$

Here T_{1H} is the proton spin–lattice relaxation time in zero magnetic field.

The difference $\Delta\Theta_f^{-1}$ between the final inverse proton spin temperature with and without the strong r.f. magnetic field irradiation is a maximum for[11]

$$\tau = T_{1H} \ln(1+x)/x \qquad (V.16)$$

where

$$x = \varepsilon W_{NH} T_{1H}/3 \qquad (V.17)$$

In this case the difference $\Delta\Theta_f^{-1}$, which is proportional to the double resonance signal, is equal to

$$\Delta\Theta_f^{-1} = \Theta_i^{-1} x/(1+x)^{(1+1/x)} \qquad (V.18)$$

The magnitude of the double resonance signal increases with x. For $x \gg 1$, we can thus measure the S spectra with the same sensitivity as the I spectra. For the assumed Gaussian form of the cross-relaxation rate W_{NH}, the maximum double resonance signals are obtained when $\omega_M = \omega_L$. Since the cross-relaxation rates W_{NH} are usually for all three nitrogen quadrupole transitions approximately equal, the sensitivities of the detection of different nitrogen quadrupole resonance frequencies by the multiple coupling technique are also approximately equal.

E. Results

The new technique has been experimentally tested in four cases.

(a) In the case of zero magnetic field nuclear double resonance between ^{1}H and ^{14}N nuclei in thymine and polyglycine. The double resonance lineshape of the highest frequency ^{14}N quadrupole transition in thymine was measured at two different r.f. magnetic field amplitudes. The single coupling technique was used. The results are shown in Fig. 19. The measured lineshape

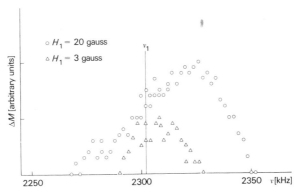

Fig. 19.   Measured ^1H–^{14}N double resonance lineshape of the highest frequency ^{14}N NQR transition in thymine.

qualitatively agrees with the calculated one. When an r.f. magnetic field intensity of 30 gauss was used, the cross-relaxation time was found to be approximately 10 ms, which is well below our shortest experimentally obtainable time for which the sample stays in zero magnetic field.

In a continuation of our quadrupole resonance studies of biologically interesting molecules, we also determined the ^{14}N quadrupole coupling of the peptide nitrogens in polyglycine.[17] We believe that this is the first observation of nuclear quadrupole resonance of ^{14}N in a large polypeptide. The polycrystalline sample of polyglycine was obtained from Sigma Chem. Company. The approximate molecular weight was 6 000, so that the molecule

$$
\begin{array}{ccccccc}
H & H & H & H & H & H & O^- \\
\diagdown & | & | & | & | & | & \diagup \\
H\text{–}N^+(1)\text{—} & C\text{—}C\text{—}N(2)\text{—} & C & \cdots\cdots & C\text{—}N(n)\text{—}C\text{—}C & \\
\diagup & | & \| & | & \| & | & \diagdown \\
H & H & O & H & O & H & O
\end{array}
$$

contained about $n = 105$ glycine subunits. The spectra were measured at 77 K. The corresponding ^{14}N quadrupole coupling constant is $e^2qQ/h = 3.097$ MHz and the asymmetry parameter $\eta = 0.76$. This value is much higher[2] than the quadrupole coupling constant of glycine, $e^2qQ/h = 1.25$ MHz, $\eta = 0.51$. It is also higher[2] than the quadrupole coupling constant of the terminal nitrogens N(I) in diglycine ($e^2qQ/h = 1.28$ MHz, $\eta = 0.41$) and triglycine ($e^2qQ/h = 1.18$ MHz, $\eta = 0.46$) but of the same order as the values obtained for the peptide nitrogens in these two compounds.[2] In diglycine, $e^2qQ/h(N2) = 3.03$ MHz, $\eta(N2) = 0.41$ whereas in triglycine $e^2qQ/h(N2) = 3.01$ MHz, $\eta(N2) = 0.48$ and $e^2qQ/h(N3) = 3.08$ MHz, $\eta(N3) = 0.76$. Thus it seems safe to assign the observed ^{14}N NQR spectrum in polyglycine to the peptide backbone nitrogens. The spectrum of the terminal nitrogens should be too weak to be observable.

107

(b) In the case of nuclear double resonance between purely magnetic ^{23}Na and ^{19}F spins in NaF in a strong static magnetic field. The double resonance lineshape at the frequency $\omega_F - \omega_{Na}$ in a single coupling process was measured. The result is shown in Fig. 20(a). The r.f. field amplitude was approximately 80 gauss and the irradiation time was 1 s. The Larmor frequency of ^{19}F was 18 MHz. In this case the cross-relaxation time was found to be 150 ms.

(c) In the case of ^{1}H–^{14}N double resonance in an NH$_4$H$_2$PO$_4$ powder sample. The double resonance lineshape at the frequency $\omega_H - \omega_N$ was measured. The proton Larmor frequency was 18 MHz, the irradiation time was 1 s and the r.f. magnetic field amplitude was 80 gauss. The results are shown in Fig. 20(b). The unusual double resonance lineshape is due to the nuclear quadrupole broadening of the ^{14}N resonance lines.

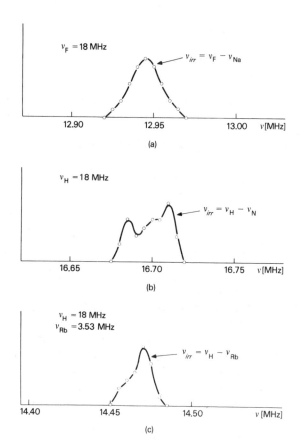

Fig. 20. Measured nuclear double resonance spectra in a strong static magnetic field. (a) ^{19}F–^{23}Na double resonance in NaF; (b) ^{1}H–^{14}N double resonance in powdered NH$_4$H$_2$PO$_4$; (c) ^{1}H–^{87}Rb double resonance in RbH$_2$PO single crystal.

(d) In the case of ^1H–^{87}Rb double resonance in a RbH_2PO_4 single crystal. The experimental conditions were the same as in (c). The ^{87}Rb resonance line at the frequency 3.53 MHz was chosen. The results are shown in Fig. 20(c).

In the case of zero-field nuclear double resonance detection of ^{14}N NQR spectra through the proton signal, the sensitivity of the new technique is higher than the sensitivity of the level-crossing technique which can be used for the same purpose. The new technique is of particular importance in some cases when level-crossing signals cannot be obtained. This for example is the case when

(i) the spin–lattice relaxation time of the S spin system is much shorter than the time for which the sample stays in zero static magnetic field,

(ii) when the level-crossing times are too short for sufficient energy exchange between the two spin systems, and

(iii) when the quadrupole resonance frequencies of the S nuclei are too high for the level crossing to occur.

VI. DOUBLE RESONANCE RELAXATION MEASUREMENTS

A. Double Resonance Relaxation Measurements in the Rotating Frame for $T_{1B} \leq T_{1A}$

The double resonance spin–lattice relaxation problem[18,19] in the rotating frame is equivalent to the thermodynamic problem of energy exchange between two heat reservoirs, A and B, the thermal coupling of which to the surrounding heat bath is characterized by time constants $T_{1\rho A}$ and $T_{1\rho B}$ respectively. When thermal contact between the two reservoirs is established, it is characterized by a time constant τ_{AB} which is short compared to $T_{1\rho A}$ or $T_{1\rho B}$. For the sake of simplicity, we shall assume that the energy of a given reservoir can be described by a spin temperature T_s

$$E_A = C_A\beta_A, \qquad E_B = C_B\beta_B, \qquad \text{where } \beta = 1/kT_s \qquad (VI.1)$$

which is in thermal equilibrium when T_s is equal to the lattice temperature. $T_{1\rho B}$ is determined by observing E_A or β_A as a function of the time of thermal contact to the B reservoir. The mathematical problem is the determination of the time development of the inverse spin temperature of the A system, β_A, when in thermal contact with the B system. The change in the inverse spin temperature is described by

$$\frac{d\beta_A}{dt} = -\frac{\beta_A}{T_{1A}} - \frac{\beta_A - \beta_B}{\tau_{AB}} \qquad (VI.2)$$

$$\frac{d\beta_B}{dt} = -\frac{\beta_B}{T_{1B}} - \frac{\beta_B - \beta_A}{\tau_{BA}} \qquad (VI.3)$$

with

$$\tau_{BA} = \tau_{AB}/\varepsilon \tag{VI.4a}$$

where ε is the ratio of the magnetic heat capacities of the two systems:

$$\varepsilon = C_B/C_A \tag{VI.4b}$$

For given initial conditions $\beta_A = \beta_A(0)$ and $\beta_B = 0$ we find

$$\beta_A(t) = \beta_A(0) \exp\left(-\frac{t}{\varepsilon\tau_{AB}}\right)$$
$$+ (1-\varepsilon) \exp\left(-\frac{t}{(1+\varepsilon)T_{1\rho A}}\right) \exp\left(-\frac{t}{(1+\varepsilon)T_{1\rho B}}\right) \tag{VI.5}$$

provided that $\varepsilon\tau_{AB} \ll T_{1\rho A}, T_{1\rho B}$.

If there is no thermal contact between the two systems ($\tau_{AB} = 0$), the inverse spin temperature of the A system on the other hand varies as

$$\beta_A^{(0)}(t) = \beta_A^{(0)}(0) \cdot \exp\left(-t/T_{1\rho A}\right) \tag{VI.6}$$

As the magnetization of the A system M_A is proportional to β_A, all time constants of the problem can be determined by observing the time development of β_A via the corresponding free induction decay signals.

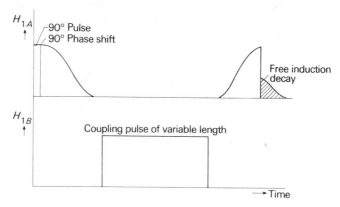

Fig. 21. Pulse sequence for double resonance relaxation measurements of $T_{1\rho B}$ in the rotating frame for $T_{1\rho B} < T_{1\rho A}$.

The actual experiment is performed as follows (Fig. 21). First a low spin temperature is generated in the A spin system by adiabatic demagnetization in the rotating frame (ADRF). Then a B r.f. pulse of variable length is turned on for a time $t_{Br.f.}$ to establish thermal contact between the two spin systems by equalizing the transition frequencies in the two rotating frames:

$$\omega_{eff}(A) = \omega_{eff}(B) \qquad (VI.7)$$

where $\omega_{eff} = \gamma H_{eff}$.

The remaining A spin magnetization is finally measured via a free induction signal (or a 'dipolar' signal after a 45° pulse) at a fixed time $t \approx T_{1\rho A}$, thus making the term $\exp[-t/(1+\varepsilon)T_{1\rho A}]$ in Eq. (VI.5) constant. The logarithm of the A signal amplitude as a function of $t_{Br.f.}$ is a straight line with a slope $[\varepsilon/(1+\varepsilon)](1/T_{1\rho B})$, from which $T_{1\rho B}$ can be determined.

B. Double Resonance Relaxation Measurements in the Rotating Frame for $T_{1B} > T_{1A}$

If T_{1B} is long compared to T_{1A}, another procedure has to be used for the determination of T_{1B}. First the A spin system is cooled by ADRF. Then thermal contact between the A and B systems is established by applying a large effective field

$$H_{eff} = (1/\gamma)[(\Delta\omega)^2 + (\gamma H_1)^2]^{1/2} \qquad (VI.8)$$

(which is off resonance by $\Delta\omega$) to the B system, thus cooling it down too. After that, $H_{eff}(B)$ is turned off, so that the thermal contact is disconnected and the B

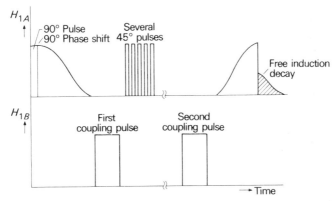

Fig. 22. Pulse sequence for double resonance relaxation measurements of T_{1B} in the rotating frame for $T_{1B} > T_{1A}$.

magnetization relaxes along the laboratory field H_0 with the time constant T_{1B}. The A system is now first saturated by a series of 45° pulses and then remagnetized by reconnection to the B system. The time constant T_{1B} is then determined by measuring the magnetization of the A system as a function of the time t between the two thermal contacts (Fig. 22). The time development of the magnetization of the A system is here given by

$$M_A(t) \approx M_A(0) \cdot \varepsilon \cdot \exp(-t/T_{1B}) \qquad (VI.9)$$

C. Double Resonance Relaxation Measurements of Quadrupolar Relaxation Times by Level Crossing in the Laboratory Frame

A disadvantage of the double resonance methods in the rotating frame lies in the fact that for the determination of the NQR spectra and relaxation times of nuclei with a non-zero quadrupole coupling single crystals are required. This problem can be avoided by multiple level crossing and spin mixing in the laboratory frame, where powder samples can be used as well as single crystals.

The A system is first magnetized in a strong external magnetic field, and then adiabatically demagnetized to zero-field by moving the sample out of the magnet. At an intermediate field the Zeeman splitting of the A system and the Zeeman perturbed electric quadrupole splitting of the B system are equal, and resonance energy transfer takes place. The system is kept for a time t in zero-field and then adiabatically remagnetized, resulting in a second level crossing.

T_{1B} is simply determined by measuring the A signal after remagnetization as a function of the time the sample stays in zero-field (Fig. 23)

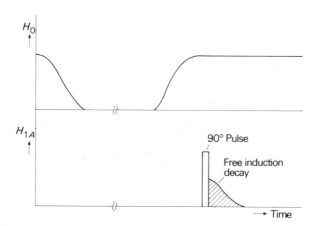

Fig. 23. Pulse sequence and variation of the Zeeman magnetic field with time for relaxation measurements via level crossing in the laboratory frame.

$$M_A(t) = A \exp(-t/T_{1A}) + B \exp(-t/T_{1B})$$

The relaxation time T_{1B} can be assigned to a given spin species by saturating the energy levels of the corresponding species or by an inspection of the relative magnitude of the two coefficients, A and B.

The above method works well if $T_{1A} \ll T_{1B}$. If T_{1A} is not short, the A spin system is, after the demagnetization, saturated by a weak pulse H_{1A}. The A spin Zeeman signal after adiabatic remagnetization is thus completely due to level crossing with the B spin system. By varying the time between the adiabatic demagnetization and remagnetization, the spin–lattice relaxation time of the B spin system in zero magnetic field T_{1B} is obtained.

D. Double Resonance Relaxation Measurements in the Laboratory Frame via the Solid Effect

If level crossing between the A and the B system is not possible, or if T_{1B} is shorter than the time between the adiabatic demagnetization and remagnetization, one can measure T_{1B} via the solid effect with the help of the measuring cycles shown in Fig. 24.

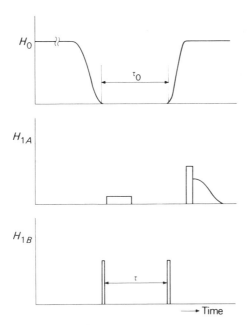

Fig. 24.   Pulse sequence and variation of the Zeeman magnetic field with time for relaxation measurements via the solid effect.

After adiabatic demagnetization, the two spin systems are coupled with a strong radiofrequency pulse of appropriate frequency. After the coupling pulse is turned off, the A spin system is saturated with a weak r.f. field H_{1A}. In a time τ after the first coupling pulse, a second coupling pulse is applied. The spin temperature of the A system is thus completely determined by the relaxation time T_{1B} of the B spin system in zero magnetic field. In a time τ_0 after the demagnetization, the system is adiabatically remagnetized and the spin temperature of the A system is determined from the free induction decay signal. T_{1B} is determined by varying the time between the two coupling pulses τ and keeping the time τ_0 between the adiabatic demagnetization and remagnetization fixed.

E. Measurement of Short Quadrupolar Relaxation Times by ADRF

If the cross-relaxation rate W_{AB} between the A and B spin systems in the rotating or dipolar frames is so large that

$$W_{AB}T_{1\rho B} \gg 1$$

and if $T_{1\rho B}$ is much shorter than $T_{1\rho A}$

$$T_{1\rho A} \gg T_{1\rho B}$$

$T_{1\rho B}$ can be determined in a very simple way by spin mixing in the rotating or dipolar frame. Here W_{AB} is given by

$$W_{AB} = \hbar^{-2}|V_{AB}|^2 g(\omega_{AB}), \qquad \omega_{AB} = \omega_A - \omega_B$$

where V_{AB} is the matrix element of the dipolar interaction between the A and B spins and

$$g(\omega_{AB}) = \iint g(\omega'_A)g(\omega''_B)\delta(\omega'_A - \omega''_B)\, d\omega'_A \, d\omega''_B$$

is the density of the final overlapping states.

$T_{1\rho B}$ is simply determined by adiabatic demagnetization of the A spins into the rotating frame, allowing the A spins to cross-relax with the B spins for a time τ, remagnetizing the A spins and measuring the A spin magnetization as a function of the time τ (Fig. 24). Assuming that the spin heat capacity of the A system is larger than the heat capacity of the 'mixed' reservoir, which is in turn larger than the heat capacity of the B dipolar system, i.e.

$$C_{DA} > C_{DAB} > C_{DB}$$

$T_{1\rho B}$ is obtained from the measured relaxation rate $(T_{1\rho})^{-1}$ of the coupled system as

$$(T_{1\rho})^{-1} \simeq (T_{1\rho B})^{-1} \frac{C_{DAB}}{C_{DA}}$$

REFERENCES

1 S. R. Hartmann and E. L. Hahn, *Phys. Rev.*, **128**, 2042 (1962).
2 F. M. Lurie and C. P. Slichter, *Phys. Rev.*, **133**, A1108 (1964).
3 A. Hartland, *J. Phys. C*, **2**, 246 (1969).
4 R. Blinc, M. Mali, R. Osredkar, A. Prelesnik, I. Zupančič, and L. Ehrenberg, *J. Chem. Phys.*, **55**, 4843 (1971); *Acta Chem. Scand.*, **25**, 2403 (1971); *Chem. Phys. Letters*, **9**, 85 (1971). R. Blinc, M. Mali, R. Osredkar, A. Prelesnik, J. Seliger, and I. Zupančič, *ibid.*, **14**, 49 (1972).
   Y. Tsutsumi, M. Kunitomo, T. Terao, and T. Hashi, *J. Phys. Soc. Jap.*, **26**, 16 (1969).

5 R. E. Slusher and E. L. Hahn, *Phys. Rev.*, **166,** 332 (1968).
6 J. Koo, Ph.D. Thesis, University of California, Berkeley (unpublished).
7 R. Blinc, J. Seliger, R. Osredkar, and M. Mali, *Phys. Lett.*, **47A,** 131 (1974).
8 G. W. Leppelmeier and E. L. Hahn, *Phys. Rev.*, **141,** 724 (1966).
9 D. T. Edmonds and P. A. Speight, *Phys. Lett.*, **34A,** 325 (1971).
10 D. T. Edmonds, M. J. Hunt, and A. L. Mackay, *J. Magn. Resonance*, **9,** 66 (1973).
11 D. T. Edmonds, M. J. Hunt, A. L. Mackay, and C. P. Summers, *Advances in Nuclear Quadrupole Resonance*, Vol. 1, Heyden, London, 1974, p. 145.
12 R. Blinc, M. Mali, R. Osredkar, A. Prelesnik, J. Seliger, I. Zupančič, and L. Ehrenberg, *J. Chem. Phys.*, **57,** 5087 (1972).
13 J. Seliger, R. Blinc, M. Mali, R. Osredkar, and A. Prelesnik, *Phys. Rev.* (to be published).
14 M. Goldman, *Spin Temperature and Nuclear Magnetic Resonance in Solids*, Chapter 7, Oxford University Press, London, 1971.
15 A. Abragam and W. G. Proctor, *Compt. Rend.*, **246,** 2253 (1958).
16 A. Landesman, *J. Phys. Solids*, **18,** 210 (1961).
17 R. Blinc, M. Mali, R. Osredkar, J. Seliger, and L. Ehrenberg, *Chem. Phys. Lett.* **28,** 158 (1974).
18 E. L. Hahn, *Magnetic Resonance and Relaxation*, Ed.: R. Blinc, North Holland Publishing Company, 1967, p. 14.
19 D. Stehlik, *Pulsed Magnetic and Optical Resonance*, Ed.: R. Blinc, 'J. Stefan' Institute, 1972, p. 63.

4. NUCLEAR QUADRUPOLE RESONANCE OF IODINE IN POLYIODIDES

Daiyu Nakamura and Masaji Kubo

Department of Chemistry, Nagoya University, Chikusa, Nagoya, Japan

I. INTRODUCTION

Although it had been well known for a long time that iodine dissolves in an aqueous solution of ammonium or alkali iodides, ammonium triiodide was obtained as crystals for the first time by Johnson[1] in 1878. Since then, various triiodides and other polyiodides have been prepared by several investigators.[2-4] However, it was not until 1935 that the geometric structure of polyiodide ions was elucidated by Mooney, who carried out an X-ray analysis on ammonium triiodide.[5] She found that almost linear I_3^- units exist in crystals with the two I–I distances different from each other by about 0.3 Å. In a subsequent paper,[6] she reported that the dichloroiodide ion, ICl_2^-, has a linear symmetric structure in crystals. This confirmed the linear structure of trihalide ions, and suggested that the asymmetric structure might not be a universal characteristic of triiodide ions.

The determination of the structure of the triiodide ion led to theoretical investigations on the nature of chemical bonds between iodine atoms. Since both an iodine molecule and an iodine ion have completed octets, the formation of an additional covalent bond in the triiodide ion must involve electrons in excess over those filling the outermost $5p$ orbitals of iodine. Therefore, the electronic state must be described by introducing at least one orbital of higher energy. Kimball[7] considered the $5d$ or $6s$ orbital for the bivalency of the central iodine atom, while Pauling[8] and others[9] interpreted the linear structure in terms of the trigonal–bipyramidal hybridized orbitals of the central atom. These authors attempted to explain the structure and stability of this ion by means of a covalent bond model. On the other hand, Pimentel[10] treated bond formation in trihalide ions by a simple molecular orbital theory without introducing atomic orbitals higher than the $5p$ orbitals of the central iodine atom. According to his

117

approximation, the ground electronic configuration of the I_3^- ion is expressed by $\phi_1^2\phi_2^2$, in which the lower molecular orbital, ϕ_1, is bonding whereas the second one, ϕ_2, is formally non-bonding and accommodates two electrons between terminal iodine atoms without introducing any node. This indicates that the charge density on each terminal iodine atom is higher than that on the central atom.

Since the nuclear quadrupole resonance of halogens permits us to evaluate the electronic charge distribution among the valence orbitals of halogens, it provides a powerful means for testing various theories of bond formation. A number of papers[11-19] has been published on the NQR of halogens in polyhalides. In fact, polyhalides are amenable to NQR measurements, because the naturally abundant isotopes of all halogens except fluorine have a nuclear spin equal to or greater than $3/2$. Cornwell and Yamasaki[13,14] carried out NQR studies on chloroiodides and reported complete NQR data on both chlorine and iodine. Kurita *et al.*[15] reported the NQR of chlorine in some additional chloroiodides. These authors concluded that the chlorine–iodine bond in chloroiodides could be interpreted adequately by an MO bonding scheme proposed by Pimentel rather than by a covalent bond model involving the hybridized orbitals of the central iodine atom.

Although various polyiodide ions are known to exist in crystals up to the enneaiodide ion,[5,20-30] the NQR of iodine has been investigated only in some triiodides and caesium octaiodide. Therefore, we will focus our attention on the electronic structure of triiodide and octaiodide ions as elucidated by the results of NQR investigations. Because various non-equivalent iodine atoms exist in polyiodide crystals and also because resonant iodine atoms show two frequencies, ν_1 and ν_2, the assignment of the observed frequencies to non-equivalent iodine atoms in crystals presents an important problem to be solved.

II. NQR FREQUENCIES, ASSIGNMENT, AND CRYSTAL STRUCTURE

Since iodine-127 has a nuclear spin equal to $5/2$, one can observe two resonance frequencies given by

$$\nu_1 = \tfrac{3}{20}(e^2qQ/h)(1 + 1.0926\eta^2 - 0.634\eta^4 + \cdots) \tag{1}$$

$$\nu_2 = \tfrac{3}{10}(e^2qQ/h)(1 - 0.2037\eta^2 + 0.162\eta^4 - \cdots) \tag{2}$$

Here e^2qQ/h and η denote the quadrupole coupling constant and the asymmetry parameter, respectively. It is obvious that the relation, $2\nu_1 \geq \nu_2$, holds between these frequencies.

The nuclear quadrupole resonance of iodine in polyiodides was observed for the first time in 1955 by Kojima *et al.*,[11] who studied ammonium triiodide and found two resonance frequencies, 366.23 and 732.13 MHz at 25°. They assumed the existence of symmetric triiodide ions in crystals and tentatively

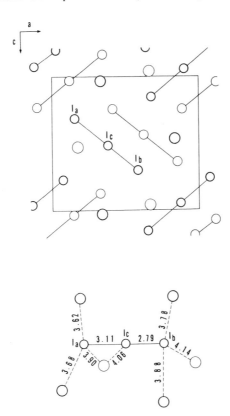

Fig. 1.   Projection of the structure of ammonium triiodide on (010) and the environ-ment of a triiodide ion. Small and large circles represent iodine atoms and ammonium ions respectively. Heavy circles lie in the plane of the drawing, while light circles lie in a plane at height $b/2$. The figures are redrawn on the basis of refined values of the cell dimensions and positional parameters determined by Cheesman and Finney.[30]

assigned these lines to ν_1 and ν_2 of end iodine atoms. However, Mooney[5] had already carried out the X-ray crystal structure analysis of this compound. It was an early paper (1935), so that the accuracy of measurement was not very high, but the following conclusions were reached. Ammonium triiodide forms orthorhombic crystals belonging to the space group *Pnma*, with four crystallo-graphically equivalent triiodide ions in the unit cell. Each triiodide ion is almost linear and asymmetric with interiodine distances of 2.82 and 3.10 Å (in her paper, she states that this difference might be due to inaccuracies in locating atomic positions). Figure 1 shows a projection of the structure on (010) and the surroundings of a triiodide ion. If Mooney's structure is correct, six frequencies should be observed in iodine quadrupole resonance. In fact, Sasane *et al.*[16] and Bowmaker and Hacobian[18] have independently observed four lines, in addition to the aforementioned two lines. As shown in Table 1, the six resonance

Table 1. Nuclear quadrupole resonance frequencies of ^{127}I in ammonium triiodide

	Temp., °C	Frequency, MHz	
		ν_1	ν_2
a	23	87.34±0.03	172.21±0.06
	Liquid N_2	70.55±0.01	138.07±0.06
b	21	239.2±0.1	475.86±0.03
	Liquid N_2	265.1±0.1	515.10±0.03
c	29	366.21±0.02	732.13±0.02[a]
	Liquid N_2	368.81±0.02	

[a] Observed by Kojima *et al.*[11] at 25°.

frequencies are distributed over a fairly wide frequency range, indicating that the three types of non-equivalent iodine atoms are different from one another to a considerable extent, in agreement with Mooney's structure. A recent reinvestigation carried out by Cheesman and Finney[30] has confirmed interatomic distances reported by Mooney, but the triiodide ion was found to be linear rather than slightly bent.

Five, five, and six resonance frequencies have been reported for thallium(I), rubidium, and caesium triiodides, respectively.[16,18,19] The expected highest frequency line has not been detected for thallium(I) and rubidium triiodides because of experimental difficulties. Mention should be made here of the fact that thallium(I) triiodide shows a quadrupole resonance pattern different from those of gallium(III) and indium(III) iodides,[31-33] and that the pattern bears a striking resemblance to those of the other triiodides mentioned above. Moreover, a preliminary X-ray analysis carried out by Hazell[34] has indicated that thallium(I) triiodide is isomorphous with ammonium triiodide. Consequently, it is evident that the compound, TlI_3, obtained by adding iodine to thallous iodide is not thallium(III) iodide, but thallium(I) triiodide, $Tl^+I_3^-$. Rubidium and caesium triiodides also are isomorphous with ammonium triiodide.[16,26] Being isomorphous with one another, all these triiodides show similar quadrupole resonance patterns as illustrated in Fig. 2.

Now we will assign the observed five or six resonance frequencies to the ν_1 and ν_2 of individual iodine atoms in a triiodide anion. Fortunately, the assignment can be made easily and unequivocally for the triiodides having markedly asymmetric I_3^- ions.

For ammonium triiodide, the lowest frequency line is surely ν_1. The corresponding ν_2 line is required to satisfy the relation, $2\nu_1 \geq \nu_2$. Then, the second lowest line surely represents the corresponding ν_2 frequency and the third must be the ν_1 of another iodine atom in an I_3^- ion. The corresponding ν_2 cannot be determined uniquely from the theoretical requirement alone, because the frequencies of both the fourth and the fifth lines are lower than

120

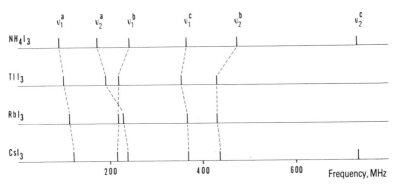

Fig. 2.   NQR spectra of some triiodides.

twice the frequency of the third line. Therefore, a clue to an unequivocal assignment must be looked for elsewhere. In view of the linear structure of triiodide ions in crystals, the asymmetry parameters of all iodine atoms in the anion should be small, i.e., $2\nu_1$ should be close to ν_2. This indicates that only the combination of the third and the fifth lines are allowed as the ν_1 and ν_2 of the second iodine atom. If the third and the fourth lines are combined, the asymmetry parameter amounts to as much as 50%. Two remaining lines are attributable to the third iodine atom. The resonance lines of thallium(I), rubidium, and caesium triiodides can be interpreted in a similar manner. The results are shown in Fig. 2.

The X-ray analysis of ammonium and caesium triiodides[26,30] has indicated that the shortest interiodine distance, I_b–I_c in Fig. 1, is fairly close to the internuclear distance 2.68 Å of an iodine molecule in crystals.[35,36] On the other hand, the I_a–I_c distance is considerably longer than the interiodine distance of the iodine molecule. As an approximation, let an I_3^- ion be assumed to be formed from an iodine molecule and an iodine ion approaching each other along the molecular axis, the ion thereby polarizing the molecule to stabilize the system by electrostatic interaction. Then, it is expected that I_a bears the largest negative charge, I_b is charged to a lesser extent, and I_c has a fractional positive charge. This qualitative expectation is supported by the following fact. In the crystals of ammonium and caesium triiodides, each I_a atom is surrounded by four cations with a mean separation equal to 3.78 and 3.82 Å respectively, whereas each I_b has four neighbouring cations separated by 3.98 and 4.12 Å respectively. The closer distances suggest the presence of a stronger attractive centre having a greater net negative charge. On the other hand, each I_c atom in ammonium and caesium triiodides has only two neighbouring cations at a distance of 4.06 and 4.10 Å respectively. Accordingly, the three ν_1 as well as the corresponding ν_2 frequencies are attributable to I_a, I_b, and I_c in the order of increasing frequency. Thus the assignment of observed frequencies is achieved for these compounds.

121

Table 2. Nuclear quadrupole resonance frequencies of ^{127}I in tetraethylammonium triiodide

Frequency, MHz		
Liquid N_2 temp.	$-77\,°C$	Assignment[a]
167.38 ± 0.08	166.28 ± 0.07	$\nu_1\,(At)$
$175.3\ \pm0.1$	$174.0\ \pm0.1$	$\nu_1\,(Bt)$
334.69 ± 0.03	$332.3\ \pm0.3$	$\nu_2\,(At)$
350.34 ± 0.03		$\nu_2\,(Bt)$
356.75 ± 0.02	$356.8\ \pm0.3$	$\nu_1\,(Bc)$
386.88 ± 0.03	$383.6\ \pm0.3$	$\nu_1\,(Ac)$

[a] Terminal atoms are indicated by t, while c stands for central.

Bowmaker and Hacobian[18] reported five resonance frequencies for tetra-ethylammonium triiodide. Later, Harada *et al.*[19] found another line in addition to those reported by the previous workers. Accordingly, six frequencies are known for this compound as given in Table 2. Since the two closely-spaced lines appearing below 180 MHz are certainly ν_1, the corresponding ν_2 frequencies should be among four other lines appearing above 330 MHz. Consequently, the two remaining lines must be the ν_1 frequencies of positively charged iodine atoms. This consideration leads to the conclusion that two crystallographically non-equivalent I_3^- ions having a symmetric structure exist in this crystal.

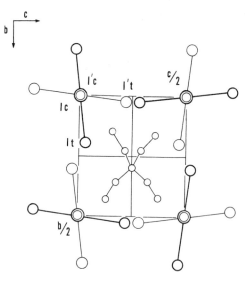

Fig. 3. The structure of modification I of tetramethylammonium triiodide. The iodine atoms at heights 0 and $a/2$ are shown by heavy and light circles respectively.

It is known from an X-ray crystal analysis carried out by Migchelsen and Vos[28] that tetraethylammonium triiodide forms two crystal modifications, I and II. Two kinds of non-equivalent symmetric I_3^- ions exist in the former crystal, which belongs to the space group *Cmca*, as shown in Fig. 3, whereas two kinds of almost linear asymmetric I_3^- ions are found in the latter modification. Harada *et al.*[19] have taken X-ray powder patterns and confirmed that the crystals of modification I yielded the six NQR frequencies referred to above. Thus, the lowest and the second lowest frequencies are undoubtedly the ν_1 frequencies of terminal iodine atoms belonging to $(I_3^-)_A$ and $(I_3^-)_B$ ions respectively. From the theoretical requirement, the third and the fourth lines are attributed to the corresponding ν_2 frequencies of the I_3^- ions A and B respectively. The highest and the second highest lines represent the ν_1 frequencies of central iodine atoms. Since the total charge of a triiodide ion should be $-e$, the large negative charge on the terminal iodine atoms requires a large positive net charge on the central iodine atom. This means that if the ν_1 and ν_2 frequencies of the terminal iodine atoms in one kind of triiodide ions are lower than those in the other kind of the ions, the higher ν_1 frequency of central iodine atoms is expected for the former kind of triiodide ions. Therefore, the highest frequency line can be assigned to the ν_1 frequency of the central iodine atom in an $(I_3^-)_A$ ion and the second highest to the ν_1 frequency of the central atom in an $(I_3^-)_B$ ion as shown in Table 2.

Tetra-*n*-propylammonium triiodide gives rise to six resonance lines as does tetraethylammonium triiodide, indicating the existence of two kinds of non-equivalent symmetric I_3^- ions in crystals.[19] However, the crystals of these two compounds are not isomorphous with each other, as confirmed by Harada *et al.* from X-ray powder pattern experiments.

Two resonance frequencies, 168.1 and 379.78 MHz at liquid nitrogen temperature, have been reported for dimethylammonium triiodide.[19] The low-frequency line can be attributed to ν_1 of terminal iodine atoms. The corresponding ν_2 should appear below $2\nu_1 = 336.2$ MHz. However, no resonance lines have been detected in the anticipated frequency range. Therefore, the line at 379.78 MHz must be assigned to ν_1 of the central iodine atoms. Since the line at 168.1 MHz is weak, the corresponding ν_2 is probably too weak to be observed by the usual NQR measurements. The fact that terminal and central iodine atoms each show a single ν_1 line indicates the presence of one kind of symmetric I_3^- ion in these crystals.

Di-*n*-propylammonium triiodide shows two ν_1 frequencies from terminal iodine atoms and two ν_1 frequencies from central iodine atoms.[19] The spectrum bears a close resemblance to that of tetraethylammonium triiodide, although ν_2 frequencies have not been observed. This indicates that two kinds of symmetric I_3^- ion exist in these crystals.

Tetra-*n*-butylammonium triiodide shows ten resonance lines at liquid nitrogen temperature, as shown in Table 3.[18,19] Four frequencies appearing in a rather wide frequency range below 200 MHz are attributable to ν_1 of

Table 3. Nuclear quadrupole resonance frequencies of ^{127}I in tetra-*n*-butylammonium triiodide at liquid nitrogen temperature

Frequency, MHz		
ν_1	ν_2	Assignment
156.63±0.07	312.13±0.05	1a
167.41±0.08	334.6 ±0.3	2a
185.7 ±0.1	371.1 ±0.3	2b
195.5 ±0.1	390.2 ±0.3	1b
359.63±0.03		3c
361.10±0.03		4c

terminal iodine atoms. Two lines at 312.13 and 334.6 MHz surely represent ν_2 corresponding to the lowest and the second lowest ν_1 frequencies, respectively. Two lines at 359.63 and 361.10 MHz are attributed to ν_1 of the central iodine atoms, because if they are assumed to be ν_2, the asymmetry parameter becomes too large. The two highest frequencies are interpreted adequately as ν_2. Therefore there are two kinds of asymmetric I_3^- ion in these crystals. The results are shown in Table 3.

Tetramethylammonium triiodide yields fifteen resonance lines in a frequency range of 140–400 MHz, indicating the presence of three kinds of non-equivalent asymmetric I_3^- ion in the crystals.[18,19] Although Migchelsen and Vos attempted to determine the structure of this crystal,[28] they failed to determine accurate positional parameters, because a superstructure occurred and also because the compound was unstable on exposure to X-rays. The crystal structure of tetra-*n*-butylammonium triiodide has also not been solved as yet. In view of the fact that asymmetric I_3^- ions exist, as revealed by the results of NQR measurements, despite the presence of considerably large cations in these crystals, it will be a very interesting problem to determine the extent of asymmetry of the I_3^- ions by accurate X-ray crystal analysis.

Caesium octaiodide forms monoclinic crystals belonging to the space group $P2_1/a$ and having two equivalent octaiodide ions in the unit cell.[23] An octaiodide ion has a centre of symmetry and is Z-shaped, as shown in Fig. 4. Accordingly, there are four kinds of crystallographically non-equivalent iodine atoms in crystals. This indicates that eight resonance frequencies, i.e. four ν_1 and four ν_2, are expected to be observed in a fairly wide frequency range. Iodine atoms, I_a, I_c, and I_b form an almost linear structural unit. The interiodine distances agree closely with those in ammonium and caesium triiodides. Two iodine atoms, I_d, form a diatomic unit having the shortest I–I distance and an almost linear arrangement of $I_a \cdots I_d–I_d \cdots I_a$. Because the chain of iodine atoms is bent at I_a, a fairly large asymmetry parameter is anticipated for I_a, whereas all other iodine atoms are expected to have small asymmetry parameters.

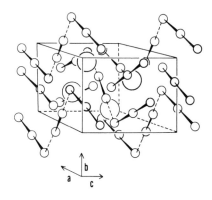

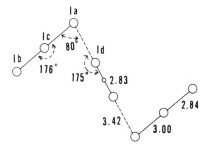

Fig. 4. Crystal structure of caesium octaiodide and the molecular structure of an octaiodide ion.

Seven resonance lines have been reported for this compound[18,19] as given in Table 4. From a comparison with the frequencies of ammonium and caesium triiodides, lines at 373.04 and 737.4 MHz are assigned to ν_1 and the corresponding ν_2 of I_c. Similarly, lines at 214.56 and 426.61 MHz are attributed to ν_1

Table 4. Nuclear quadrupole resonance frequencies of ^{127}I in caesium octaiodide

Frequency, MHz			
Liquid N_2 temp.	$-77\,°C$	Room temp. (°C)	Assignment
133.83±0.05	133.26±0.07	132.80±0.08 (20)	$\nu_1(a)$
187.57±0.07	185.4 ±0.1	183.42±0.08 (20)	$\nu_2(a)$
214.56±0.03	213.4 ±0.1	212.3 ±0.1 (20)	$\nu_1(b)$
320.45±0.03	318.27±0.03	316.06±0.04 (18)	$\nu_1(d)$
373.04±0.03	371.75±0.03	370.29±0.03 (18)	$\nu_1(c)$
426.61±0.03	424.8 ±0.3		$\nu_2(b)$
		739.4 ±0.1 (25)	$\nu_2(c)$

125

and ν_2 of I_b atoms. By comparing the frequency data of caesium octaiodide with $\nu_1 = 333.941$ and $\nu_2 = 643.298$ MHz of iodine crystals,[37,38] a line at 320.45 MHz is surely attributable to the ν_1 frequency of I_d atoms. The corresponding ν_2 line has not been detected as yet. Accordingly, two remaining frequencies at 133.83 and 187.57 MHz can be assigned to ν_1 and ν_2 of I_a. As expected, they yield a large asymmetry parameter amounting to 61%.

III. QUADRUPOLE COUPLING CONSTANT AND ASYMMETRY PARAMETER

Tables 5 and 6 show the quadrupole coupling constants and the asymmetry parameters of iodine in some triiodides and caesium octaiodide at liquid nitrogen temperature.

The quadrupole coupling constants of terminal iodine atoms in symmetric I_3^- ions, as in tetraethylammonium triiodide, fall in the very narrow frequency range of 1 110–1 170 MHz. This is a reasonable result, because the electronic charge of a symmetric I_3^- ion is distributed on both terminal iodine atoms almost equally and because the effect of the alkylammonium cations on the electric field gradient of the iodine atoms in the crystals of these compounds is expected to be small, provided that strong intermolecular interactions such as hydrogen bonding are absent. In fact, the variation of NQR frequencies due to different cations is known to be small (less than about 10%) in hexa-

Table 5. Quadrupole coupling constants and asymmetry parameters of [127]I in some alkylammonium triiodides at liquid nitrogen temperature. The assignments of observed frequencies are shown in parentheses.

Compound	e^2qQ/h,MHz		η^c	
$(CH_3)_4NI_3$	964.3(1t)	1147.1(2t)	0.010(1t)	0.053(2t)
	1 176.8(3t)	1 181.5(4t)	0.038(3t)	0.052(4t)
	1 216[a](5, 6t)	2 414(7c)	0.14[a](5t, 6t)	0(7c)
	2 540(8c)	2 597(9c)	0(8c)	0(9c)
$(C_2H_5)_4NI_3$	1 115.0(At)[b]	1 168(Bt)	0.013(At)	0.02(Bt)
	2 378(Bc)	2 579(Ac)	0(Bc)	0(Ac)
$(n-C_3H_7)_4NI_3$	1 145.9(1t)	1 153.3(2t)	0.009(1t)	0.044(2t)
	2 515(3c)	2 515(4c)	0(3c)	0(4c)
$(n-C_4H_9)_4NI_3$	1 041.0(1a)	1 115(2a)	0.053(1a)	0.02(2a)
	1 237(2b)	1 301(1b)	0.03(2b)	0.04(1b)
	2 398(3c)	2 407(4c)	0(3c)	0(4c)
$(CH_3)_2NH_2I_3$	1 121(At)	2 532(Ac)	0(At)	0(Ac)
$(n-C_3H_7)_2NH_2I_3$	1 109(1t)	1 148(2t)	0(1t)	0(2t)
	2 480(3c)	2 539(4c)	0(3c)	0(4c)

[a] Averages are taken.
[b] A and B are used when the assignment is complete. Otherwise, different (or possibly different) kinds of iodine atoms are numbered consecutively.
[c] When the asymmetry parameter cannot be evaluated, it is assumed to be zero.

Table 6. Quadrupole coupling constants and asymmetry parameters of ^{127}I in some triiodides having markedly asymmetric I_3^- ions and caesium octaiodide at liquid nitrogen temperature. The assignments of observed frequencies are given in parentheses.

Compound	e^2qQ/h, MHz			η^a		
NH$_4$I$_3$	466.8(a)	1 725.0(b)	2 458.7(c)	0.130(a)	0.151(b)	0(c)
TlI$_3$	632.7(a)	1 453.0(b)	2 382(c)	0.188(a)	0.120(b)	0(c)
RbI$_3$	774.2(a)	1 449.8(b)	2 465.7(c)	0.056(a)	0.058(b)	0(c)
CsI$_3$	819.0(a)	1 436.6(b)	2 477.5(c)	0.02(a)	0.041(b)	0(c)
Cs$_2$I$_8$	664.6(a)	1 423.4(b)	2 487(c)	0.610(a)	0.068(b)	0(c)
	2 136(d)			0(d)		

^{a} When the asymmetry parameter cannot be evaluated, it is assumed to be zero.

halometallates(IV).[39-42] In the case of the asymmetric I_3^- ions in ammonium, thallium(I), rubidium, and caesium triiodides, the quadrupole coupling constants of the terminal iodine atoms I_a, separated to a greater extent from the central iodine atom, are as small as 470 to 820 MHz, whereas those of iodine atoms I_b separated to a lesser extent amount to 1 430 to 1 730 MHz. The extent of asymmetry is very large, as expected from the asymmetric structure of the triiodide ions in these compounds. In the case of the tetramethylammonium and tetra-n-butylammonium triiodides having asymmetric I_3^- ions, the nuclear quadrupole coupling constant of I_a ranges from 960 to 1 180 MHz, while that of I_b falls in the range 1 180 to 1 300 MHz, revealing a rather small extent of asymmetry. Even if a complete assignment is unfeasible for the former compound, the lowest and the highest e^2qQ/h values of the terminal iodine atoms can be attributed to I_a and I_b, respectively. Although the structure of these crystals has not been clarified as yet, it is highly probable that there exist nearly symmetric I_3^- ions as found in modification II of tetraethylammonium triiodide.[28] The quadrupole coupling constants of central iodine atoms I_c are almost the same (2 380 to 2 600 MHz) regardless of whether the anions are symmetric or not.

By examining the quadrupole coupling constants of terminal iodine atoms, one can classify triiodides into the following three groups: (i) triiodides having symmetric I_3^- ions, (ii) those having nearly symmetric I_3^- ions, and (iii) those having markedly asymmetric I_3^- ions. As discussed in the previous section, asymmetric I_3^- units are discernible in caesium octaiodide. The large asymmetry of the I_3^- unit is approximately equal to that found in I_3^- ions in triiodides belonging to group (iii). Although no NQR study has been performed as yet, tetraphenylarsonium triiodide, in which symmetric I_3^- ions were found for the first time by Mooney-Slater,[27] is presumed to belong to group (i). Triiodide ions of group (i) must be located in a unit cell in such a way that the two terminal iodine atoms in an I_3^- ion are crystallographically equivalent. Therefore the central iodine atoms are required to lie at special positions with appropriate symmetry properties, such as a centre of symmetry, a twofold axis, or a mirror plane perpendicular to the axis of the ion. On the other hand, triiodide ions in compounds of groups (ii) and (iii) can occupy general positions in a unit cell. These conclusions will be helpful when one attempts to determine the crystal structure of triiodides already investigated by NQR.

The existence of symmetric, nearly symmetric, and considerably asymmetric I_3^- ions in crystals suggests that the variation of the structure is due to an intrinsic property of the I_3^- ion rather than to the packing effect in crystal lattices. From an analogy to the H_3 molecule, Slater has proposed a possible explanation for the occurrence of symmetric and asymmetric forms of the triiodide ion.[43] When the distance between terminal iodine atoms is below a certain critical value, there is a single minimum for the energy of the I_3^- ion as a function of the position of the central iodine atom, this minimum lying midway between the two terminal atoms. However, when the distance between the terminal atoms is longer, there are two energy minima and an asymmetric I_3^- ion is formed. Until Mooney-Slater showed the existence of symmetric I_3^- ions in tetraphenylarsonium triiodide,[27] the latter case had been believed to be common for actual triiodide ions in crystals.

Hach and Rundle[20] and Mooney-Slater[27] have attributed the observed asymmetry of I_3^- ions to the crystal environment which leads to an unbalance of small positive ions distributed around a terminal iodine atom. They concluded that symmetric triiodide ions might be found only in crystals having sufficiently large cations, as in tetraphenylarsonium triiodide. However, Migchelsen and Vos[28] have found symmetric and asymmetric I_3^- ions existing in modifications I and II respectively of tetraethylammonium triiodide. Furthermore, the existence of asymmetric I_3^- ions in tetra-n-butylammonium triiodide as revealed by NQR indicates clearly that the cationic size is not the single factor determining the structure of I_3^- ions in crystals. In crystals belonging to group (ii), the arrangement of ions surrounding an I_3^- ion might play an important role in giving rise to asymmetric triiodide ions.

The asymmetry parameter of iodine is large in the ammonium and thallium(I) triiodides, which have relatively small cations. In particular, the asymmetry parameter of 18.8% for I_a in thallium(I) triiodide is among the largest ever reported for I_3^- ions. As discussed below, electrostatic interactions between an iodine atom and neighbouring small cations surrounding it in an asymmetric manner might be responsible for a large asymmetry parameter. As expected from the linear or almost linear structure of I_3^- ions, iodine atoms in other triiodides have reasonably small asymmetry parameters which are capable of being explained in terms of the asymmetry of the electric field gradient originating from ionic charges distributed in the crystal.

IV. CHARGE DISTRIBUTIONS IN TRIIODIDE AND OCTAIODIDE IONS

The ionicity i of iodine atoms in polyiodides can be calculated in accordance with the Townes–Dailey procedure.[44] For a negatively charged iodine atom, one has

$$e^2qQ/h = (1-i)(e^2qQ/h)_{atom} \tag{3}$$

whereas the following relation holds for a positively charged iodine atom, i.e., one whose observed e^2qQ/h is greater than the absolute value of the quadrupole coupling constant, $(e^2qQ/h)_{atom}$, of atomic ^{127}I.

$$e^2qQ/h = (1+i+2\varepsilon i)(e^2qQ/h)_{atom} \tag{4}$$

Here an empirical constant ε equal to 0.12 takes into account the contraction of the p orbitals in a partially positive iodine atom. Table 7 shows the ionicity of the iodine atoms calculated from these equations and the total calculated charge on the polyiodide ions.

The charge on the terminal iodine atoms varies to a considerable extent owing to the difference in the structure of the I_3^- ions, whereas the central iodine atom, I_c, carries a small positive fractional charge regardless of whether the

Table 7. Charge distribution in I_3^- and I_8^{2-} ions in some polyiodides

Compound (classification)	Type of I_3^-	Iodine atom	Charge, e	Total calculated charge, e
NH_4I_3(iii)	Asym.	a	−0.80	
		b	−0.25	
		c	+0.05	−1.00
TlI_3(iii)	Asym.	a	−0.72	
		b	−0.37	
		c	+0.03	−1.06
RbI_3(iii)	Asym	a	−0.66	
		b	−0.37	
		c	+0.06	−0.97
CsI_3(iii)	Asym.	a	−0.64	
		b	−0.37	
		c	+0.06	−0.95
$(CH_3)_4NI_3$(ii)	Asym.	t(av)	−0.50	
		c(av)	+0.08	−0.92
$(C_2H_5)_4NI_3$(i)	Sym.	t(av)	−0.50	
		c(av)	+0.07	−0.93
$(n-C_3H_7)_4NI_3$(i)	Sym.	t(av)	−0.50	
		c(av)	+0.08	−0.92
$(n-C_4H_9)_4NI_3$(ii)	Asym.	a(av)	−0.53	
		b(av)	−0.45	
		c(av)	+0.04	−0.94
$(CH_3)_2NH_2I_3$(i)	Sym.	t	−0.51	
		c	+0.08	−0.94
$(n-C_3H_7)_2NH_2I_3$(i)	Sym.	t(av)	−0.51	
		c(av)	+0.08	−0.94
Cs_2I_8		a	−0.71	
		b	−0.38	
		c	+0.07	
		d	−0.06	−2.16

129

triiodide ions are symmetric or not. The total calculated charges of the triiodide and octaiodide ions agree very well with the theoretical values of $-e$ and $-2e$ respectively, indicating the adequacy of the evaluation of the ionicity.

In order to interpret the linear structure of triiodide ions, Pimentel[10] as well as Hach and Rundle[20] proposed a molecular orbital scheme for bonding in an I_3^- ion using only the p_σ orbitals of iodine atoms. According to this model, only one MO is bonding in the triiodide ion and hence there is only one bonding electron pair for two bonds. The non-bonding MO is confined to the terminal iodine atoms, indicating the accumulation of high electron density on them in agreement with the charge distribution determined by NQR. Although Rundle[45] suggested some minor contribution from outer d orbitals to the p_σ bonding orbitals as well as the participation of π bonding from some outer d orbitals, recent theoretical calculations[46,47] indicate that the contribution from outer d orbitals to the bonding in the I_3^- ion is negligible.

In order to understand the electronic state of the I_3^- ion, it is instructive to consider the resonance structures, I, II, and III. The structure III takes into

$$
\begin{array}{ccc}
\text{I} & \text{II} & \text{III} \\
I_a^- \quad I_c{-}I_b & I_a{-}I_c \quad I_b^- & I_a^- \quad I_c^+ \quad I_b^-
\end{array}
$$

account the polarization of an iodine molecule by an approaching negative iodine ion. From the charge distribution in the I_3^- ions evaluated above, it is concluded that contributions from the resonance structures, I, II, and III amount to 43, 43, and 8% respectively for symmetric I_3^- ions, 49, 41, and 4% for nearly symmetric I_3^- ions in compounds belonging to group (ii), and 66, 29, and 5% for asymmetric I_3^- ions in the triiodides of group (iii). The data for tetramethylammonium triiodide are not included in this evaluation because the complete assignment of the observed frequencies has not been performed. The values of the ionicity of I_a atoms in ammonium and thallium(I) triiodides involve some errors because of the neglect of the strong electrostatic interactions between iodine atoms and neighbouring small cations. It is therefore rather fortuitous that the sum of the contributions from the three resonance structures amounts exactly to 100%.

V. ANOMALOUS TEMPERATURE DEPENDENCE OF THE RESONANCE FREQUENCIES

Although quadrupole resonance frequencies normally decrease with increasing temperature,[48,49] resonance lines showing a positive temperature coefficient have been reported for ammonium, thallium(I), and tetramethylammonium triiodides.[16,19] In the case of tetramethylammonium triiodide, only the lowest frequency ν_1 line shows a positive temperature coefficient, while both the ν_1 and ν_2 frequencies of I_a in ammonium and thallium(I) triiodides show positive temperature coefficients, as shown in Figs. 5 and 6.

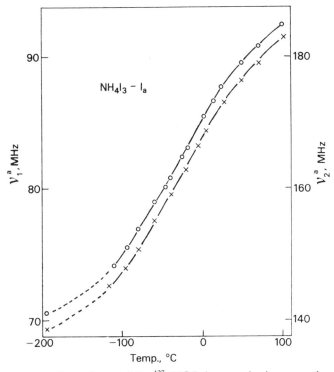

Fig. 5. Temperature dependence of the ^{127}I NQR frequencies in ammonium triiodide. Circles and crosses represent the ν_1(a) and ν_2(a) frequencies respectively.

The positive temperature coefficient has been interpreted in terms of a decrease in the $d_\pi-p_\pi$ character of metal–ligand bonds due to thermal vibration in some octahedral complex ions[39,40,50] or by the formation of OH $\cdots$ Cl hydrogen bonds in sodium tetrachloroaurate(III) and sodium tetrachloroiodate(III) dihydrates.[51-53] However, these explanations are not applicable to the present case. Firstly, hydrogen bonding is not conceivable in thallium(I) triiodide. Secondly, only a limited number of triiodides show the anomaly. It is hardly believable that the nature of I–I bonds in a triiodide ion is responsible for the positive temperature coefficient. Surely the present case represents another type of anomalous temperature dependence.

The lattice constants of ammonium and thallium(I) triiodides are smaller than those of the isomorphous rubidium and caesium triiodides. In compounds having small cations, a terminal iodine atom I_a has four neighbouring cations at close distances in directions almost perpendicular to the bond axis as shown in Fig. 1. In the absence of neighbouring ions, the field gradient eq_{zz} at I_a is given by

$$eq_{zz} = N_x eq_x + N_y eq_y + N_z eq_z \tag{5}$$

131

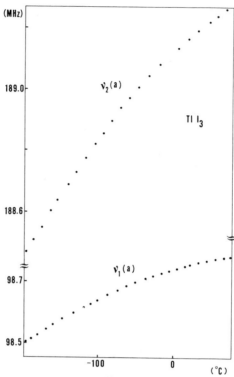

Fig. 6. Temperature dependence of the ^{127}I NQR frequencies in thallium(I) triiodide.

where N_x, N_y, and N_z denote the number of electrons in the $5p_x$, $5p_y$, and $5p_z$ orbitals of the iodine atom respectively, and eq_x, eq_y, and eq_z are the contributions of a single electron in these orbitals to the field gradient eq_{zz}, the z-axis being directed towards I_c. Both eq_x and eq_y are positive, while eq_z is negative. Since the $5p_x$ and $5p_y$ orbitals of I_a are almost fully occupied, one can write

$$eq \equiv eq_{zz} = 2eq_x + 2eq_y - N_z|eq_z| \tag{6}$$

In a fictitious, vibrationless, crystal of ammonium or thallium(I) triiodide, neighbouring cations attract and deform the electron cloud of the $5p_x$ and $5p_y$ orbitals thereby decreasing eq_x and eq_y. Alternatively, the strong electrostatic field of the neighbouring cations promotes electrons in the $5p_x$ and $5p_y$ orbitals to higher atomic orbitals of iodine thereby decreasing N_x and N_y from the maximum of two. Accordingly eq_{zz} decreases. Thermal vibrations partially quench this effect and hence the observed e^2qQ/h increases with increasing temperature, provided that the normal negative temperature coefficient is masked. The rotation of ammonium ions and the possible existence of weak hydrogen bonding must be taken into account for ammonium triiodide, because it shows an extraordinarily large positive temperature coefficient with its maximum at about $-40°$.

132

REFERENCES

1 G. S. Johnson, *J. Chem. Soc.*, **33**, 397 (1878).
2 J. W. Mellor, *A Comprehensive Treatise on Inorganic and Theoretical Chemistry*, Vol. II, Longmans, Green and Co., London, 1931, p. 233.
3 *Gmelins Handbuch der anorganischen Chemie*, Part 8, Verlag Chemie, Berlin, 1933, p. 403.
4 A. I. Popov, *Halogen Chemistry*, Vol. 1, V. Gutmann, editor, Academic Press, London and New York, 1967, p. 225.
5 R. C. L. Mooney, *Z. Krist.*, **90**, 143 (1935).
6 R. C. L. Mooney, *Z. Krist.*, **100**, 519 (1939).
7 G. E. Kimball, *J. Chem. Phys.*, **8**, 188 (1940).
8 L. Pauling, *The Nature of the Chemical Bond*, 2nd Ed., Cornell University Press, Ithaca, New York, 1948, p. 111.
9 E. Cartmell and G. W. A. Fowles, *Valency and Molecular Structure*, Butterworth and Co., Ltd., London, 1956, p. 177.
10 G. C. Pimentel, *J. Chem. Phys.*, **19**, 446 (1951).
11 S. Kojima, A. Shimauchi, S. Hagiwara, and Y. Abe, *J. Phys. Soc. Japan*, **10**, 930 (1955).
12 S. Kojima, K. Tsukada, S. Ogawa, and A. Shimauchi, *J. Chem. Phys.*, **23**, 1963 (1955).
13 C. D. Cornwell and R. S. Yamasaki, *J. Chem. Phys.*, **27**, 1060 (1957).
14 R. S. Yamasaki and C. D. Cornwell, *J. Chem. Phys.*, **30**, 1265 (1959).
15 Y. Kurita, D. Nakamura, and N. Hayakawa, *J. Chem. Soc. Japan* (*Pure Chem. Sect.*), **79**, 1093 (1958).
16 A. Sasane, D. Nakamura, and M. Kubo, *J. Phys. Chem.*, **71**, 3249 (1967).
17 G. L. Breneman and R. D. Willett, *J. Phys. Chem.*, **71**, 3684 (1967).
18 G. A. Bowmaker and S. Hacobian, *Aust. J. Chem.*, **21**, 551 (1968).
19 H. Harada, D. Nakamura, and M. Kubo, *J. Magn. Resonance*, **13**, 56 (1974).
20 R. J. Hach and R. E. Rundle, *J. Amer. Chem. Soc.*, **73**, 4321 (1951).
21 J. Broekema, E. E. Havinga, and E. H. Wiebenga, *Acta Cryst.*, **10**, 596 (1957).
22 E. E. Havinga and E. H. Wiebenga, *Acta Cryst.*, **11**, 733 (1958).
23 E. E. Havinga, K. H. Boswijk, and E. H. Wiebenga, *Acta Cryst.*, **7**, 487 (1954).
24 W. J. James, R. J. Hach, D. French, and R. E. Rundle, *Acta Cryst.*, **8**, 814 (1955).
25 E. H. Wiebenga, E. E. Havinga, and K. H. Boswijk, *Advan. Inorg. Chem. Radiochem.*, **3**, 133 (1961).
26 H. A. Tasman and K. H. Boswijk, *Acta Cryst.*, **8**, 59 (1955).
27 R. C. L. Mooney-Slater, *Acta Cryst.*, **12**, 187 (1959).
28 T. Migchelsen and A. Vos, *Acta Cryst.*, **23**, 796 (1967).
29 T. Bernstein and F. H. Herbstein, *Acta Cryst.*, **B24**, 1640 (1968).
30 G. H. Cheesman and A. J. T. Finney, *Acta Cryst.*, **B26**, 904 (1970).
31 R. G. Barnes and S. L. Segel, *J. Chem. Phys.*, **25**, 180 (1956).
32 S. L. Segel and R. G. Barnes, *J. Chem. Phys.*, **25**, 578 (1956).
33 G. W. Ludwig, *J. Chem. Phys.*, **25**, 159 (1956).
34 A. C. Hazell, *Acta Cryst.*, **16**, 71 (1963).
35 P. M. Harris, E. Mack, Jr. and F. C. Blake, *J. Amer. Chem. Soc.*, **50**, 1583 (1928).
36 I. I. Kitaigorodskii, V. Khotsyanova, and M. Struchkov, *Zhur. Fiz. Khim.*, **27**, 780 (1953).
37 R. V. Pound, *Phys. Rev.*, **82**, 343 (1951).
38 H. G. Dehmelt, *Z. Physik*, **130**, 356 (1951).
39 M. Kubo and D. Nakamura, *Advan. Inorg. Chem. Radiochem.*, **8**, 257 (1966).
40 T. L. Brown and L. G. Kent, *J. Phys. Chem.*, **74**, 3572 (1970).

41 T. B. Brill, R. C. Gearhart, and W. A. Welsh, *J. Magn. Resonance*, **13**, 27 (1974).
42 T. B. Brill and W. A. Welsh, *J. Chem. Soc., Dalt.*, 357 (1973).
43 J. C. Slater, *Acta Cryst.*, **12**, 197 (1959).
44 B. P. Dailey and C. H. Townes, *J. Chem. Phys.*, **23**, 118 (1955).
45 R. E. Rundle, *Acta Cryst.*, **14**, 585 (1961).
46 E. H. Wiebenga and D. Kracht, *Inorg. Chem.*, **8**, 738 (1969).
47 W. Gabes and M. A. M. Nijman-Meester, *Inorg. Chem.*, **12**, 589 (1973).
48 H. Bayer, *Z. Physik*, **130**, 227 (1951).
49 T. Kushida, *J. Sci. Hiroshima Univ.*, **A19**, 327 (1955).
50 T. L. Brown, W. G. McDugle, Jr., and L. G. Kent, *J. Amer. Chem. Soc.*, **92**, 3645 (1970).
51 C. W. Fryer and J. A. S. Smith, *J. Chem. Soc., A*, 1029 (1970).
52 A. Sasane, T. Matuo, D. Nakamura, and M. Kubo, *J. Magn. Resonance*, **4**, 257 (1971).
53 A. Sasane, D. Nakamura, and M. Kubo, *J. Magn. Resonance*, **8**, 179 (1972).

5. THE *A PRIORI* CALCULATION OF THE ELECTRONIC CONTRIBUTION TO THE NUCLEAR QUADRUPOLE COUPLING CONSTANT: theoretical aspects and recent results

R. Moccia and M. Zandomeneghi

Istituto di Chimica Fisica della Università, Via Risorgimento 35, 56100 Pisa, Italy

INTRODUCTION

The development of sophisticated experimental techniques has afforded the possibility of very accurate measurements of the nuclear quadrupole coupling constants (NQCC) for a wide variety of nuclei in a large number of systems.[1] These data, apart from their experimental usefulness, represent a valuable source of information either to probe the quality of the computed electronic wave functions (w.f.) for known nuclear quadrupole moments or to evaluate these quadrupole moments when their value may be uncertain.[2-4] To this purpose, one of the main problems lies in the electronic contribution to the experimental quadrupole coupling constant, assuming, as will be done throughout this review, that the dynamics of the nucleons will not be affected by external fields. Thus no nuclear polarization effects will be considered.[5] As far as we know, this represents quite an acceptable approximation and in any case it would involve problems beyond the scope of this article. Since there is no evidence in our case against its correctness, the Born–Oppenheimer approximation will be assumed throughout.

The attention here will be focused upon the problem of defining and calculating the electronic contributions to the NQCC, mainly in molecules. The dramatic progress made in recent years in the available computer facilities has spurred great activity in the calculation of accurate electronic w.f., beyond the MOSCF level, using a variety of theoretical approaches. Unfortunately several of the theoretical and methodological points involved in the actual calculations of the NQCC have received little attention.

The main purpose of this article is then to present in a compact way some of the methods available at present to compute accurately the electronic contributions to the NQCC. The relative merits of these methods, for the particular problem in hand, cannot be assessed in general from a numerical viewpoint because of the lack of sufficient computations.

After defining in Section I what are these electronic contributions to NQCC, the first-order effect will be considered in Section II in some detail, looking at it both as an expectation value of an observable and as a first-order property. While in the former technique, the accuracy depends in a general way on the quality of the w.f., in the latter there exists the possibility of improving judiciously the w.f. The more delicate effects which may affect the NQCC, always within the approximation stated above, such as the vibrational corrections and second-order effects due to the nuclear quadrupole moment and the finite size of the nuclei, will be briefly reviewed in Sections III and IV in order to assess their relative importance. In the last Section V, a collection of the most recent *a priori* calculations of the electronic contribution to the NQCC is presented. Finally the relationships between the first-order electronic contribution to the NQCC and the molecular force constants are sketched in Appendix A and the problem of the electronic core polarization effects is briefly reviewed in Appendix B.

I. THEORETICAL BACKGROUND

With the assumption that the nuclear forces are the only ones which rule the dynamics of the nucleons (as far as the internal degree of freedom are concerned) and the adoption of the Born–Oppenheimer approximation, the electrostatic electronic Hamiltonian of a molecule or of an atom may be written as[*]

$$\mathcal{H} = \sum_j \left(-\tfrac{1}{2}\Delta_j\right) + \sum_\alpha \mathcal{V}^\alpha_{N_e} + \sum_{i<j} \frac{1}{r_{ij}} + \sum_{\alpha<\beta} \mathcal{V}^{\alpha\beta}_{NN} \tag{I-1}$$

where the indices i, j run over all the electrons and the indices α, β over all the nuclei. The symbol $\mathcal{V}^\alpha_{N_e}$ stands for the electrostatic interaction between all electrons and the nucleons pertaining to the nucleus α, while $\mathcal{V}^{\alpha\beta}_{NN}$ stands for the electrostatic interaction between the nucleons pertaining to the nucleus α and those pertaining to the nucleus β.

Indicating by q_ν, etc., the electric charge of the nucleon ν of the nucleus α, we may write these two terms as

$$\mathcal{V}^\alpha_{N_e} = -\sum_j \sum_{\nu_\alpha} \frac{q_{\nu_\alpha}}{|\bar{r}_j - \bar{r}_{\nu_\alpha}|}; \qquad \mathcal{V}^{\alpha\beta}_{NN} = \sum_{\nu_\alpha} \sum_{\nu_\beta} \frac{q_{\nu_\alpha} q_{\nu_\beta}}{|\bar{r}_{\nu_\alpha} - \bar{r}_{\nu_\beta}|} \tag{I-2a,b}$$

---

[*] a.u. will be used throughout unless otherwise specified.[6]

By (1) using the Laplace expansion, (2) retaining only the most important terms (according to the extremely small size of the nuclei), (3) neglecting the contribution arising from the electrons within the nuclear volumes, and (4) using the result that the expectation values of nucleon operators of odd parity are zero, we have

$$\mathcal{V}_{N_e}^\alpha + \sum_{\beta(\neq\alpha)} \mathcal{V}_{NN}^{\alpha\beta} = -\sum_i \frac{Z_\alpha}{|\vec{r}_i - \vec{r}_\alpha|} + \sum_{\beta(\neq\alpha)} \frac{Z_\alpha Z_\beta}{|\vec{r}_\alpha - \vec{r}_\beta|} + \hat{h}_{quadr} \qquad \text{(I-3a)}$$

$$\hat{h}_{quadr} = \sum_{m=-2}^{+2} (-1)^m \hat{V}_{2,-m}^\alpha \hat{Q}_{2,m}^\alpha \qquad \text{(I-3b)}$$

where $\vec{r}_\alpha$, $\vec{r}_\beta$ are the position vectors of the nuclear centres of mass and the quantities appearing in the second term $\hat{h}_{quadr}$ of Eq. (I-3a), the electrostatic quadrupolar interaction, are

$$\hat{V}_{2,m}^\alpha = \left(\frac{4\pi}{5}\right)^{1/2} \left\{ \sum_{\beta(\neq\alpha)} Z_\beta Y_{2,m}(\hat{r}_\beta) \frac{1}{|\vec{r}_\alpha - \vec{r}_\beta|^3} - \sum_i Y_{2,m}(\hat{r}_i) \frac{1}{|\vec{r}_i - \vec{r}_\alpha|^3} \right\} \qquad \text{(I-4a)}$$

$$\hat{Q}_{2,m}^\alpha = \left(\frac{4\pi}{5}\right)^{1/2} \sum_{\nu_\alpha} q_{\nu_\alpha} Y_{2,m}(\hat{r}_{\nu_\alpha}) r_{\nu_\alpha}^2 \qquad \text{(I-4b)}$$

The nucleus-electron w.f. (considering for simplicity only the nucleus α), neglecting the quadrupolar interaction $\hat{h}_q$, is

$$\Psi_0\rangle = |IM\rangle_\alpha |\Phi_n(\mathbf{R})\rangle \qquad \text{(I-5)}$$

where I and M are the quantum numbers associated with the square and the projection, upon some quantization axis, of the total nuclear angular momentum, n is the collection of the possible electronic quantum numbers, and $\mathbf{R}$ stands for the set of nuclear coordinates. The other nuclear quantum numbers do not appear because it is assumed that no nuclear excitation processes will take place.

The energy will be $E_{nucleus} + E_n$ and, due to the nuclear spin, it will be at least $2I+1$ times degenerate. This degeneracy may be removed by the quadrupole interaction $\hat{h}_q$ and the subsequent splitting may be computed by perturbation theory. Among the many formulations possible,[7] we have chosen that described at length by Löwdin[8] because it is very neat and concise.

If Ψ represents the set (row) of the $2I+1$ states [Eq. (I-5)] degenerate at zero order and distinguished only by the value of M, the energy levels in the presence of a perturbation corresponding to the unperturbed ones E_0 are given exactly by the roots E of the equation

$$\det \langle \Psi(\mathbf{1}E_0 + \hat{h}_q + \hat{h}_q \hat{P}\hat{T}_E \hat{P}\hat{h}_q - \mathbf{1}E)\Psi\rangle = 0 \qquad \text{(I-6)}$$

In the above formula, the operator

$$\hat{P} = \mathbf{1} - \Psi\rangle\langle\Psi\Psi\rangle^{-1}\langle\Psi$$

is a projection operator which eliminates any possible component lying in the space spanned by $\boldsymbol{\Psi}$ and the operator $\hat{T}_E$, the reduced resolvent, is defined as

$$\hat{T}_E = \hat{P}(1E - \hat{P}\hat{\mathscr{H}}_0\hat{P} - \hat{P}\hat{h}_q\hat{P})^{-1}\hat{P} \tag{I-7}$$

By using an expansion in the perturbation $\hat{h}_q$ for $\hat{T}_E$, the following terms are easily derived

$$\hat{T}^{(0)} = \hat{P}(1E_0^{(0)} - \hat{P}\hat{\mathscr{H}}_0\hat{P})^{-1}\hat{P} \tag{I-8a}$$

$$\hat{T}^{(1)} = -\hat{T}^{(0)}(1E_0^{(1)} - \hat{P}\hat{h}_q\hat{P})\hat{T}^{(0)}, \quad \text{etc.} \tag{I-8b}$$

By substitution in Eq. (I-6), the energy eigenvalues up to zero order will be the roots of the equation

$$\det \langle \boldsymbol{\Psi}(1E_0^{(0)} - 1E)\boldsymbol{\Psi}\rangle = 0 \tag{I-9a}$$

up to first-order

$$\det \langle \boldsymbol{\Psi}(1E_0^{(0)} + \hat{h}_q - 1E)\boldsymbol{\Psi}\rangle = 0 \tag{I-9b}$$

up to second-order

$$\det \langle \boldsymbol{\Psi}(1E_0^{(0)} + \hat{h}_q + \hat{h}_q\hat{P}(1E_0^{(0)} - \hat{P}\hat{\mathscr{H}}_0\hat{P})^{-1}\hat{P}\hat{h}_q - 1E)\boldsymbol{\Psi}\rangle = 0, \quad \text{etc.} \tag{I-9c}$$

The elements of the matrices to be diagonalized are readily obtained using the w.f. [Eq. (I-5)] and the Wigner–Eckart theorem.[9] They are:

up to zero order

$$\boldsymbol{H}_{M,M'}^{(0)} = E_0^{(0)} \delta_{MM'} \tag{I-10a};$$

up to first-order

$$\boldsymbol{H}_{M,M'}^{(0)+(1)} = E_0^{(0)} \delta_{MM'} + \langle\|Q\|\rangle(-1)^{M-M'} V_{M'-M}^{00} C(I2I; M'M - M') \tag{I-10b};$$

up to second-order

$$\boldsymbol{H}_{M,M'}^{(0)+(1)+(2)} = E_0^{(0)} \delta_{MM'} + \langle\|Q\|\rangle(-1)^{M-M'} V_{M'-M}^{00} C(I2I; M'M - M')$$

$$+ \langle\|Q\|\rangle^2 \sum_{M''} C(I2I; M'', M - M'')C(I2I; M'M'' - M')$$

$$\sum_n \frac{V_{M''-M}^{0n} V_{M'-M''}^{n0}}{E_0^{(0)} - E_n^{(0)}}, \quad \text{etc.} \tag{I-10c}$$

In Eqs. (I-1) to (I-10) $C(j_1j_2j_3M_1M_2)$ indicate the $3-j$ symbols (whose properties automatically exclude $|M'-M| > 2$), $\langle\|Q\|\rangle$ is the reduced matrix element of the nuclear contribution, and the $V_m^{nn'}$ elements are defined as

$$V_m^{nn'} = \frac{\langle\Phi_n^{(0)} \hat{V}_{2,m} \Phi_{n'}^{(0)}\rangle}{\langle\Phi_n^{(0)}\Phi_n^{(0)}\rangle^{1/2}\langle\Phi_{n'}^{(0)}\Phi_{n'}^{(0)}\rangle^{1/2}} \tag{I-11}$$

Since the $\Phi_n\rangle$'s depend upon $\mathbf{R}$, the nuclear coordinates, the matrix elements (I-11) will also be dependent upon them.

The reduced matrix element $\langle\|Q\|\rangle$ is related to the more usual symbol Q of the nuclear quadrupole moment[2-4] by

$$Q = 2\langle\|Q\|\rangle C(I2I; I0I) \tag{I-12}$$

To find the relation between these quantities and the more usual ones q and η, it is convenient to define the real operators $\hat{q}_m$ which are expressed using the cartesian basis instead of the spherical basis employed to define the $\hat{V}_{2,m}$.

$$\hat{q}_{|m|} = \frac{1}{\sqrt{2}}[\hat{V}_{2,-|m|} + (-1)^m \hat{V}_{2,|m|}]$$

$$\hat{q}_0 = \hat{V}_{2,0} \tag{I-13}$$

$$\hat{q}_{-|m|} = \frac{i}{\sqrt{2N}}[\hat{V}_{2,-|m|} - (-1)^m \hat{V}_{2,|m|}]$$

It is easily found that for the principal axes

$$q_1^{00} = q_{-1}^{00} = q_{-2}^{00} = 0 \tag{I-14}$$

$$q_0^{00} = \tfrac{1}{2}q \tag{I-15}$$

$$\sqrt{3}\frac{|q_2^{00}|}{q_0^{00}} = \eta \tag{I-16}$$

In addition, if the electronic w.f. is totally symmetric under the symmetry operations of the local point group (the sub-group of the symmetry point group of the molecules, including all operations which do not move the nucleus under consideration), it may be easily seen that

$q_2^{00} = 0$ if there is a n-fold axis with $n \geq 3$, and
$q_0^{00} = 0$ if there is more than one n-fold axis with $n \geq 3$ (groups like
$\qquad$ T, T_d, T_h, O, O_h, R_3).

II. FIRST-ORDER TREATMENT

A. Electronic Contribution as an Expectation Value

Generally the first-order treatment, which reduces to the diagonalization of the matrix (I-10b), is widely accepted as a satisfactory approximation. To this level of accuracy, the theory relates simply the experimental data directly and only to $\langle\|Q\|\rangle$, q_0^{00} and q_2^{00}. The quantities q_0^{00} q_2^{00} by their definitions are, apart from the contribution of the other nuclei, the expectation values of the (spinless) one-electron operators (the origin is located upon the nucleus under consideration)

$$\hat{q}_{0_{\text{elec}}} = -\sum_i \frac{2z_i^2 - x_i^2 - y_i^2}{2r_i^5} \qquad \hat{q}_{2_{\text{elec}}} = -\sum_i \frac{3^{1/2}}{2} \frac{x_i^2 - y_i^2}{2r_i^5} \qquad \text{(II-1a,b)}$$

Their accuracy may be discussed within the general theoretical aspects of the calculation of one-electron operators. There is no doubt that the problems inherent in their accurate calculation are different from those regarding other one-electron operators, but it is certainly worthwhile to look at them from a general stand point.

The accuracy of the expectation values of operators (not only one-electron operators) has been investigated by several authors and several bounds have been established for them.[10] Unfortunately most of these methods are based upon the possibility of obtaining estimates of the projection of the approximate w.f. φ upon the true eigenfunction ψ. These estimates require the knowledge[11] either of the expectation value of the squared Hamiltonian $\langle \varphi \mathscr{H}^2 \varphi \rangle$ or the experimental knowledge of a number of consecutive energy levels of the system belonging to the same symmetry (the greater the numbers, the better the estimates). The difficulties inherent in fulfilling these requirements, among other reasons, have restricted the practical application of the method of bounds to exceedingly simple systems. Thus these methods will not be described here and the interested reader may find all the details in the above references.[10,11,13]

The calculation of the spinless one-electron operators requires in principle the knowledge of only the spinless first-order density matrix $P(\bar{r}, \bar{r}')$,[12] and for operators like $\hat{q}_0$ and $\hat{q}_2$ simply the diagonal part $P_{00}(\bar{r})$, which is defined as

$$P_{00}(\bar{r}) = N \int \psi_0(r\sigma; x_2 \dots x_N)\psi_0^*(r\sigma; x_2 \dots x_N)\, d\sigma\, dx_2 \dots dx_N \quad \text{(II-2)}$$

where x_i are the spin–spatial coordinates $(\sigma_i, \bar{r}_i)$ of the electron j, N is the total number of electrons and ψ_0 the electronic w.f. At first sight, it might appear that because in the expectation value expression

$$q^{00}_{m_{\text{elec}}} = \int \hat{q}_{m_{\text{elec}}}(\bar{r})P_{00}(\bar{r})\, dv \qquad \text{(II-3)}$$

the operator weights heavily the density matrix in the immediate neighbourhood of the nucleus, it would be necessary to worry about the quality of $P(\bar{r})$ only in that region of space; the requirements for the quality of $P(\bar{r})$ in the rest of the space should not be so exacting. Unfortunately the situation is far from being so simple. In fact, in the immediate neighbourhood of a nucleus the dominant part of the electronic density is the spherical one which, for symmetry reasons, gives zero contribution. The component of quadrupolar symmetry of $P(\bar{r})$ will be more important in region of space not so close to the nucleus. The electronic contribution arising from the part of space far from the nucleus, while being important in general, will be heavily counterbalanced by the contributions of other nuclei (for a neutral molecule). It must be recognized therefore that the situation is rather complicated and a simple qualitative approach may easily give a wrong picture. Perhaps it may be said that the region of space around the nucleus up to the directly bonded atoms is probably the most important one. For these reasons also it seems convenient to proceed in a more formal way and we begin by examining the general problems inherent in the calculation of accurate electronic charge distributions.

It seems appropriate to start with the SCFMO type w.f., both because of their practical availability and their theoretical importance as an extremely convenient starting point of more sophisticated approaches. It is a widely held belief that the electronic charge distribution given by SCFMO w.f. is highly accurate. This contention is based upon a consequence of the Brillouin theorem[14] which states that *if*

$$\Phi_0 = \sqrt{N!} \mathcal{A} \psi_1(x_1) \ldots \psi_N(x_N) \tag{II-4}$$

is the best single detor built with the first N of a complete set of spin orbitals ψ which must then satisfy the HF equations

$$[\hat{h}(x_1) + \hat{\mathcal{G}}(x_1)] \psi_r(x_1) = \varepsilon_r \psi_r(x_1) \tag{II-5}$$

where $\hat{h}$ is the one-electron operator and

$$\mathcal{G}(x_1)\psi_j(x_1) = \sum_{K=1}^{N} \int \frac{dx_2}{|\vec{r}_1 - \vec{r}_2|} \psi_K^*(x_2)(1 - P_{12})\psi_k(x_2)\psi_j(x_1) \tag{II-6}$$

is the HF effective electron-repulsion operator, *then*

$$\langle \Phi_0 \mathcal{H}_0 \Phi_j^\nu \rangle = 0 \qquad (j \leqslant N; \nu > N) \tag{II-7}$$

In the above equation, Φ_j^ν stands for the detor obtainable from Φ_0 by substituting the spin orbital ψ_j with the spin orbital ψ_ν, i.e. the energy matrix elements between the best single detor and any single excitation detor are zero. All possible N-electron detors Φ, obtainable from the complete set Ψ, which may conveniently be indicated by

$$\Phi_{j_1 j_2 \cdots j_n}^{\nu_1 \nu_2 \cdots \nu_n} \qquad (n \leqslant N) \tag{II-8}$$

are eigenfunctions of an N-electron Hamiltonian $\mathcal{H}_{HF}(x_1 \ldots x_N)$

$$\mathcal{H}_{HF} \Phi_{j_1 \cdots j_n}^{\nu_1 \cdots \nu_n} = \left[\sum_{j=1}^{N} \varepsilon_j + \sum_{m=1}^{n} (\varepsilon_{\nu_m} - \varepsilon_{j_m}) \right] \Phi_{j_1 \cdots j_n}^{\nu_1 \cdots \nu_n} \tag{II-9}$$

with

$$\mathcal{H}_{HF}(x_1 \ldots x_N) = \sum_{j=1}^{N} [\hat{h}(x_j) + \hat{\mathcal{G}}(x_j)] \tag{II-10}$$

By considering $\hat{\mathcal{H}}_{HF}$ as an unperturbed Hamiltonian closely resembling the true $\hat{\mathcal{H}}$ (at least for the ground state) and by defining a fictitious perturbation $\hat{h}'$

$$\hat{h}' = \hat{\mathcal{H}} - \hat{\mathcal{H}}_{HF} = \sum_{i<j} 1/r_{ij} - \sum_j \hat{\mathcal{G}}(x_j) \tag{II-11}$$

it is possible to improve systematically, by perturbation methods, the quality of the electronic w.f. and then of the first-order density matrix. If $\Phi^{(1)}$, $\Phi^{(2)}$, etc., are the successive corrections to Φ_0, the corrections to the first-order density matrix ρ, once the normalization $\langle \Phi_0 \Phi^{(l)} \rangle = 0$ ($l = 1, 2 \ldots$) is assumed, are

$$\rho^{(1)} = \rho_{\Phi^{(1)}\Phi_0} + \rho_{\Phi_0\Phi^{(1)}} \tag{II-13a}$$

$$\rho^{(2)} = \rho_{\Phi^{(2)}\Phi_0} + \rho_{\Phi_0\Phi^{(2)}} + \rho_{\Phi^{(1)}\Phi^{(1)}} - \langle \Phi^{(1)}\Phi^{(1)} \rangle \rho_{\Phi_0\Phi_0} \tag{II-13b}$$

In the above equations, the symbol $\rho_{\Phi^{(l)}\Phi^{(m)}}$ stands for

$$\rho_{\Phi^{(l)}\Phi^{(m)}}(xx') = N \int \Phi^{(l)*}(x'x_2 \ldots x_N)\Phi^{(m)}(xx_2 \ldots x_N) \, dx_2 \ldots dx_N.$$

Straight-forward application of Schrödinger perturbation theory gives as the lth-order correction

$$\Phi^{(l)} = \sum_{n=1}^{N} \sum_{j_1<j_2<\ldots<j_n} \sum_{\nu_1<\nu_2<\ldots<\nu_n} C_{j_1\ldots j_n}^{(l)\nu_1\ldots\nu_n} \Phi_{j_1\ldots j_n}^{\nu_1\ldots\nu_n} \tag{II-15}$$

with

$$C_{j_1\ldots j_n}^{(1)\nu_1\ldots\nu_n} = \frac{\langle \Phi_{j_1\ldots j_n}^{\nu_1\ldots\nu_n} h' \Phi_0 \rangle}{\sum_{m=1}^{n} (\varepsilon_{j_m} - \varepsilon_{\nu_m})} = \frac{\langle \Phi_{j_1\ldots j_n}^{\nu_1\ldots\nu_n} \mathcal{H}\Phi_0 \rangle}{\sum_{m=1}^{n} (\varepsilon_{j_m} - \varepsilon_{\nu_m})} \tag{II-16a}$$

[where the last equation holds because of Eq. (II-9)], and

$$C_{j_1\ldots j_n}^{(2)\nu_1\ldots\nu_n} = \frac{\langle \Phi_{j_1\ldots j_n}^{\nu_1\ldots\nu_n}(h' - E_0^{(1)})\Phi^{(1)} \rangle}{\sum_{m=1}^{n} (\varepsilon_{j_m} - \varepsilon_{\nu_m})}, \quad \text{etc.} \tag{II-16b}$$

By exploiting the Brillouin theorem, Eq. (II-7), Eq. (II-16a) says that no single excitations contribute to $\Phi^{(1)}$. On the other hand because no more than two-electron operators are contained in $\mathcal{H}$, it is easily concluded that only the double excitations contribute to $\Phi^{(1)}$. The conclusion is that the first-order correction $\rho^{(1)}$, as given by Eq. (II-13a), vanishes.

The contention that the HF first-order density matrix is very accurate is therefore based upon the assumption that the second-order effects, which are readily seen to be generally different from zero, are negligible. There is little doubt that this is the case for many closed-shell situations, but it is advisable to exercise some care before taking this for granted in all situations. The need to be careful is emphasized by situations in which different possible occupation schemes of the spin orbitals describing states of a given symmetry lead to HF energies very close to each other. An extreme case of this is the occurrence of degeneracy. In such an instance, while the first-order correction to ρ vanishes, the second-order correction may be quite noticeable. This is because when different occupation schemes give almost equal HF energies, the single-detor approximation breaks down because the true w.f. projects heavily upon different detors.

The Brillouin theorem holds for a single-detor w.f. which, as is well known, cannot in general satisfy the requirements of being pure spin states (we are considering purely electrostatic Hamiltonians) and of providing irreducible representations for the spatial symmetry group of the B. and O. Hamiltonian. Although it has been argued that from an energy viewpoint the spin-and-symmetry requirements may act as additional unnecessary constraints (the symmetry dilemma),[15] it appears desirable to have them satisfied when other observables are of interest. Now it happens that a w.f. expressed as a linear combination of detors, in order to comply with the spin-and-symmetry requirements, does not satisfy the Brillouin theorem. In such an instance, first-order corrections should be expected for the density matrix and their importance is very difficult to assess.

Let us return now to the cases when the single-detor approximation may represent a satisfactory description of the state of affairs, i.e. in closed shells with well-defined occupation schemes. The first-order corrections are exactly zero while the second-order corrections, as given by Eqs. (II-13b), (II-15) and (II-16), are generally different from zero and are interpreted as correlation corrections. Before discussing these correlation effects, it should be remarked that all the arguments presented so far refer to truly self-consistent field orbitals, which satisfy the integrodifferential Eqs. (II-5) and then the Brillouin theorem. In practice this is hardly ever the case, because the orbitals are practically always expanded upon some convenient basis. Thus the best which may be obtained is to have the Brillouin theorem satisfied only for orbitals which lie completely in the space spanned by the basis employed. This means that variations in the first-order density matrix are to be expected upon increasing the basis size. The effect of the extension of the basis upon the expectation values of the operators we are interested in is estimated in Appendix B with special emphasis given to the inner core polarization.

The computation of the electron-correlation corrections through the formulae (II-13b), (II-15), (II-16) while being straightforward does not lend itself to simple interpretations, because of the very many terms involved. Although some attempts have been made[16] to identify the most important contributions for specific cases, it is useful to look for some simplifying arguments. To retain some of the pictorial simplicity of the orbital description, it is convenient to use localized orbitals[17] instead of the canonical ones. This permits us to visualize the charge distribution as due to contributions arising from different regions of space: those close to the nucleus α containing the localized orbitals identifiable with the K shell, the bonds directly attached to it and the lone pairs, and those farther away. It seems a reasonable assumption that the strongest effect of $\rho^{(2)}$ will be confined to the region of space containing the above-mentioned group of orbitals. The implication of this reasoning is that in the perturbation formulae those excitations involving orbitals other than the ones mentioned may be neglected. If g_a indicates a localized orbital obtained from the canonical ones $\varphi_1 \varphi_2 \ldots$ through a unitary transformation

$$g_a = \sum_j U_{ja} \varphi_j \tag{II-17}$$

the expansion coefficients $C_{ab}^{(1)\nu\mu}$ and $C_a^{(2)\nu}$ [only the single excitations of $\Phi^{(2)}$ will contribute to $\rho^{(2)}$, see Eq.(II-13b)] needed in this case are

$$C_{ab}^{(1)\nu\mu} = \sum_{i<j} \sum_{c<d} \frac{T_{ij,cd} T_{ij,ab}^*}{\varepsilon_i + \varepsilon_j - \varepsilon_\nu - \varepsilon_\mu} \langle \Phi_{ab}^{\nu\mu} h' \Phi_0 \rangle \tag{II-18a}$$

$$C_a^{(2)\nu} = \sum_i \sum_c \frac{U_{ic} U_{ia}^*}{\varepsilon_i - \varepsilon_\nu} \langle \Phi_a^\nu (h' - E^{(1)}) \Phi^{(1)} \rangle \tag{II-18b}$$

where

$$T_{ij,cd} = (U_{ic} U_{jd} - U_{id} U_{jc}) \tag{II-19}$$

It should be noted that in agreement with the reasoning presented above, only the occupied orbitals need to be localized.

In practice all the summations appearing in the perturbation formulae (which in principle are infinite sums and must include the continuum as well) are truncated because of the unavoidable use of finite bases. Thus everything reduces to the diagonalization to different order in λ of the finite matrix

$$H^\circ + \lambda h' \tag{II-20}$$

whose elements are

$$H_{\nu_1 \ldots \nu_n, \mu_1 \ldots \mu_m \atop i_1 \ldots i_n, j_1 \ldots j_m} = \langle \Phi_{i_1 \ldots i_n}^{\nu_1 \ldots \nu_n} \hat{\mathscr{H}}_{HF} \Phi_{j_1 \ldots j_m}^{\mu_1 \ldots \mu_m} \rangle \tag{II-21a}$$

$$h'_{\nu_1 \ldots \nu_n, \mu_1 \ldots \mu_m \atop i_1 \ldots i_n, j_1 \ldots j_m} = \langle \Phi_{i_1 \ldots i_n}^{\nu_1 \ldots \nu_n} \hat{h}' \Phi_{j_1 \ldots j_m}^{\mu_1 \ldots \mu_m} \rangle \tag{II-21b}$$

B. Variational Wave Function

The variational theorem has been the cornerstone of approximation methods in the quantum chemistry calculations of accurate w.f., especially with linear variational parameters. In this category falls the well known CI method, which corresponds to the exact diagonalization of the matrix (II-20) (with $\lambda = 1$). The resulting expansion coefficients are related to the perturbation series, except for a normalization constant, by

$$C_{k_1 \ldots k_n}^{\lambda_1 \ldots \lambda_n} = \sum_{l=0}^{\infty} C_{k_1 \ldots k_n}^{(l)\lambda_1 \ldots \lambda_n} \tag{II-22}$$

The CI method is straightforward but slow in converging with respect to the huge number of configurations which may easily arise. Thus a wise selection of the configurations to be included (for instance, along the line sketched above for localized orbitals) might be decisive in bringing the problem within the practical limits.[18] There have been a great number of theoretical investigations designed to overcome the intrinsic difficulties of the general CI method. Many of them were developed within the framework of the many-body theory[19] for nuclei and used extensively the second quantization formalism. Some of them, which have been applied in quantum chemistry, will be briefly discussed using a more conventional, quantum-chemical, formalism. To simplify the exposition of the methods, which do not involve the use of interelectronic coordinates, it is convenient to introduce the operators[20] $\hat{b}_{j_1 \ldots j_n}^{\nu_1 \ldots \nu_n}$ defined by

$$\hat{b}_{j_1 \ldots j_n}^{\nu_1 \ldots \nu_n} \Phi_0 = \Phi_{j_1 \ldots j_n}^{\nu_1 \ldots \nu_n}$$

The general CI-type w.f. may then be written as (II-23)

$$\Psi = \sum_{n=0}^{\infty} \sum_{j_1 < j_2 < \cdots < j_n} \left(\sum_{\nu_1 < \nu_2 < \cdots < \nu_n} C_{j_1 \ldots j_n}^{\nu_1 \ldots \nu_n} \hat{b}_{j_1 \ldots j_n}^{\mu_1 \ldots \nu_n} \right) \Phi_0 \qquad \text{(II-24)}$$

where as usual the C-coefficients are the linear variational parameters. The summation in parentheses in Eq. (II-24) may be defined as the operator $\hat{U}_{j_1 \ldots j_n}$

$$\hat{U}_{j_1 \ldots j_n} = \sum_{\nu_1 < \nu_2 \ldots < \nu_n} C_{j_1 \ldots j_n}^{\nu_1 \ldots \nu_n} \hat{b}_{j_1 \ldots j_n}^{\nu_1 \ldots \nu_n} \qquad \text{(II-25)}$$

which operating upon ϕ_0 has the effect of substituting the n spin orbitals $\psi_{j_1} \ldots \psi_{j_n}$ with a n-electron antisymmetric function. It should be noted that the product of two (or more) operators $\hat{U}_{j_1 \ldots j_n} \hat{U}_{k_1 \ldots k_m}$ will be zero if they have one or more indices in common, because it would imply the substitution of spin orbitals which have been already substituted. The expansion (II-24), once the definition (II-25) is used, reads

$$\Psi = \sum_{n=0}^{N} \sum_{j_1 < j_2 < \cdots < j_n} \hat{U}_{j_1 \ldots j_n} \Phi_0 \qquad \text{(II-26)}$$

This last expression represents what is generally called the cluster expansion of the w.f.[21] A more useful form than Eq. (II-26), which is merely another way of writing the exact w.f., may be achieved once it is recognized that the n-order cluster $\hat{U}_{j_1 \ldots j_n}$ includes the contributions arising from the products of lower-order clusters. Thus the irreducible or linked n-electron clusters are generated by the operators $\hat{T}_{j_1 \ldots j_n}$ defined by the sequence of relations

$$\hat{U}_{j_1} = \hat{T}_{j_1}$$
$$\hat{U}_{j_1 j_2} = \hat{T}_{j_1} \hat{T}_{j_2} + \hat{T}_{j_1 j_2} \qquad \text{(II-27)}$$
$$\hat{U}_{j_1 j_2 j_3} = \hat{T}_{j_1} \hat{T}_{j_2} \hat{T}_{j_3} + \hat{T}_{j_1} \hat{T}_{j_2 j_3} + \hat{T}_{j_2} \hat{T}_{j_1 j_3} + \hat{T}_{j_3} \hat{T}_{j_1 j_2} + \hat{T}_{j_1 j_2 j_3}$$

etc. It is clear that the n-electron irreducible cluster operators may be written analogously to the reducible cluster operator $\hat{U}_{j_1 \cdots j_n}$ as

$$\hat{T}_{j_1 \cdots j_n} = \sum_{\nu_1 < \nu_2 < \cdots < \nu_n} A_{j_1 \cdots j_n}^{\nu_1 \cdots \nu_n} \hat{b}_{j_1 \cdots j_n}^{\nu_1 \cdots \nu_n} \tag{II-28}$$

where the A-coefficients are derivable from the C-coefficients through the relations (II-27). The $\hat{T}$'s then generate clusters which cannot be reduced to products of lower-order clusters.

It may be proved that[21] (but for a normalization constant)

$$\Psi = \left\{ \exp \left[\sum_{n=1}^{N} \sum_{j_1 < j_2 < \cdots < j_n} \hat{T}_{j_1 \cdots j_n} \right] \right\} \Phi_0 \tag{II-29}$$

The great importance of the irreducible cluster expansion (I-29) lies in the fact that, due to the presence of no more than two-body operators in the Hamiltonian, *the most important irreducible clusters are those of order $n = 2$* and their importance will rapidly decrease with n. This is the idea underlying several of the methods proposed.

Let us examine in this framework some specific methods starting from Eq. (II-29):

(1) Limiting the terms up to $n = 2$ and expanding the exponential up to first order, we obtain

$$\Psi \cong \left(1 + \sum_{j} \hat{T}_j + \sum_{j_1 < j_2} \hat{T}_{j_1 j_2} \right) \Phi_0 \tag{II-30}$$

which represents a rather widely used method of limited CI. Within this method, special techniques may be devised in order to reduce the computational effort either by the use of approximated natural spin orbitals[23] or by the use, already mentioned, of localized spin orbitals.

(2) Dividing the electrons in pairs $j_k j_l$ according (for instance) to the chemical concepts of K shells, bonds and lone pairs and retaining the cluster up to $n = 2$, we obtain

$$\Psi \cong \left\{ \exp \left[\sum_{j_k j_l} (\hat{T}_{j_k} + \hat{T}_{j_l} + \hat{T}_{j_k j_l}) \right] \right\} \Phi_0 = \prod_{(j_k, j_l)} (1 + \hat{T}_{j_k} + \hat{T}_{j_l} + \hat{T}_{j_k j_l}) \Phi_0 \tag{II-31}$$

where the summation is carried out over all the selected exclusive pairs. Expression (II-31) represents what is called the *geminal product w.f.*[24] In order to attach to the pairs $j_k j_l$ the chemical significance indicated above, it is necessary to employ, for the space part, orbitals localized in the regions presumably occupied by the K shells, bonds and lone pairs. To simplify the computational part of this method, which is a particular example of the group function approach,[25] the strong-orthogonality constraint[19] is generally adopted. This constraint is realized by performing the summation indicated in Eq. (II-28) upon selected subsets of the empty spin orbitals, ν, different for each pair and not containing common elements. This *strong-orthogonal geminal product w.f.* may be particularly valid for systems where the localized orbitals are well separated.

(3) With a w.f. like

$$\Psi = \left\{ \exp\left[\sum_{j_1 < j_2} \hat{T}_{j_1 j_2} \right] \right\} \Phi_0 \tag{II-32}$$

the variational problem of determining the (non-linear) coefficients may become quite formidable. In order to reduce the computational efforts to maneagable size, Sinanoğlu[26] proposed the use of a variational w.f. of the type

$$(1 + \hat{T}_{j_k j_l}) \Phi_0 \tag{II-33}$$

in order to determine the coefficients $A_{j_k j_l}$ separately for *each of the possible* pairs $j_k j_l$; such a w.f. is to be considered as a satisfactory approximation to those appearing in Eq. (II-32). The w.f. (II-33) so obtained, which is clearly not the best possible one of that kind, is then used to evaluate the observables of interest. (A possible improvement may be achieved by using $\hat{T}_{j_k} + \hat{T}_{j_l} + \hat{T}_{j_k j_l}$ in place of only $\hat{T}_{j_k j_l}$.)

It must be emphasized that, in contrast to the geminal method, here all possible pairs are considered. Because of the approximation of using different variational w.f.'s for the different pairs, this method, which decouples the electron pairs, is called the *independent electron-pairs method* as well as the *Sinanoğlu many-electron theory*.

(4) By observing that

$$e^{\hat{T}_j} \Phi_0 = (1 + \hat{T}_j) \Phi_0 \tag{II-34}$$

$$e^{(\hat{T}_j + \hat{T}_k + \hat{T}_{jk})} \Phi_0 = (1 + \hat{T}_j + \hat{T}_k + \hat{T}_{jk}) \Phi_0 \tag{II-35}$$

etc., it becomes clear that the variational determination of a w.f. like (II-34) or (II-35) is equivalent to the variational determination of a w.f. like

$$(1 + \hat{U}_j) \Phi_0 \tag{II-34a}$$

$$(1 + \hat{U}_j + \hat{U}_k + \hat{U}_{kl}) \Phi_0, \quad \text{etc.} \tag{II-35a}$$

Nesbet[27] has used this approach to compute the *net increments* of some observables due to the successive inclusions of irreducible clusters of increasing order. The procedure adopted by this author is the following: if $\Delta F_{j_1 \cdots j_n}$ is the *gross increment* to some property $\hat{F}$ defined as the difference between the expectation values of $\hat{F}$ calculated with the variational w.f.

$$\{\exp[\hat{T}_{j_1} + \cdots + \hat{T}_{j_n} + \hat{T}_{j_1 j_2} + \cdots + \hat{T}_{j_{n-1} j_n} + \cdots + \hat{T}_{j_1 j_2 \cdots j_n}]\} \Phi_0 \tag{II-36}$$

and with Φ_0, then this increment may be factorized as a sum of net increments $f_{j_1\dots}$ due to the inclusion of successive higher-order clusters according to

$$\Delta F_{j_1} = f_{j_1}$$

$$\Delta F_{j_1 j_2} = f_{j_1} + f_{j_2} + f_{j_1 j_2} \tag{II-37}$$

$$\Delta F_{j_1 j_2 j_3} = f_{j_1} + f_{j_2} + f_{j_3} + f_{j_1 j_2} + f_{j_1 j_3} + f_{j_2 j_3} + f_{j_1 j_2 j_3}, \quad \text{etc.}$$

This method permits us to assess numerically the importance of the nth cluster by summing the contributions arising from all possible n-clusters considered separately, and to make some estimates of the nth-order gross increments without performing the actual computations. For instance, if the 3-body irreducible cluster contributions may be neglected, then

$$\Delta F_{j_1 j_2 j_3} \cong f_{j_1} + f_{j_2} + f_{j_3} + f_{j_1 j_2} + f_{j_1 j_3} + f_{j_2 j_3} \tag{II-38}$$

The value of this method is strictly connected with the goodness of the approximation (II-38) which considers negligible the contributions arising from high-order irreducible clusters. It should be said that the essence of this approach is an extension of the methods of Brueckner[28] and of Bethe and Goldstone,[29] which imply, analogously to Sinanoğlu, that the electron-pair problem may be solved independently for each pair.

The methods sketched above, as well as others not reported here, are all based upon the use of a fixed set of spin orbitals which may conveniently be or not be HF spin orbitals. If this imposition is relaxed (it is not a constraint if the complete set is fully employed), other methods become available. For instance, in the *multiconfiguration* SCF[12,30] method a limited CI-type w.f. is used in which the orbitals are not fixed but are optimized together with the CI expansion coefficients by an iterative SCF-type procedure. Another method[23,31] is based upon the idea that the natural spin orbitals would supply an ideal set with which to carry out a fast-converging CI-type calculation. The occupation numbers are used as the right parameters to indicate the intrinsic importance of the spin orbitals. In order to apply this idea in practice (in principle, the natural spin orbitals are known *a posteriori*), the most important double-excited configurations from an energy viewpoint are selected for a first, limited, CI-type calculation. Then the approximated natural spin orbitals which result from diagonalizing the first-order density matrix are used to start a new CI calculation and so on.

Finally it should be mentioned that besides variational w.f.'s expressed solely as linear combinations of spin-orbital products, which is essentially the language used in the many-body techniques, there are some altogether different approaches. They are based upon the introduction into the variational w.f. of the explicit interelectronic coordinates.[32] Apart from the original Hylleraas method, whose extension to molecules of even moderate size appears unlikely, one should mention the Boys' transcorrelated w.f. method which employs a variational w.f. of the type

$$\psi = F\Phi \tag{II-39}$$

where Φ is a single detor and F, the correlation factor, is a suitable symmetric function containing the interelectronic coordinates. By a deft selection of the analytical form of F and by proving that within a reasonable range of accuracy the so called "transcorrelated" Hamiltonian $F^{-1}\mathcal{H}F$ may be adopted, Boys and Handy were able to formulate a maneageable approach which appears to be successful.[33,34]

III. FIRST-ORDER OBSERVABLES

A. Perturbation Approach

In the previous section, the q_m^{00} quantities have been defined as expectation values of the one-electron operator $\hat{q}_m$ evaluated for the w.f. φ_0. The accuracy of these quantities is thus dependent upon the accuracy of φ_0 and some methods have been described which should guarantee better accuracy for q^{00} by improving the quality of φ_0.

Another way of looking at the problem, which may be convenient for exploiting a different approach, is based upon the idea of considering the expectation values of an operator $\hat{h}^\alpha$ as first-order observables.[36] These are defined as follows: if $\mathcal{H}^{00}$ is a zeroth-order Hamiltonian and $\alpha\hat{h}^\alpha$ is a perturbation of intensity α, the energy eigenvalues E_n will be analytical functions of α (barring the unlikely occurrence, for physical quantities, of non-analytical behaviour), i.e.

$$E_0(\alpha) = E_0(0) + \alpha\left(\frac{\partial E_0}{\partial\alpha}\right)_0 + \frac{1}{2}\alpha^2\left(\frac{\partial^2 E_0}{\partial\alpha^2}\right)_0 + \cdots \tag{III-1}$$

The expansion coefficients $E_0^\alpha = (\partial E_0/\partial\alpha)_0$; $E_0^{\alpha\alpha} = (\partial^2 E_0/\partial\alpha^2)_0$ etc. are called respectively first-order, second-order observables, etc. Straightforward application of perturbation theory gives

$$E_0^\alpha = \langle\varphi_0^{00}\hat{h}^\alpha\varphi_0^{00}\rangle \tag{III-2}$$

where φ_0^{00} is the unperturbed w.f. The need for the two superscripts will become apparent in the following.

Equation (III-2) establishes that the expectation value of an operator may also be considered as the first derivative of the energy with respect to some suitable perturbation parameter. This viewpoint, extremely useful when we are dealing with an approximate variational w.f., also affords a novel approach in the perturbation scheme. Let us consider, for instance, the eigenvalue problem

$$[\mathcal{H}_{HF} + \lambda_1(\mathcal{H} - \mathcal{H}_{HF}) + \lambda_2\hat{q}_m - E_0(\lambda_1, \lambda_2)]\varphi_0(\lambda_1, \lambda_2) = 0 \tag{III-3}$$

similar to the one considered in the previous section [see Eqs (II-8); (II-16)] but for the addition of the fictitious perturbation $\lambda_2 \hat{q}_m$. It must be noted that in Eq. (III-3), for simplicity, only one of the possible component $\hat{q}_m$ has been written. The result can be generalized by a consideration of all components. Assuming that a Taylor expansion around $\lambda_1 = \lambda_2 = 0$ is legitimate, both for φ_0 and E_0, the looked-for quantity q_m will be expressed as

$$q_m = \left(\frac{\partial E_0}{\partial \lambda_2}\right)_{\lambda_1 = \lambda_2 = 0} = \sum_{j=0}^{\infty} \frac{\lambda^j}{j!} E_0^{j,1} \tag{III-4}$$

with

$$E_0^{j,1} = \left[\left(\frac{\partial}{\partial \lambda_1}\right)^j \frac{\partial}{\partial \lambda_2} E_0\right]_{\lambda_1 = \lambda_2 = 0} \tag{III-5}$$

Eq. (III-4) shows how to include the correlation corrections systematically. The $E_0^{j,1}$ corrections are derived through the solution of the perturbation equations which may be obtained formally by taking the partial derivative of Eq. (III-3) with respect to λ_1 and λ_2. For instance, Eq. (III-3) is rewritten compactly as

$$[H(\lambda_1, \lambda_2) - E_0(\lambda_1, \lambda_2)]\varphi(\lambda_1, \lambda_2) = 0 \tag{III-6}$$

with

$$H(\lambda_1, \lambda_2) = \mathcal{H}_{HF} + \lambda_1(\mathcal{H} - \mathcal{H}_{HF}) + \lambda_2 \hat{q}_m \tag{III-7}$$

The $n + m$th-order perturbation equation (of order n in λ_1 and m in λ_2) is obtained as

$$\sum_{j=0}^{n} \sum_{k=0}^{m} [H - E_0]^{j,m} \varphi^{n-j, m-k} = 0 \tag{III-8}$$

where the notation should be clear from Eq. (III-5). Since in our case the only $H^{j,k}$ different from zero are

$$H^{0,0} = \mathcal{H}_{HF}; \qquad H^{1,0} = \mathcal{H} - \mathcal{H}_{HF}; \qquad H^{0,1} = \hat{q}_m \tag{III-9}$$

the first corrections $E^{j,k}$, once we have adopted the normalization condition

$$\langle \varphi_0^{00} \varphi_0^{j,k} \rangle = \delta_{j0} \delta_{k0} \tag{III-10}$$

are easily seen to be[37]

$$E_0^{0,1} = \langle \varphi_0^{00} H^{0,1} \varphi_0^{00} \rangle \tag{III-11a}$$

$$E_0^{1,1} = \langle \varphi_0^{00} H^{0,1} \varphi_0^{1,0} \rangle + \langle \varphi_0^{00} H^{1,0} \varphi_0^{0,1} \rangle \tag{III-11b}$$

$$E_0^{2,1} = \langle \varphi_0^{00} H^{0,1} \varphi_0^{2,0} \rangle + 2\langle \varphi_0^{00} H^{1,0} \varphi_0^{1,1} \rangle, \quad \text{etc.} \tag{III-11c}$$

The required corrections to the w.f. φ_0 may be conveniently expressed as

$$\varphi_0^{1,0} = -(H^{00} - E_0^{00})^{-1}\hat{P}H^{1,0}\varphi_0^{00} \tag{III-12a}$$

$$\varphi_0^{0,1} = -(H^{00} - E_0^{00})^{-1}\hat{P}H^{0,1}\varphi_0^{00} \tag{III-12b}$$

$$\varphi^{2,0} = -2(H^{00} - E_0^{00})^{-1}\hat{P}(H^{1,0} - E_0^{1,0})\varphi_0^{1,0} \tag{III-12c}$$

$$\varphi^{1,1} = -(H^{00} - E_0^{00})^{-1}\hat{P}[(H^{0,1} - E_0^{0,1})\varphi_0^{1,0} - (H^{1,0} - E_0^{1,0})\varphi^{0,1}], \quad \text{etc.} \tag{III-12d}$$

where the projection operator $\hat{P} = (1 - \varphi_0^{00})\langle\varphi_0^{00}\varphi_0^{00}\rangle^{-1}\langle\varphi_0^{00}\rangle$ annihilates any components along φ^{00}.

By substitution of expressions (III-12) into Eqs. (III-11), the following alternative forms for the perturbation corrections $E_0^{1,1}$, $E_0^{2,1}$, etc. may be obtained

$$E_0^{1,1} = \begin{cases} (\langle\varphi_0^{00}H^{1,0}\varphi_0^{0,1}\rangle + \text{c.c.}); & \text{(III-13a)} \\ (\langle\varphi_0^{00}H^{0,1}\varphi_0^{1,0}\rangle + \text{c.c.}); & \text{(III-13b)} \end{cases}$$

$$E_0^{2,1} = \begin{cases} (\langle\varphi_0^{00}H^{0,1}\varphi_0^{2,0}\rangle + \text{c.c.}) + 2\langle\varphi^{1,0}(H^{0,1} - E^{0,1})\varphi^{1,0}\rangle; & \text{(III-14a)} \\ 2(\langle\varphi_0^{1,0}(H^{1,0} - E_0^{1,0})\varphi_0^{0,1}\rangle + \text{c.c.} + 2\langle\varphi^{1,0}(H^{0,1} - E^{0,1})\varphi^{1,0}\rangle); & \text{etc.} \\ & \text{(III-14}b\text{)} \end{cases}$$

Inspection of formulae (III-13) and (III-14) reveals that:

(1) The first-order correction (in the correlation perturbation $H^{1,0}$) to $(\partial E/\partial \lambda_2)$ may be evaluated in two different ways by exploiting the interchange theorem.[37,38] In the case of an HF unperturbed w.f., the first-order correction $E_0^{1,1}$ is zero, in agreement with what has been said in the previous section.

(2) The second-order corrections (III-14a) are identical to those obtainable through Eq. (II-13b), but Eq. (III-14b) shows a formally different method of computation.

The advantages of having different ways of calculating this correction lie in the unavoidable necessity of approximating the perturbed functions $\varphi_0^{0,1}$, $\varphi_0^{1,0}$, $\varphi_0^{2,0}$, $\varphi_0^{1,1}$, etc. (as well as the unperturbed one φ_0^{00}) by suitable means. Thus it may be much easier to approximate the first-order corrections $\varphi^{1,0}$ and $\varphi^{0,1}$ instead of the second-order corrections $\varphi^{0,2}$, etc. As a matter of fact, the first-order perturbed w.f.'s are obtainable by using variational techniques to look for the stationary conditions $\delta I = 0$ under arbitrary variations of $\varphi^{0,1}$ within the Hylleraas[37,39] functional

$$I = [\langle\varphi^{0,1}(H^{0,1} - E^{0,1})\varphi^{00}\rangle + \text{c.c.}] + \langle\varphi^{0,1}(H^{00} - E^{00})\varphi^{0,1}\rangle \tag{III-15}$$

In addition, if φ^{00} refers to the ground state of the system, the function (III-15) will also be a minimum at the stationary point. Thus we have the possibility of approximating $\varphi^{0,1}$ in the most convenient way by looking for the minimum of (III-15). It must be pointed out, however, that if a *fixed basis* is used to approximate any of the perturbed w.f.'s, the results will be identical whether one uses expression (III-14a) or (III-14b). The convenience arises from the freedom of using variational forms most suitable to describe the perturbation at hand.

The simultaneous perturbation approach may also prove very useful in calculating the effect of the environment upon the expectation value of $\hat{q}_m$ for a molecule in a crystal. In such an example, without going outside the HF approximation, the $H^{1,0}$ perturbation may be assimilated into the external potential due to the surrounding molecules. Formulae (III-13) and (III-14) (in this case the first-order correction will be different from zero) may be used to estimate the external field effect upon q_m^{00}. It should be observed that within the approximation of a perturbing field $\hat{V}(\bar{r})$ whose origins are all external, the HF method automatically implies a perturbation which includes, besides the direct term $\hat{V}(\bar{r})$, an indirect contribution due to the polarization of the orbitals. If such an effect is properly taken into account, $H_{HF}^{1,0}$ will be

$$H_{HF}^{1,0} = \sum_j [\hat{V}(\bar{r}_j) + \hat{\mathcal{G}}^{1,0}(\bar{r}_j)] \tag{III-16}$$

where $\hat{\mathcal{G}}^{1,0}$ indicates the first-order correction to the HF effective electron field. The inclusion of this contribution leads to the so-called *perturbed coupled* HF method[49] (PCHF), while its omission leads to the uncoupled perturbed HF method (PUHF).

Most of the considerations and most of the formulae presented so far, which are based upon the HF approximation, assume that it is always possible to achieve the solutions of integrodifferential equations like (III-3), (III-12). In practice, this is certainly not the case, because the best which may be accomplished is the use of large basis sets. The effect of this limitation upon quantities like q_m^{00} is not easy to assess because, as is widely known, basis functions of hardly any importance from an energetical viewpoint may be very important for other observables. It has been claimed that the quadrupole polarization of the K shells, while contributing a negligible amount to the energy, is of great importance for an accurate value of q_m^{00}.[40] This aspect has been considered in some detail in Appendix A.

B. Variational Approach

The idea that q_m^{00} may be considered as a first-order observable may be conveniently exploited[36] to obtain alternative ways to compute it when arbitrary variational wave functions (v.w.f.) are used.

Let

$$\varphi[D_1, D_2, \ldots, D_n] = \varphi[\mathbf{D}] \tag{III-17}$$

be a v.w.f. (the electronic coordinates are not indicated for simplicity) which depends upon *n-independent* variational parameters $D_1, D_2, \ldots, D_n$. The expectation value of the energy for a Hamiltonian $\mathcal{H}_0 + \alpha \hat{h}^\alpha$ is

$$E(\alpha) = \frac{\langle \varphi[\mathbf{D}(\alpha)](\mathcal{H}_0 + \alpha \hat{h}^\alpha)\varphi[\mathbf{D}(\alpha)]\rangle}{\langle \varphi[\mathbf{D}(\alpha)]\varphi[\mathbf{D}(\alpha)]\rangle} \tag{III-18}$$

where the dependence of $\mathbf{D}$ upon α will be established by asking that $E(\alpha)$ must be stationary with respect to all D's for any value of α, i.e. (for all real quantities)

$$\left(\frac{\partial E}{\partial D_M}\right) = 2\frac{\langle \varphi^M Q(\mathcal{H}_0 + \alpha \hat{h}^\alpha)\varphi\rangle}{\langle \varphi\varphi\rangle} = 0 \quad \text{(any } \alpha) \tag{III-19}$$

where

$$\varphi^M = \frac{\partial \varphi}{\partial D_M} \quad \text{and} \quad Q = (1 - \varphi)\langle \varphi\varphi\rangle^{-1}\langle \varphi\rangle$$

Now from (III-18)

$$\left(\frac{\partial E}{\partial \alpha}\right)_{\alpha=0} = \sum_M \left(\frac{\partial D_M}{\partial \alpha}\frac{\partial E}{\partial D_M}\right)_{\alpha=0} + \left(\frac{\langle \varphi \hat{h}^\alpha \varphi\rangle}{\langle \varphi\varphi\rangle}\right)_{\alpha=0} \tag{III-20}$$

Thus if a v.w.f. is optimized at $\alpha = 0$ with respect to all the variational parameters, we find that the first-order property is given as the expectation value of $\hat{h}^\alpha$ (Helmann–Feynman theorem[41]). In practice, no v.w.f. is optimized with respect to all possible variational parameters. In fact, even where all variational parameters explicitly appearing in the v.w.f., have been fully optimized, we may still imagine that we are working with a more flexible v.w.f. in which some of the variational parameters are arbitrarily kept fixed to convenient numerical values.[36] For instance, if in a multiconfiguration SCF v.w.f. all mixing coefficients, except that corresponding to the leading config-uration, are put equal to zero, we reduce to the single-detor SCF v.w.f. Thus the usual SCF w.f. may be considered as a v.w.f. optimized only with respect to a select set of the possible variational parameters, while the rest of them are arbitrarily kept fixed to the value zero.

This way of thinking gives the possibility of improving the accuracy of the first-order observables by the following approach. We may consider that the set of values $\mathbf{D} = \mathbf{D}_0$ differs from the best possible set $\mathbf{D}_b$ [i.e. $(\partial E/\partial D_M)_{\mathbf{D}_0} \neq 0$, $(\partial E/\partial D_M)_{\mathbf{D}_b} = 0$, $M = 1, 2 \ldots$] by an amount $x_b = \mathbf{D}_b - \mathbf{D}_0$. If x_b is small, which appears to be quite a reasonable assumption if we work with accurate v.w.f., the variation of the energy E with x will be satisfactorily given by a series expansion up to second order,

$$E(\mathbf{D}_0 + x) = E(\mathbf{D}_0) + \tilde{A}x + \tfrac{1}{2}\tilde{x}Bx + 0(x^3) \tag{III-21}$$

where x has been considered as a (column) vector and the vector elements A_M and matrix elements B_{MN} are defined by (all quantities assumed real and $\langle\varphi\varphi\rangle_{D_0}=1$)

$$A_M = \left(\frac{\partial E}{\partial D_M}\right)_{D_0} = 2\langle\varphi^M Q\mathcal{H}\varphi\rangle_{D_0} \tag{III-22}$$

$$B_{MN} = 2\langle\varphi^{MN}Q\mathcal{H}\varphi\rangle_{D_0} + 2\langle\varphi^M(\mathcal{H}-E)\varphi^N\rangle_{D_0}$$
$$-4A_M\langle\varphi^N\varphi\rangle_{D_0} - 4A_N\langle\varphi^M\varphi\rangle_{D_0} \tag{III-23}$$

Neglecting the residue $0(x^3)$, x_b is obtained by asking for the minimum of (III-21) which occurs at

$$x_b = -B^{-1}A \tag{III-24}$$

The improved value of the energy within the above-mentioned approximation is given from (III-21) by

$$E(D_0+x_b) = E(D_0) - \tfrac{1}{2}\tilde{A}B^{-1}A \tag{III-25}$$

This last equation yields the following expression for the first-order property E^α:

$$E^\alpha = \langle\varphi\hat{h}^\alpha\varphi\rangle_{D_0} - 2\tilde{A}_0 B_0^{-1} A_0^\alpha + 0(A_0^2) \tag{III-26}$$

where

$$A_{0M}^\alpha = \langle\varphi^M Q\hat{h}^\alpha\varphi\rangle_0 \tag{III-27}$$

and the subscript 0 indicates that all quantities are evaluated at $D=D_0$, $\alpha=0$. Equation (III-26) is an approximation because the terms in A_0^2 and higher are neglected. This should be an acceptable simplification if D_0 is reasonably close to D_b, because from (III-24) it is seen that x_b is proportional to A.

The second term of Eq. (III-26) is the correction to be added to the expectation value of $\hat{h}^\alpha$ calculated with the v.w.f. evaluated at $\alpha=0$ *not* perfectly optimized. As an example, let us consider a v.w.f. like that of Eq. (II-30), which may be written as

$$\Phi = \Phi_{HF} + \sum_j\sum_\nu D_j^\nu\Phi_j^\nu + \sum_{j<k}\sum_{\nu<\mu} D_{jk}^{\nu\mu}\Phi_{jk}^{\nu\mu} \tag{III-28}$$

with $\Phi(D_0)=\Phi_{HF}$ (i.e., $D_j^\nu=D_{jk}^{\nu\mu}=0$ at $D=D_0$). Making explicit use of the Brillouin theorem, we may write the elements of the vector A and of the matrix B as

$$A_{j\nu} = 0; \qquad A_{j\nu k\mu} = \langle jk\nu\mu\rangle - \langle jk\mu\nu\rangle$$

$$A_{j\nu}^\alpha = \langle \nu\hat{h}^\alpha j\rangle; \qquad A_{j\nu k\mu}^\alpha = 0$$

$$B_{j\nu;l\lambda} = \delta_{jl}\delta_{\nu\lambda}(\varepsilon_\lambda - \varepsilon_l) + \langle\nu lj\mu\rangle - \langle\nu l\mu j\rangle$$

$$B_{j\nu;l\lambda m\rho} = (1-\hat{P}_{\lambda\rho})\delta_{\nu\lambda}(\langle lmj\rho\rangle - \langle lm\rho j\rangle) + (1-\hat{P}_{lm})\delta_{jm}(\langle l\nu\lambda\rho\rangle - \langle l\nu\rho\lambda\rangle)$$

154

$$\mathbf{B}_{j\nu k\mu;l\lambda mp} = \delta_{jl}\delta_{km}\delta_{\nu\lambda}\delta_{\mu p}(\varepsilon_\lambda + \varepsilon_p - \varepsilon_m - \varepsilon_l) + \delta_{jl}\delta_{km}(\langle\nu\mu\lambda p\rangle$$

$$-\langle\nu\mu p\lambda\rangle) + \delta_{\nu\lambda}\delta_{\mu p}(\langle lmjk\rangle - \langle lmjk\rangle) + (1-\hat{P}_{jk})$$

$$\times(1-\hat{P}_{\nu\mu})(1-\hat{P}_{l,m})(1-\hat{P}_{\lambda p})\delta_{\nu\lambda}\delta_{jm}(\langle l\nu k p\rangle - \langle l\nu p k\rangle)$$

where $\hat{P}_{jk}$ stands for the transposition operator.

A connection with the more standard perturbation approach in this specific example may be established by writing the matrix $\mathbf{B}$ as

$$\mathbf{B} = \boldsymbol{\varepsilon} + \boldsymbol{\Delta}$$

where $\boldsymbol{\varepsilon}$ stands for the diagonal matrix arising from the ε differences of the diagonal elements and is supposed to be the dominant part of $\mathbf{B}$. Since

$$\mathbf{B}^{-1} = \boldsymbol{\varepsilon}^{-1} - \boldsymbol{\varepsilon}^{-1}\boldsymbol{\Delta}\boldsymbol{\varepsilon}^{-1} + \boldsymbol{\varepsilon}^{-1}\boldsymbol{\Delta}\boldsymbol{\varepsilon}^{-1}\boldsymbol{\Delta}\boldsymbol{\varepsilon}^{-1} - \cdots \tag{III-29}$$

it may be seen that while the first term of the expansion (III-29) does not give any contribution (from the Brillouin theorem!), the second term gives the second-order perturbation correction (III-14). The analysis of the remaining terms is better accomplished by using the diagram technique, but this type of analysis will not be pursued here being beyond the scope of this article.

The variational approach may also be applied to investigate the effect of an external field upon the field gradient as happens in the case already mentioned of a molecule in a perturbing environment. In this case, it is necessary to look for the second-order properties $\partial^2 E/\partial\lambda_1\,\partial\lambda_2$ from a variational viewpoint.[36]

IV. VIBRATIONAL AND OTHER CORRECTIONS

In the Born–Oppenheimer approximation, the complete w.f. $\Psi\rangle$ for a (non-rotating) molecule may be written as a product of the w.f. (I–5) times the vibrational part $\theta_v(R)$

$$\Psi\rangle = |IM\rangle_\alpha|\Phi_n(R)\rangle|\theta_v\rangle \tag{IV-1}$$

where v stands for the collection of quantum numbers which specify the vibrational state and R, as usual, stands for the collection of internal nuclear coordinates. Since the vibrational motion is much faster than the nuclear reorientation, all matrix elements appearing in Eqs. (I-10b) and (I-10c) should be vibrationally averaged. The implication is that to first order [Eq. (I-10b)], the matrix elements

$$V_m^{00}(R) = \frac{\langle\Phi_0(R)\hat{V}_{2,m}(R)\Phi_0(R)\rangle}{\langle\Phi_0(R)\Phi_0(R)\rangle} \tag{IV-2}$$

which depend upon R both directly through the operator $\hat{V}_{2,m}$ and indirectly through the R-dependence of $\Phi_0(R)$, should be replaced by

$$\langle V_m^{00}\rangle_v = \frac{\langle\theta_v V_m^{00}(R)\theta_v\rangle}{\langle\theta_v\theta_v\rangle} \tag{IV-3}$$

The evaluation of this last quantity involves:

(i) The determination of the electronic w.f. Φ_0 for a set of geometrical nuclear arrangements which should permit us, through Eq. (IV-2), to achieve a sufficiently detailed knowledge of the property as a function of the nuclear coordinates.

(ii) The determination of the vibrational w.f. and the subsequent evaluation of the expectation value (IV-3).

This type of approach is straightforward but from a numerical viewpoint rather difficult to fulfil. Thus there are very few numerical examples in the literature of this type of calculation.[42,43]

A more practical approach, which reduces the calculation involved in part (ii), is based upon the commonly used expansions of the property (IV-3) as a power series in the normal coordinates $q_1, q_2 \ldots$, i.e.

$$V_m^{00}(q_1 \ldots) = V_m^{00}(0, 0 \ldots) + \sum_j \alpha_j q_j + \tfrac{1}{2} \sum_{j,k} \beta_{jk} q_j q_k + \tfrac{1}{6} \sum_{j,k,l} \gamma_{jkl} q_j q_k q_l + \cdots \qquad \text{(IV-4)}$$

and upon the use of perturbation theory to obtain vibrational w.f.'s corrected for the anharmonicity terms in the potential $V(q_1 q_2 \ldots)$ appearing in the nuclear vibrational Schrödinger equation. Kern and his collaborators[44] have developed all the formulae which lead to relations expressing the vibrational average as a power series in the vibrational quantum numbers characterizing the unperturbed harmonic problem, i.e.

$$\langle V_m^{00} \rangle_v = V_m^{00}(0, 0, \ldots) + \sum_{j=1} A_j(v_j + \tfrac{1}{2}) + \sum_{j<k} B_{jk}(v_j + \tfrac{1}{2})(v_k + \tfrac{1}{2}) + \cdots$$

$$\text{(IV-5)}$$

Table 1. Vibrational corrections to the electric field gradient

Nucleus	Ref.	$q_{(\text{equil.})}/\text{a.u.}$	$q_{\text{vib.aver.}}/\text{a.u.}$	Observation
Li (LiH)	61	−0.0440	−0.0430 ($v = 0$)	SCFMO, w.f. extended basis $R_e = 3.015$ a.u.
Li (LiH)	63	−0.0399	−0.0383 ($v = 0$)	MCSCF ($R_e = 3.05$ a.u.)
H (LiH)	63	0.0493	0.0488 ($v = 0$)	MCSCF ($R_e = 3.05$ a.u.)
H (H_2O)	44	0.548	0.542 ($v_1 = v_2 = v_3 = 0$)	SCFMO+CI (8 STO)
H (H_2O)	44	−0.564[a]	−0.569 ($v_1 = v_2 = v_3 = 0$)	SCFMO (double-ζ + polarization)
O (H_2O)	44	−1.858	−1.822 ($v_1 = v_2 = v_3 = 0$)	SCFMO (double-ζ + polarization
N (NH_3)	62	−0.792	−0.802 ($\{v\} = 0,0^+$)	OCESCFMO 77 STO

[a] The sign indicated here is as is reported in the original reference.

where the summations run over all possible internal degrees of freedom and the quantities A_j, B_{jk} are complicated functions of the quantities α, β, γ as well as of the quadratic and cubic expansion coefficients of the vibrational potential $V(q_1 \ldots)$. It should be observed that this treatment as it stands cannot be applied to molecules which are not in pure vibrational states, as may happen in the case of molecules with minima like ammonia.

The quantities α_j, β_{jk} etc., which are generally found from the calculations indicated in (i), might be also obtained from experimental measurements of the dependence of the NQCC on the rotational state, as shown by Kukolich.[35]

Unfortunately the derivation of these quantities from the experimental data involves a knowledge of the force field and the final numerical values depend critically upon the type of force field adopted.[43] Luckily these vibrational corrections, according to the available data reported in the literature, are rather small, as may be inferred from Table 1.

V. ELECTRONIC POLARIZATION EFFECT

In Section I we have shown how to handle the problem of finding the electronic contribution to the NQCC up to second order. These second-order terms may be interpreted as the interaction between the field gradient arising from electronic polarization produced by the field of the nuclear quadrupole moment and the nuclear quadrupole moment itself. The summations appearing in Eq. (I-10c) may be considered as the elements of the tensor $\mathbf{R}$

$$R_{m,m'} = \sum_n \frac{V_m^{0n} V_m^{n0'}}{E_0^{(0)} - E_n^{(0)}} \tag{V-6}$$

describing the electronic quadrupole polarization due to the nuclear quadrupole field. In the specific case of the ^{14}N nucleus in ammonia ($I = 1$, C_{3v} point group), it may be shown that the polarization effect will produce simply the level shifts (the degeneracy partially removed by the first-order effect will not be decreased by the second-order effect) given by[43]

$$\Delta E_{M=0} = \langle \|Q\| \rangle^2 [(\tfrac{4}{10}) R_{0,0} + (\tfrac{3}{10})(R_{1,-1} + R_{-1,1})] \tag{V-7a}$$

$$\Delta E_{M=\pm 1} = \langle \|Q\| \rangle^2 [\tfrac{1}{10} R_{0,0} + \tfrac{3}{10}(R_{2,-2} + R_{-2,2}) + \tfrac{3}{20}(R_{1,-1} + R_{-1,1})] \tag{V-7b}$$

It is clear that if the electron cloud polarizes isotropically, i.e. if

$$R_{0,0} = Re R_{1,-1} = R_l R_{2,-2} \tag{V-8}$$

then the energy level shifts will be equal and this effect cannot be derived from transition-energy data as is the case for the NQCC. The actual calculations of the tensor $\mathbf{R}$ involves many subtle points but the overall evidence strongly suggests that while the level shifts may be noticeable, as compared to the first-order effects, the second-order level splitting will be negligible, i.e. the electron polarizes isotropically.[43]

This isotropic behaviour arises from the fact that the polarization effects are confined to a very small region of space around the nucleus where the unperturbed electron distribution displays practically spherical symmetry. Thus, unless strong non-spherical fields are present, these second-order effects upon the NQCC are expected to be negligible. Finally we must mention the finite nuclear size, as well as all other possible effects which may arise when we improve the model Hamiltonian and all of which may influence the NQCC.[45] The treatment of these small terms is best accomplished by perturbation theory. In fact if $\hat{V}_{Nl}^\alpha(\bar{r})$ is the electrostatic potential arising from a finite nuclear-size model which one wishes to consider, the perturbation $\hat{h}'$ to be taken into account is simply

$$\hat{h}' = \hat{V}_{Nl}^\alpha(\bar{r}) - \frac{Z_\alpha}{r} \tag{IV-8}$$

The available calculations[43] show that in this case there is also no detectable effect upon the NQCC.

From the foregoing, it seems proper to conclude that the calculations available at the present have not reached as yet the stage at which these finer effects should be taken into account.

VI. SUMMARY OF THE RESULTS OF SOME RECENT CALCULATIONS

In this section, the results of the most recent accurate *a priori* calculations of q and η are reported, together with a few other data and information regarding the type of calculation. The definitions of the quantities q and η comply with Eqs. (I-15), (I-16), (II-1a) and (II-1b).

The nuclear contribution to $\hat{q}_0$ and $\hat{q}_2$ is explicitly given by

$$q_{0\,nuc}^\alpha = \sum_{\beta(\neq\alpha)} Z_\beta \frac{2z_\beta^2 - x_\beta^2 - y_\beta^2}{r_{\alpha\beta}^5}; \qquad q_{2\,nuc}^\alpha = \sum_{\beta(\neq\alpha)} Z_\beta \frac{3^{1/2}}{2} \frac{x_\beta^2 - y_\beta^2}{2r_{\alpha\beta}^5} \tag{VI-1a,b}$$

All results refer to principal axes, even when the original reference did not give them. There is not general agreement about the sign to be assigned to q because some authors, differently from the convention adopted here, define q as the field gradient expectation value, i.e.

$$q_{elect} = \left\langle \Phi \sum_i \frac{2z_i^2 - x_i^2 - y_i^2}{r_{\alpha i}^5} \Phi \right\rangle$$

In the following tables, the signs reported agree with the definition (II-1a) (II-1b), (V-1a), (V-1b). In a few cases, which are properly indicated, it has not been possible to identify the definition adopted, and the signs have been left as in the original paper. All data are given in atomic units (a.u.).* For the case when the NQCC is given in hertz in the original reference, the transformation factor adopted is

$$(eq)_{\text{a.u.}} = \frac{0.425\,583}{Q\ \text{cm}^2} \cdot 10^{-32} \nu\,(\text{Hz})$$

and the value of Q used is that given in the original reference;

The tables, where all the data referring to a given chemical element are collected together, report in columns 6 and 7 the values of g and η respectively; in addition, they also give:

first column: the molecule;

second column: the method of calculation, indicated by standard notation;

third column: the basis set employed, characterized by a widely; used notation;[46]

fourth column: the total molecular energy;

fifth column: the geometry to which the calculation refers;

eighth column: a few comments regarding some relevant features; and

ninth column: the reference(s).

In addition, it must be indicated that all the cases for which it was not possible to identify the definition adopted for q (see above) have been marked by crosses. To avoid ambiguities in some cases, the chemical element to which the reported q refers has been underlined.

---

* $(eq)_{\text{Volts/m}^2} = (eq)_{\text{a.u.}} \times 9.717\,54 \times 10^{21}$.

Table 2. Hydrogen

Molecule	Method	Basis	Energy	Geometry	q	η	Comments	Ref.
LiH	SCFMO	STO, very large set	−7.9873	Experimental ($R = 3.015$)	0.0563	—	H.F. Level	71
LiH	VALENCE CI	7 STO + 8 Elliptic orb.	−8.0561	$R = 3.046$	0.0516	—		42
LiH	MCSCF	STO $23\sigma + 8\pi + 4\delta$	−8.0213	$R = 3.05$	0.0493 0.0488	— —	no vibr. correct. with vibr. correct. ($v = 0$)	63
LiH	SCFMO + CI	11 STO	−8.0555	$R = 3.015$	0.0595	—		56
BeH	SCFMO	STO, very large set	−15.1531	Experimental	0.1286	—	H.F. level	50
BH	SCFMO	STO, very large set	−25.1314	Experimental	0.1686	—	H.F. level	50
CH	SCFMO	STO, very large set	−38.279	Experimental	0.2522	—	H.F. level	50
CH_4	SCFMO	18 STO	−40.185	$R = 2.067$	−0.349†	—		68
CH_4	SCFMO	39 STO	−40.2041	$R = 2.067$	0.316	—		55
CH_4	SCFGF	39 STO − 34 Localized MO	−40.3035	$R = 2.067$	0.324	—	Strongly orthogonal geminals	55
CH_4	SCFMO + CI	39 STO − 39 MOs	−40.4040	$R = 2.067$	0.3205	—	All single and double excited configurations (832) of pure $^1A_1$ symmetry	75
CH_3^-	SCFMO	GTO(10s6p1d/8s1p) [6s2p1d/3s1p]	−39.5129	$R = 2.090$ Out of plane angle = 23.5°	0.3219	0.1519		59
CH_3^-	SCFMO + CI	GTO(10s6p1d/8s1p) [6s2p1d/3s1p]	−39.6645		0.3189	0.1482	931 CI	59

Table 2. *Cont.*

Molecule	Method	Basis	Energy	Geometry	q	η	Comments	Ref.
NH_3	SCFMO	GTO(13s7p1d/8s1p) [6s2p1d/3s1p]	−56.2117	R=1.913 Out of plane angle=22.14°	0.4331	0.133		59
NH_3	SCFMO+CI	29 STO	−56.3747	{ R=1.912 HN̂H=106.67°	0.4368	0.128	948 CI	59
NH_3	SCFMO		−56.1827		0.4286	0.152		55
NH_3	SCFGF	28 Localized M.O.	−56.2581		0.4234	0.149	Strongly orthogonal geminals	55
NH_3	SCFMO	STO Double zeta	−56.167	R=1.9233 Out of plane angle=22°	−0.457†	0.120		68
NH_3	SCFMO	13 STO for bond GTO Double-zeta quality		Close to exp.	0.4977	0.09		70
HCN			−92.829		0.365			51
H_2O	SCFMO	STO minimal set +2p	−75.7661		0.5918	0.1044	without vibr. corr.	57
H_2O	SCFMO+CI	8 STO (1s Double zeta)			0.548	0.120	with vibr. corr. (v=0)	44a
					0.542	0.121		
H_2O	SCFMO	STO-Double zeta	−76.005	Close to exp.	−0.576†	0.111		68
H_2O	SCFMO	GTO	−76.015	Close to exp.	−0.547†	0.115		68
H_2O	SCFMO	GTO(9s5p3d/4s) [4s3p2d/2s]	−76.0420		0.5060	0.124		58
H_2O		(9s5p2d/4s1p) [4s3p2d/2s1p]	−76.0501		0.5089	0.137		58
H_2O	SCFMO	27 STO	−76.0374	{ R=1.8103 HÔH=105°	0.5413	0.1183	Strongly orthogonal geminals	55
H_2O	SCFGF	27 localized M.O.	−76.0997		0.5520	0.1145		55
H_2O	SCFMO	GTO(9s5p2d/4s1p) [4s3p2d/2s1p]	−76.0504	Exp.	−0.502†	0.138	no vibr. corr.	44
			−76.0510	Theoretical equilib.	−0.564†	0.127	with vibr. corr. (v=0)	44
				Exp	−0.569	0.129		
HCHO	SCFMO	80 GTO→54 CGF	−113.8917	Exp	0.262	0.0157	Comparison with other ab initio values	52

Table 2. Cont.

Molecule	Method	Basis	Energy	Geometry	q	η	Comments	Ref.
Pyridine	SCFMO	GLO Double-zeta quality	-246.3265	Near exp.	H_α 0.382, H_β 0.382, H_γ 0.390	0.026, 0.037, 0.026		69
Pyrazine	SCFMO	GLO Double-zeta quality	-262.2547	Near exp.	0.433	0.07		69
C_6H_6	SCFMO	STO minimal	-230.213		0.340			60
	SCFMO	GTO(9s5p1d/4s1p)[4s2p1d/2s1p]	-230.749	$R_{CC}=2.63622$ $R_{CH}=2.05039$	0.3156	0.0762		60
$CH_3\underline{Si}H_3$	SCFMO	GTO(9s5p1d/4s1p/12s9p1d)[4s2p1d/7s4p1d]	-330.2993	Close exp.	-0.307	0.0080	effect of polarization orbitals reported	66
$CH_3Si\underline{H}_3$					0.3133	—		
NaH	SCFMO	STO, very large set	-162.393	Exp.	-0.186	0.015		50
MgH	SCFMO		-200.157	Exp.	0.0426	—		50
AlH			-242.463	Exp.	0.0733	—		50
SiH			-289.436	Exp.	0.0947	—		50
PH			-341.293	Exp.	0.1431	—		50
SH			-398.101	Exp.	0.1855	—		50
ClH			-460.110	Exp.	0.2445	—		50
MnH	SCFMO	GTO, 8s5p6d2f/5s2p	-1150.423	R=3.35, R=3.25	-0.078, -0.092	—, —	$^7\Sigma^+$ state, $^7\Pi$ state	73

Table 3. Carbon

Molecule	Method	Basis	Energy	Geometry	q	η	Comments	Ref.
CO	SCFMO	62GTO→40CGF	-112.7622	Experimental	-1.135	—		52
HCHO	SCFMO	80GTO→54CGF	-113.8917	Experimental	0.648	0.710		52
CH$_3^-$	SCFMO	67GTO→36CGF	-39.5129	$R = 2.090$	-0.3687	—		59
	SCFMO + CI	67GTO→36CGF	-39.6641	Out of plane angle 23.5°	-0.3800	—	931 CI (single and double excitations)	59
CF$_2$	SCFMO	GTO(9s5p1d/9s5p1d) [4s2p1d/4s2p1d]	-236.7252	Experimental	0.993	0.776	Double-zeta plus polarization quality	72
$\underline{C}$H$_3$NC	SCFMO	GTO(9s5p/9s5p/4s) [4s2p/4s2p/2s]	-131.8507		-0.46	—	Isomerization of CH$_3$NC is studied along a possible reaction coordinate	64
$\underline{C}$H$_3$CN	SCFMO	GTO(9s5p/9s5p/4s) [4s2p/4s2p/2s]	-131.8785		-0.28		Double-zeta quality	64
C$_6$H$_6$	SCFMO	GTO(9s5p1d/4s1p) [4s2p1d/2s1p]	-230.749	$R_{CC} = 2.63622$ $R_{CH} = 2.05039$	0.4483	0.0998	Double-zeta plus polarization quality	60
C$_5$NH$_5$ (Pyridine)	SCFMO	GLO of Double zeta quality	-246.3265	Close to experimental	C$_\alpha$ -0.295 C$_\beta$ -0.260 C$_\gamma$ -0.272	0.885 0.423 0.081		69
C$_4$N$_2$H$_4$ (Pyrazine)	SCFMO	GLO of Double zeta quality	-262.2547	Close to experimental	-0.293	0.915		69
CH$_3$SiH$_3$	SCFMO	GTO(9s5p1d/12s9p/ 4s1p)[4s2p1d/ 7s4p1d/2s1p]	-330.2993	Close to experimental Staggered conformation	0.1540		30% variation of q if polarization functions are suppressed	66

163

Table 4. Nitrogen

Molecule	Method	Basis	Energy	Geometry	q	η	Comments	Ref.
N_2	SCFMO	STO minimal	-108.57		-0.310	—		51
	SCFMO	best limited	-108.63		-1.341	—		51
	SCFMO	GTO Double-zeta quality	-108.80		-1.080	—		51
	SCFMO	GTO, (9s5p2d)[4s3p2d]	-108.9732	$R = 2.068$ (exp.)	-1.3574	—	Various basis contractions	54
	SCFMO	[5s3p]	-108.8890		-1.2474	—		
	SCFMO	[4s3p]	-108.8877		-1.2480	—		
	SCFMO	STO, 4s3p	-108.8865		-1.2433	—		
NH_3	SCFMO	OCE 66 STO	-56.1182	$R = 1.912$ HN̂H = 112.17°	-0.758	—	Vibrational correction 1.8%; other corrections evaluated	62
		OCE 77 STO	-56.1227		-0.786	—		
	SCFMO	32 STO	-56.1860	$R = 1.912$ HN̂H = 106.67° (Close to exp.)	-0.802	—	Strongly orthogonal geminals	62
	SCFGF	28 Localized M.O.	-56.2581		-0.797	—		62
	SCFMO	STO Double zeta	-56.167	Close to exp.	1.315†	—		68
		GTO Double-zeta quality	-56.003	Close to exp.	0.920†	—		68
	SCFMO	GLO Double-zeta quality	-56.142	$R = 1.9089$	-1.960	—	Sensitivity of q to small angle variation and to basis is shown	51
		GLO Extended	-56.159	HN̂H = 107°	-1.045	—	Analysis of individual MO contributions	51
		GLO, Double-zeta quality		$R = 1.9089$ HN̂H = 115° Out of plane angle 22.14°	-2.356	—		51
	SCFMO	GTO(13s7p1d/8s1p) [6s2p1d/3s1p]	-56.2117		-1.0994	—		59
	SCFMO + CI	948 CI	-56.3747	$R = 1.913$	-1.1664	—		59

Table 4. *Cont.*

Molecule	Method	Basis	Energy	Geometry	q	η	Comments	Ref.
HCN	SCFMO	GTO, Double-zeta quality	−92.829	Close to exp.	−0.854	—		51
	SCFMO	54 STO	−92.9147	Exp.	−1.1948	—		53
FCN	SCFMO	48 STO	−191.7667	$R_{CF} = 2.38109$, $R_{CN} = 2.20156$	−0.7114	—		53
ClCN	SCFMO	56 STO	−551.8247	$R_{CCl} = 3.0784$, $R_{CN} = 2.1978$	−0.9417	—		53
NCCN	SCFMO	64 STO	−184.6465	$R_{CN} = 2.186$, $R_{CC} = 2.608$	−1.1908	—		53
NCCCH	SCFMO	60 STO	−168.5784	$R_{CN} = 2.1864$, $R_{CC} = 2.6116$, $R_{CC} = 2.2734$, $R_{CH} = 1.9975$	−1.0940	—		53
CN⁻	SCFMO	26 STO	−167.2563	$R = 2.1791$	−1.0848	—		53
OCN⁻	SCFMO	48 STO	−489.9107	Exp.	−0.1819	—		53
SCN⁻	SCFMO	56 STO		$R_{SC} = 2.95$, $R_{CN} = 2.3$	−0.5552	—		53
NF₂	SCFMO	GTO(9s5p1d/9s5p1d) [4s2p1d/4s2p1d]	−253.2235	Exp.	−1.258	0.6184		72
NH₂OH	SCFMO	GLO, Double-zeta quality		At calculated equilibrium	−2.038	0.495	For details on computed geometry see W. Fink and L. C. Allen, *J. Chem. Phys.*, **47**, 895 (1967)	56
NH₂NH₂	SCFMO	GLO, Double-zeta quality		equilibrium	−2.390	0.561		56
CH₃NH₂	SCFMO	GLO, Double-zeta quality	−95.1127, −95.1088	Staggered, Eclipsed	−1.918, −1.943	0.177, 0.188		56
C₅NH₅	SCFMO	GLO, Double-zeta quality	−246.3265	Close to exp.	−1.887	0.137	Individual contributions from MOs	69
C₄N₂H₄ (Pyrazine)	SCFMO	GLO, Double-zeta quality	−262.2547	Close to exp.	−1.942	0.049		69

Table 5. Oxygen

Molecule	Method	Basis	Energy	Geometry	q	η	Comments	Ref.
H_2O	SCFMO	STO minimal+2p		Experimental	-2.62	0.78	0.1 a.u. above H.F. limit of energy	57
H_2O	SCFMO	STO Double-zeta	-76.005	Close to exp.	2.288†	0.7928		68
H_2O	SCFMO	GTO Double-zeta quality	-76.015		2.017†	0.7809		68
H_2O	SCFMO	GTO(9s5p2d/4s1p) [4s3p2d/2s1p]	-76.0504	Experimental	-1.922	0.796	Vibrational corrections no vibr. corr.	44
			-76.0510	At calculated equilibrium $R = 1.7789$ HÔH = 106.11°	-1.858	0.861	with vibr. corr. ($v = 0$)	44
					-1.822	0.833		
H_2O	SCFMO	GTO(9s5p3d/4s) [4s3p2d/2s]	-76.042		-1.9711	0.7913	Effect of basis contractions	58
H_2O	SCFMO	GTO(9s5p2d/4s1p) [4s3p2d/2s1p]	-76.0501		-1.9245	0.7968		58
H_2O	SCFMO	27 STO	-76.0374		-1.8190	0.7833	Strongly orthogonal geminals	55
H_2O	SCFGF	27 Localized MO's	-76.0997		-1.7906	0.7806		55
OF_2	SCFMO	GTO(9s5p1d) [4s2p1d]	-273.5294	Experimental	4.995	0.3802	Double-zeta+ polarization quality	72
CO	SCFMO	62 GTO → 40 CGF	-112.7622	Experimental	-0.697	—	MO's individual contributions reported; comparison with other computations	52
HCHO	SCFMO	80 GTO → 54 CGF	-113.8917	Experimental	-2.270	0.644		52

166

Table 6. Be, B, Li, F, Cl, Zn, Fe and Kr

Molecule	Method	Basis	Energy	Geometry	q	η	Comments	Ref.
LiH	SCFMO	STO, H.F. level	−7.9873	experimental (R = 3.015)	−0.0440 −0.0430	— —	without vibr. corr. with vibrational correction (v = 0)	61
LiH	Valence CI	7 STO on Li + 8 elliptic orbitals	−8.0561	R = 3.046	−0.0346	—	inner-shell polarized	42
LiH	MCSCF	STO 2sσ + 8π + 4δ	−8.0213	R=3.05	−0.0399 −0.0383	— —	without vibr. corr. with vibrational correction (v=0)	63
LiH	SCFMO + CI	STO, CI from a H.F. level	−8.0591	R = 3.015	−0.0413	—		61
LiH	SCFMO + CI	11 STO	−8.0555	R=3.015	−0.0429	—	All single and double CI	56
FCl	SCFMO	GTO, for C, N, O, F (9s5p)[4s2p] for Cl (12s9p)[6s4p]	−558.7975	experimental	7.343	—	inner-shell contributions are 10% of total q values	71
HOCl	SCFMO		−534.7097	experimental	6.151	0.016		71
NH2Cl	SCFMO	for H (5s)[2s]	−514.9931	experimental	5.125	0.030		71
CH3Cl	SCFMO		−499.0511	experimental	4.219	—		71
BeF2 BeF2	SCFMO SCFMO	GTO, for Be (9s2p1d)[4s2p1d] for B, C, N, O and F (9s5p1d) [4s2p1d]	−213.7351	experimental	−0.678 0.162	— —	Double-zeta + polarization quality Energy not more than	72
BF2 BF2	SCFMO SCFMO		−223.6744	R = 2.702 FBF = 120°	−1.363 −0.358	0.640 0.330	0.08 a.u. from HF level	72

Table 6. *Cont.*

Molecule	Method	Basis	Energy	Geometry	q	η	Comments	Ref.
CF₂	SCFMO		−236.7252	experimental	−2.281	0.711		72
NF₂	SCFMO		−253.2235	experimental	−3.948	0.077		72
OF₂	SCFMO		−273.5294	experimental	−5.56	0.078		72
Z̲nF₂	SCFMO	GTO(14s10p5d/9s5p)[9s5p2d/4s2p]	−1976.710	Linear configuration. At calculated equilibrium R = 3.315898	−1.253	—		67
Z̲nF₂	SCFMO				3.175	—		
F̲eF₃	SCFMO	GTO(14s9p5d/9s5p)[9s6p2d/4s2p]	−1560.871	At calculated equilibrium R = 3.59048	1.17	0.1282	⁶A′₁ state	74
F̲eF₃	SCFMO				2.21	—		
K̲rF₂	SCFMO	STO 8s6p5d2f/4s3p2d1f	−2950.734	Near experimental	1.515	—	Correct dissociation only with CI; 993 configurations employed	65
K̲rF₂	SCFMO + CI		−2950.865	Linear configuration R = 3.50	1.570	—		
K̲rF₂	SCFMO	STO 8s6p5d2f/4s3p2d1f		Near experimental	9.431	—		
K̲rF₂	SCFMO + CI			Linear configuration	7.243	—		

Note added in proof. Von Niessen, Kraemer, Dierckson and Cederbaum have recently calculated the following hydrogen q and η values (personal communication); for geometries, see ref. 76.

Molecule	Basis	$q_\alpha(\eta_\alpha)$	$q_\beta(\eta_\beta)$	$q_\gamma(\eta_\gamma)$	$q.\pi(\chi H)$
Pyridine	GTO[9s5p/4s][4s2p/2s]	0.3409(0.034)	0.3494(0.057)	0.3552(0.032)	—
Furan	GTO[9s5p/4s][4s2p/2s]	0.3584(0.073)	0.3619(0.053)	—	—
Thiophene	GTO[9s5p/4s][4s2p/2s]	0.3429(0.076)	0.3670(0.048)	—	—
Pyrrole	GTO[9s5p/4s][4s2p/2s]	0.3625(0.081)	0.3656(0.058)	—	—
*Cyclo*pentadiene	GTO[9s5p/4s][4s2p/2s]	0.3462(0.039)	0.3713(0.036)	—	0.3691(0.157)
C₅H₅P	GTO[12s9p1d/9s5p/4s][6s4p1d/4s2p/2s]	0.3065(0.066)	0.3041(0.053)	0.3230(0.015)	—
C₄H₄PH	GTO[12s9p1d/9s5p/4s][6s4p1d/4s2p/2s]	0.3559(0.074)	0.3552(0.055)	0.3036(0.065)	0.1650(0.218)
C₄H₄SiH₂	GTO[12s9p1d/9s5p/4s][6s4p1d/4s2p/2s]	0.3381(0.021)	0.3352(0.030)	—	0.1641(0.019)

APPENDIX A

Connection between q_0^{00}, q_2^{00} and the Force Constants

The relationships which relate the quantities q_0^{00}, q_2^{00} to the experimental force constants derivable from an analysis of the molecular vibrational spectra may be obtained by the following arguments.[47] First of all, to simplify the final formulae and to make easy their extension to polyatomic molecules, we consider here the force constants referred to the cartesian coordinates $x_\alpha y_\alpha z_\alpha$ of the nuclei. These force constants are always obtainable through a known linear transformation (for infinitesimal displacements from the equilibrium), whatever the force field employed. A diagonal force constant $K_{z_\alpha z_\alpha}$ is defined by

$$K_{z_\alpha z_\alpha} = \left[\frac{\partial^2}{\partial z_\alpha^2} \frac{\langle \Phi \mathscr{H} \Phi \rangle}{\langle \Phi \Phi \rangle} \right]_{R_0} \tag{A-1}$$

where R_0 indicates the geometrical configuration at equilibrium.

For true eigenfunctions and for variational w.f. Φ *fully optimized*[43] with respect to all variational parameters appearing in Φ, the Helmann–Feynman theorem holds. Thus for these cases, the terms containing the second derivative of Φ with respect to z_α vanish. Eq. A-1 becomes

$$K_{z_\alpha z_\alpha} = \left\{ \int_{\varepsilon_\alpha} P_1(\bar{R}_1 \ldots \bar{R}_\alpha \ldots ; \bar{r}) \frac{\partial^2}{\partial z_\alpha^2} \left(-\frac{Z_\alpha}{|\bar{r} - \bar{R}_\alpha|} \right) dv \right.$$

$$+ Z_\alpha \tfrac{4}{3} \pi P_1(\bar{R}_1 \ldots \bar{R}_\alpha ; \bar{R}_\alpha) + Z_\alpha \sum_{\beta(\neq\alpha)} Z_\beta \frac{\partial^2}{\partial z_\alpha^2} \frac{1}{|\bar{R}_\beta - \bar{R}_\alpha|}$$

$$\left. + \int \frac{\partial P_1(\bar{R}_1 \ldots \bar{R}_\alpha ; \bar{r})}{\partial Z_\alpha} \frac{\partial}{\partial Z_\alpha} \left(-\frac{Z_\alpha}{|\bar{r} - \bar{R}_\alpha|} \right) dv \right\}_{R_0} \tag{A-2}$$

In this last equation, P_1 stands for the first-order spinless density matrix and the contribution to the first integral arising from the infinitesimal volume ε_α surrounding the nucleus α, i.e. $\tfrac{4}{3} \pi P_1(\bar{R}_1 \ldots \bar{R}_\alpha \ldots ; \bar{R}_\alpha)$, has been written explicitly. Thus the ε_α subscript to the first integral means that the nuclear volume of the nucleus α has been excluded. By this, the sum of the first and the third terms of Eq. (A-2) gives the quantity $2q_0^{00}$ if z_α is directed along the major principal axis. Thus Eq. (A-2) may also be written as

$$K_{z_\alpha z_\alpha} = Z_\alpha q + Z_\alpha \tfrac{4}{3} \pi P_1(\bar{R}_0 ; \bar{R}_\alpha)$$

$$- Z_\alpha \left\{ \int \frac{\partial P_1(\bar{R}_1 \ldots \bar{R}_\alpha ; \bar{r})}{\partial z_\alpha} \frac{\partial}{\partial z_\alpha |\bar{r} - \bar{R}_\alpha|} \frac{1}{} dv \right\}_{R_0} \tag{A-3}$$

Equation (A-3) is the required relationship which might be used either to evaluate the force constant exploiting the knowledge of q or to obtain a value of q exploiting the knowledge of $K_{z_\alpha z_\alpha}$. By the same reasoning, it is possible to obtain for the diagonal force constants, relative to the other two principal axes, the relationships

$$K_{x_\alpha x_\alpha} = Z_\alpha \{\sqrt{3} q_2^{00} - q_0^{00}\} + Z_\alpha \frac{4}{3} \pi \, P_1(\mathbf{R}_0; \bar{R}_\alpha)$$

$$- Z_\alpha \left\{ \int \frac{\partial P_1(\bar{R}_1 \dots \bar{R}_\alpha; \bar{r})}{\partial x_\alpha} \frac{\partial}{\partial x_\alpha} \frac{1}{|\bar{r} - \bar{R}_\alpha|} dv \right\}_{\mathbf{R}_0}$$

$$K_{y_\alpha y_\alpha} = Z_\alpha \{-\sqrt{3} q_2^{00} - q_0^{00}\} + Z_\alpha \frac{4}{3} \pi \, P_1(\mathbf{R}_0; \bar{R}_\alpha)$$

$$- Z_\alpha \left\{ \int \frac{\partial P_1(\bar{R}_1 \dots \bar{R}_\alpha; \bar{r})}{\partial y_\alpha} \frac{\partial}{\partial y_\alpha} \frac{1}{|\bar{r} - \bar{R}_\alpha|} dv \right\}_{\mathbf{R}_0} \qquad \text{(A-4a,b)}$$

The other off-diagonal force-constant expressions are easily obtainable by the same technique. Their expressions will contain, if not referred to principal axes, the quantities q_0^{00}, q_{-1}^{00}, and q_{-2}^{00} but will not include any contribution from the nuclear volume.

The three terms of Eq. (A-3) are interpreted as follows: the first two, multiplied by Δz_α, are the electrostatic forces, exerted on a charge Z_α, arising from the external non-spherical part and from the internal spherical part, respectively, of the fixed charge distribution. The last one, called the relaxation time, accounts for the polarization of the charge distribution, $\partial P/\partial z_\alpha \cdot \Delta z_\alpha$ being its variation and $(\partial/\partial z_\alpha)(Z_\alpha/|\bar{r} - \bar{R}_\alpha|)$ the force operator.

The use of Eq. (A-3), along the lines mentioned above, is not very convenient because the relaxation term is the most difficult one to evaluate. It has been observed[47] that this term, to a large extent, is compensated by the internal electrostatic spherical-charge contribution $\frac{4}{3}\pi Z_\alpha P_1(\bar{R}_0, \bar{R}_\alpha)$.

To see how this comes about, let us consider $P_1(\mathbf{R}, \bar{r})$ as a sum of a spherical part, $P^{f\alpha}$ centred at the nucleus α which follows the motion of the nucleus and the rest, ΔP_1, i.e.

$$P_1(\bar{R}_1 \dots \bar{R}_\alpha \dots; \bar{r}) = P_1^{f\alpha}(\bar{R}_1 \dots \bar{R}_{\alpha-1}, \bar{R}_{\alpha+1} \dots; |\bar{r} - \bar{R}_\alpha|)$$
$$+ \Delta P_1(\bar{R}_1 \dots \bar{R}_\alpha \dots; \bar{r}) \qquad \text{(A-5)}$$

Since $P_1^{f\alpha}$, similarly to $1/|\bar{r} - \bar{R}_\alpha|$, depends only upon the modulus of $\bar{r} - \bar{R}_\alpha$ and not upon the two vectors $\bar{r}$ and $\bar{R}_\alpha$, the following equations hold

$$\frac{\partial P_1^{f\alpha}}{\partial z_\alpha} = -\frac{\partial P_1^{f\alpha}}{\partial z}; \qquad \frac{\partial}{\partial z_\alpha} \frac{1}{|\bar{r} - \bar{R}_\alpha|} = -\frac{\partial}{\partial z} \frac{1}{|\bar{r} - \bar{R}_\alpha|}.$$

By partial integration of the relaxation term, after noting that (i) the surface term vanishes for an acceptable P_1 arising from stationary states, (ii) the term

$$\int_{\varepsilon_\alpha} P_1^{f\alpha} \frac{\partial^2}{\partial z_\alpha^2} \frac{1}{|\bar{r} - \bar{R}_\alpha|} dv$$

vanishes due to symmetry, then

$$-Z_\alpha \left\{ \int \frac{\partial P_1}{\partial z_\alpha} \frac{\partial}{\partial z_\alpha} \frac{1}{|\bar{r}-\bar{R}_\alpha|} dv \right\}_{R_0} = -Z_\alpha \left\{ \int \frac{\partial \Delta P_1}{\partial z_\alpha} \cdot \frac{\partial}{\partial z_\alpha} \frac{1}{|\bar{r}-\bar{R}_\alpha|} dv \right\}_{R_0}$$

$$-Z_\alpha \frac{4}{3}\pi\, P_1^{f\alpha}(\bar{R}_0, \bar{R}_\alpha) \qquad (A\text{-}6)$$

and (iii) by observing that $P_1^{f\alpha}(\bar{R}_0, \bar{R}_\alpha) = P_1(\bar{R}_0, \bar{R}_\alpha)$, the following expression for $K_{z_\alpha z_\alpha}$ is finally obtained

$$K_{z_\alpha z_\alpha} = Z_\alpha q - Z_\alpha \left\{ \int \frac{\partial \Delta P_1}{\partial z_\alpha} \frac{\partial}{\partial z_\alpha} \frac{1}{|\bar{r}-\bar{R}_\alpha|} dv \right\}_{R_0} \qquad (A\text{-}7)$$

This last equation has been derived explicitly taking into account the compensation mentioned above. If, as expected, $\partial \Delta P_1/\partial z_\alpha$ is very small in the neighbourhood of nucleus α, Eq. (A-7) shows the existence of a simple proportionality between $K_{z_\alpha z_\alpha}$ and q.

As a matter of fact, the numerical values obtained for some cases involving the hydrogen nucleus support this simple relationship.[48,50] Unfortunately, the situation is far from being so simple, as will be realized if we point out two simple consequences which rule out, both from a theoretical viewpoint as well as from a numerical viewpoint, the possibility of neglecting the partial relaxation term appearing in Eq. (A-7). First of all, since we know that $K_{g_\alpha g_\alpha} \geq 0$ ($g = x, y, z$), then negative values of q are automatically ruled out. Secondly, considering the two nuclei at the ends of a bond, the following equality

$$Z_\alpha q_\alpha = Z_\beta q_\beta \qquad (A\text{-}8)$$

should hold. For simplicity it has been supposed that the z_α major axis, common to both nuclei, lies along the bond. This relation is clearly not satisfied for several cases where accurate data are available. Therefore, unless there is a concourse of fortunate coincidences which make the partial relaxation term negligible, equations (A-3) and (A-8), while certainly useful for many purposes, do not represent a convenient method to avoid difficult computational efforts.

APPENDIX B

Effect of Perturbing Fields and of the Finiteness of the Basis Set upon q in the SCFMO Approximation.

In the presence of an external electrostatic potential, $\hat{g}(\bar{r})$, the matrix element $\langle \Phi_0 \hat{q}_{2,0} \Phi_0 \rangle$ will be modified through the polarization of Φ_0. If $\hat{g}(\bar{r})$ may be considered small as compared to the internal fields, the perturbation theory will give reasonable results. Indicating by $\Phi_0^{0,1}$ the first-order correction due to the perturbation $\hat{g}(\bar{r})$, the first-order corrections to the matrix elements are

$$E^{1,1} = 2\langle\Phi_0^{00}\hat{q}_{2,0}\Phi_0^{0,1}\rangle \tag{B-1}$$

By exploiting the double perturbation approach sketched in Section III, we know that because of the interchange theorem [Eq. (III-13a,b)], the following equality holds:

$$\langle\Phi_0^{00}\hat{q}_{2,0}\Phi_0^{0,1}\rangle = \langle\Phi_0^{00}\hat{g}(\bar{r})\Phi_0^{1,0}\rangle \tag{B-2}$$

where $\Phi_0^{1,0}$ stands now for the first-order correction to the w.f. due to the perturbation $\hat{q}_{2,0}$. There is thus the possibility of evaluating the correction due to $\hat{g}(\bar{r})$ using either $\Phi_0^{0,1}$ or $\Phi_0^{1,0}$.

This fact has been exploited by Sternheimer who obtained an approximate form for the first-order correction to the w.f. due to the electrostatic nuclear quadrupole field. Thus, through the use of Eq. (B-2), he made estimates of the electronic polarization correction to:

(a) spherical ions in the perturbation field of the other ions of a crystal, and

(b) spherical atomic inner cores, arising from the *central field approximation*, for systems in non S states.

This author has given several numerical tables where these polarization corrections are reported for a variety of atomic systems in different states. The corrections may be quite large and the sign may be positive (antishielding effect) as well as negative (shielding effect). It is perhaps appropriate, however, to remember that these corrections have been obtained by:

(i) adopting the one-electron approximation,

(ii) assimilating the first-order perturbation equations to those corresponding to a hydrogenic system, and

(iii) considering the perturbation potential $\hat{g}(\bar{r})$ due to the external electrons of case (b) as due also to solely external sources.

These approximations are all serious ones and their effects may be very hard to estimate.

Remaining within the one-electron w.f. approximation, a more rigorous approach should use the PCHF method[49] which would imply the solution of equations such as

$$(\hat{h}^{00} + \hat{\mathcal{G}}^{00} - \varepsilon_j^{00})\varphi_j^{0,1} + (\hat{g} + \mathcal{G}^{0,1} - \varepsilon_j^{0,1})\varphi_j^{00} = 0 \tag{B-3a}$$

$$(\hat{h}^{00} + \hat{\mathcal{G}}^{00} - \varepsilon_j^{00})\varphi_j^{1,0} + (\hat{q}_{2,0} + \mathcal{G}^{1,0} - \varepsilon_j^{1,0})\varphi_j^{00} = 0 \tag{B-3b}$$

where the φ_j indicate the orbitals (for simplicity closed-shell cases are considered) and $\mathcal{G}^{0,1}$ ($\mathcal{G}^{1,0}$) the first-order correction to the HF effective electronic field. The corrections (B-2) assume the form

$$
E_{SCF}^{1,1} = \begin{cases} 4 \sum_{j}^{occup} \langle \varphi_j^{00} \hat{g} \varphi_j^{1,0} \rangle & \text{(B-4a)} \\[30pt] 4 \sum_{j}^{occup} \langle \varphi_j^{00} \hat{q}_{2,m} \varphi_j^{0,1} \rangle & \text{(B-4b)} \end{cases}
$$

This technique is certainly suitable to evaluate the effect of the crystal field in those cases where reasonably good SCF calculations may be carried out, both for atomic ions and for molecules.

It should be observed that because the interchange theorem holds also when the first-order perturbations $\varphi^{0,1}$ and $\varphi^{1,0}$ are approximated by projecting them in a given basis set (as is practically always the case), the results will be the same no matter which of the two alternative expressions, (B-4a) or (B-4b), is used. This means that the basis set may be chosen with the simplifying criterion that it should be capable of representing accurately $\varphi^{0,1}$ ($\varphi^{1,0}$) only in the region of space where $\hat{h}^{1,0}$ ($\hat{h}^{0,1}$) is important.

It should be clear that in the absence of external fields and where no symmetry constraints are imposed upon the HF effective field (as in several open-shell cases) to achieve the full symmetry of the electrostatic many-electron Hamiltonian, the Sternheimer effect should not appear. In spite of this, the use of a limited basis set in the SCFMO calculation may bring in something very similar to the Sternheimer effect. As a matter of fact, when the basis set does not contain suitable functions the molecular orbitals corresponding to the K-shells will be almost spherical in character. This is because the p, d, etc., functions (on the same centre) usually present in the basis set are generally too diffuse to describe the polarization of the inner core. Thus we have something very similar to the Sternheimer effect, i.e. the lack of (quadrupolar) polarization of a spherical shell.

It is of some interest to investigate more closely the effect upon the value of q_0^{00} of introducing into the basis set functions particularly suitable to describe the K-shell quadrupolar polarization. Let us assume that for a given basis set $\{\chi\}$, which includes n functions, the SCF procedure has produced the set of n SCF molecular orbitals

$$
\{\varphi^0\} = \{\chi\}c \tag{B-5}
$$

m of which are (doubly) occupied (for the sake of simplicity, closed-shell cases are considered). In the basis $\{\varphi\}$ the F-matrix representation of the HF $\hat{f}$ operator is diagonal with elements equal to the SCF orbital energies E_j^0.

$$
\hat{f} = \hat{h} + \hat{\mathcal{G}} \ c_{,i} \dots c_{,m}) \tag{B-6}
$$

If a basis function χ' is adjoined to $\{\chi\}$, giving the new basis $\{\chi, \chi'\}$, we must continue the SCF procedure by beginning with the diagonalization of the pseudosecular problem of order $n-1$,

$$\boldsymbol{F}\boldsymbol{c}_j = \varepsilon_j \boldsymbol{S}\boldsymbol{c}_j \tag{B-7}$$

where now the two matrices $\boldsymbol{F}$ and $\boldsymbol{S}$ are diagonal but for the elements of the last column (row). If the added basis function χ' is of little importance from the energy viewpoint, its coefficients in the occupied molecular orbitals φ'_j must be small. This means that in such a case, the change in the $\hat{f}$ operator will be negligible and therefore it will not be necessary to pursue the SCF procedure beyond this diagonalization. Since, as assumed above, the coefficients of χ' for the occupied orbitals are small, this diagonalization may be carried out by the perturbation bordering technique, which yields for the new occupied orbitals φ'_j the approximate expressions

$$\varphi'_j \cong \varphi^0_j + \left(\frac{\boldsymbol{F}_{j,\chi'} - \varepsilon^0_j \boldsymbol{S}_{j,\chi'}}{\varepsilon^0_j - \boldsymbol{F}_{\chi'\chi'}} \right)\chi' + \cdots \tag{B-8}$$

To the same order of approximation, the first-order density matrix variation will be

$$\Delta P(\bar{r}, \bar{r}') = 2 \sum_j^{\text{occup}} \left(\frac{\boldsymbol{F}_{j,\chi'} - \varepsilon^0_j \boldsymbol{S}_{j,\chi'}}{\varepsilon^0_j - \boldsymbol{F}_{\chi'\chi'}} \right)[\varphi_j(\bar{r})\chi'(\bar{r}') + \chi'(\bar{r})\varphi_j(\bar{r}')] \tag{B-9}$$

From this last expression, the variation Δq^{00}_0 due to the introduction of the function χ' into the basis set is

$$\Delta q^{00}_0 = 4 \sum_j^{\text{occ}} \left(\frac{\boldsymbol{F}_{j,\chi'} - \varepsilon^0_j \boldsymbol{S}_{j,\chi'}}{\varepsilon^0_j - \boldsymbol{F}_{\chi'\chi'}} \right)\langle \varphi_j {}^0\hat{q}_{2,0}\chi' \rangle \tag{B-10}$$

To describe the quadrupolar polarization of a K-shell due to the quadrupolar part of the field contained in $\hat{f}$, the basis set must include d-symmetry type functions localized in the same region of space as the K-shell. Since these functions are expensive and not energetically important, it is easy to understand why they are missing in most of the bases employed in actual calculations.

Formula (B-10) may be used to evaluate the Sternheimer effect and as a matter of fact it may be proved that many of the Sternheimer calculations are analogous to the use of Eq. (B-10) where the numerical values of the needed matrix elements have been suitably approximated. To check the accuracy of Eq. (B-10) in practical cases and the reliability of the numerical estimate of the matrix elements, an actual computation was carried out for the NH_3 molecule. As may be seen from the data reported in Table 7, two SCF calculations were performed. The first one, labelled I, used a best-ζ minimal STO basis set while in the second one, labelled II, the basis set was augmented by introducing $3d$-type functions with $\zeta = 7.0$. This value of ζ was chosen by considering the model problem of a $1s$ hydrogenic electron perturbed by a quadrupolar field

Table 7. Inner core polarization—SCF calculation of the electric field gradient q (a.u.) at the ^{14}N nucleus for NH$_3$ (see text)

	1a₁ I	1a₁ II	2a₁ I	2a₁ II	3a₁ I	3a₁ II		1eₓ I	1eₓ II
$1s_0$ (6.674)	0.99618	0.99618	-0.20885	-0.20890	-0.08231	-0.08221	$2p_x$ (1.942)	0.59231	0.59234
$2s_0$ (1.942)	0.01989	0.01988	0.74277	0.74294	0.45059	0.45053	$3d_{-1}$ (7.000)	—	-0.00344
$2p_z$ (1.942)	-0.00378	-0.00375	-0.14031	-0.13914	0.89249	0.89230	$3d_{-2}$ (7.000)	—	-0.00235
$1s_{H_1}$ (1.195)	-0.00437	-0.00437	0.15223	0.15228	-0.12346	-0.12407	$1s_{H_1}$ (1.195)	0	0
$1s_{H_2}$ (1.195)	-0.00437	-0.00437	0.15223	0.15228	-0.12346	-0.12407	$1s_{H_2}$ (1.195)	-0.43408	-0.43397
$1s_{H_3}$ (1.195)	-0.00437	-0.00437	0.15223	0.15228	-0.12407	-0.12407	$1s_{H_3}$ (1.195)	0.43408	0.43397
$3d_0$ (7.000)	—	-0.00025	—	-0.00003	—	-0.00190			
ε	-15.54108	-15.54107	-1.10607	-1.10567	-0.37024	-0.37094		-0.58563	-0.58588
$\langle q \rangle_{el}$	-0.00090	0.06320	-0.05916	-0.05884	-3.20648	-3.23053		0.90402	0.90416

Geometry: $\bar{R}_N = (0; 0; 0)$; $\bar{R}_{H_1} = (0; 1.770684; -0.721400)$; $(R_{NH} = 1.912)$.

	I	II
$E_{Tot} =$	-56.00504	-56.00635
$q =$	-1.70438	-1.66373

such as $r^2 Y_{2,0}$. The first-order perturbed function, approximated by a single $3d$ STO function, was obtained by varying ζ in order to minimize the corresponding Hylleraas functional [Eq. (III-15)].

The results obtained with Eqs. (B-8) and Eq. (B-10) are not reported because they are practically coincident (to within 2 to 3%) with the exact SCF calculations. Without going into the numerous considerations which might be made about the results reported in Table 7, it is perhaps worthwhile to note that:

(i) the inner core quadrupolar polarization (orbital 1 a_1), while being clearly evident, is very small $-\Delta q_{1a_1} = 0.0630 - (-0.0009)$;

(ii) the quadrupolar polarization effects are also felt in the not-so-internal orbitals;

(iii) the overall quadrupolar polarization effect is very small; and

(iv) the energy improvement is negligible.

Since the $1a_1$ orbital without a d function has practically no quadrupolar polarization, a more sizeable Sternheimer effect was expected according to the tables reported by this author. As a matter of fact, if approximate values (based upon the same reasonings proposed by Sternheimer) are used for the $F_{i,x'}$ matrix elements, the resulting inner-core polarization effect becomes much larger. It seems proper, therefore, to emphasize the need to use much care in adopting the Sternheimer corrections, at least for molecular cases.

REFERENCES

1 See for instance the other articles of this volume and the many articles on the subject contained in *Advances in Nuclear Quadrupole Resonance*, Volume 1, Ed: J. A. S. Smith, Heyden, London, 1974.

2 E. Scrocco, *Advan. Chem. Phys.*, **5**, 319 (1963).

3 E. A. C. Lucken, *Nuclear Quadrupole Coupling Constants*, Academic Press, London, 1969.

4 T. P. Das and E. L. Hahn, *Solid State Physics*, Academic Press, New York–London, 1958.

5 G. R. Gunther-Mohr, S. Geschwind, and C. H. Townes, *Phys. Rev.*, **81**, 289L (1951).

6 For the conversion factors for other units, *see* R. McWeeny, *Nature*, **243**, 196 (1973).

7 E. C. Kemble, *The Fundamental Principles of Quantum Mechanics with Elementary Applications*, Dover, New York, 1958.

8 P. O. Löwdin, *J. Mathem. Phys.*, **3**, 969 (1962); P. O. Löwdin, *Inter. J. Quantum Chem.*, **2**, 878 (1968); P. O. Löwdin and O. Goshinski, *Inter. J. Quantum Chem.*, **5**, 685 (1971) and references therein.

9 See for instance M. E. Rose, *Elementary Theory of Angular Momentum*, John Wiley, New York, 1957.

10 P. Jennings and E. B. Wilson Jr., *J. Chem. Phys.*, **47**, 2130 (1967); A. Mazziotti, *J. Chem. Phys.*, **55**, 2657 (1971); N. W. Bazley and H. R. Fankhauser, *Chem. Phys. Lett.*, **7**, 121 (1970).

11 C. Eckart, *Phys. Rev.*, **36**, 878 (1930); H. F. Weinberg, *J. Res. Natl. Bur. Std.*, **64B**, 217 (1960); F. Weinhold, *J. Phys.*, **A1**, 305 (1968).

12 R. McWeeny and B. Sutcliffe, *Methods of Molecular Quantum Mechanics*, Academic Press, London, 1969.

13 J. Goodisman and W. Klemperer, *J. Chem. Phys.*, **38**, 721 (1963).

14 C. Möeller and M. C. Plesset, *Phys. Rev.*, **46**, 618 (1943).

15 P. O. Löwdin, *Advances in Chemical Physics*, **14**, 325 (1969); P. S. Bagus and H. F. Schaefer, *J. Chem. Phys.*, **56**, 224 (1971).

16 F. Grimaldi, *Advances in Chemical Physics*, **14**, 341 (1969); T. L. Gilbert, *Molecular Orbitals in Chemistry, Physics, and Biology*, Eds: P. O. Löwdin and B. Pullmann, Academic Press, New York, 1964.

17 S. F. Boys, *Quantum Theory of Atoms, Molecules, and Solid States*, Ed: P. O. Löwdin, Academic Press, New York, 1966; C. Edmiston and K. Ruedenberg, *ibid.*

18 S. Diner, J. P. Malrieu, F. Jordan, and M. Gilbert, *Theoret. Chim. Acta*, **15**, 100 (1969); V. Kvasnička, *Theoret. Chim. Acta*, **34**, 61 (1974).

19 P. Nozières, *Le problem à N corps*, Dunod, Paris, 1963; N. H. March, W. H. Young, and S. Sampanthar, *The Many-Body Problem in Quantum Mechanics*, Cambridge University Press, Cambridge, 1967; H. P. Kelly, *Phys. Rev.*, **182**, 84 (1969) and references therein.

20 H. Primas, *Modern Quantum Chemistry*, Part II, Ed: O. Sinanoğlu, Academic Press, New York, 1965.

21 J. Goldstone, *Proc. Roy. Soc. (London)*, **A239**, 267 (1957).

22 O. Sinanoğlu, *J. Chem. Phys.*, **36**, 706 (1963).

23 W. Meyer, *J. Chem. Phys.*, **58**, 1017 (1973).

24 A. C. Hurley, J. E. Lennard-Jones, and J. A. Pople, *Proc. Roy. Soc. (London)*, **A220**, 446 (1953).

25 R. McWeeny, *Rev. Mod. Phys.*, **32**, 335 (1960).

26 O. Sinanoğlu and K. A. Brueckner, *Three Approaches to the Electron Correlation in Atoms*, Eds: O. Sinanoğlu and E. U. Condon, Yale University Press, New Haven and London, 1970.

27 R. K. Nesbet, *Advances in Chemical Physics*, **14**, 1 (1969).

28 K. A. Brueckner, *Phys. Rev.*, **97**, 1953 (1955) and references therein.

29 H. A. Bethe and J. Goldstone, *Proc. Roy. Soc. (London)*, **A238**, 551 (1957).

30 G. Das and A. C. Wahl, *J. Chem. Phys.*, **44**, 87 (1966); E. Clementi and A. Veillard, *J. Chem. Phys.*, **44**, 3050 (1966); R. McWeeny, *Symposia of the Faraday Society*, **2**, 7 (1968); N. G. Mukherjee, *Chem. Phys. Letters*, **24**, 77 (1974).

31 P. O. Löwdin, *Phys. Rev.*, **94**, 1474 (1955); R. Alrich and W. Kutzelnigg, *J. Chem. Phys.*, **48**, 1819 (1968).

32 E. A. Hylleraas, *Z. Phys.*, **54**, 347 (1929); H. M. James and A. S. Coolidge, *J. Chem. Phys.*, **1**, 825 (1933); C. L. Pekeris, *Phys. Rev.*, **115**, 1216 (1959).

33 F. S. Boys and N. C. Handy, *Proc. Roy. Soc. (London)*, **A309**, 209 (1969); **A310**, 43, 63 (1969); and **A311**, 309 (1969).

34 N. C. Handy, *Mol. Phys.*, **26**, 169 (1973).

35 S. G. Kukolich and S. C. Wofsy, *J. Chem. Phys.*, **52**, 5477 (1970).

36 R. Moccia, *Intern. J. Quantum Chem.*, **8**, 293 (1974).

37 J. O. Hirschfelder, W. B. Brown, and S. T. Epstein, *Advances in Quantum Chemistry*, Volume 1, Ed: P. O. Löwdin, Academic Press, New York, 1964.

38 A. Dalgarno and A. L. Stewart, *Proc. Roy. Soc.*, **A247**, 245 (1958).

39 E. A. Hylleraas, *Z. Physik*, **65**, 209 (1930).

40 R. Sternheimer, *Phys. Rev.*, **80**, 102 (1950); **85**, 244 (1951); **86**, 316 (1952); **95**, 736 (1954); **A6**, 1702 (1972); **A9**, 1783 (1974); and R. F. Peierls, **A4**, 1722 (1971); R. Sternheimer and H. M. Foley, *ibid.*, **92**, 1460 (1953); H. M. Foley, R. Sternheimer,

and D. Tycko, *ibid.*, **93,** 734 (1954); R. Sternheimer, *ibid.*, **146,** 140 (1966); **159,** 266 (1967); **164,** 10 (1967).

41 See ref. 36 and references therein.

42 J. C. Browne and F. A. Matsen, *Phys. Rev.*, **135,** 1227 (1964).

43 P. Grigolini and R. Moccia, *J. Chem. Phys.*, **57,** 1369 (1972).

44 (a) C. W. Kern and R. L. Matcha, *J. Chem. Phys.*, **49,** 2081 (1968); (b) W. C. Ermler and C. W. Kern, *J. Chem. Phys.*, **55,** 4851 (1971); (c) B. J. Krohn, W. C. Ermler, and C. W. Kern, *J. Chem. Phys.*, **60,** 22 (1974).

45 R. L. Mössbauer and M. J. Clauser, *Hyperfine Interactions*, Eds: A. J. Freeman and R. B. Frankel, Academic Press, New York, 1964; P. Pyykkö and L. Linderberg, *Chem. Phys. Letters*, **1,** 34 (1970); P. Pyykkö, *ibid.*, **1,** 479 (1970).

46 J. M. Schulman, J. W. Moskowitz, and C. Hollister, *J. Chem. Phys.*, **46,** 2759 (1967).

47 L. Salem, *J. Chem. Phys.*, **38,** 1227 (1963).

48 M. Rinné and J. Depireux, *Advances in Nuclear Quadrupole Resonance*, Volume 1, Ed: J. A. S. Smith, Heyden, London, 1974, p. 357.

49 H. Peng, *Proc. Roy. Soc. (London)*, **A178,** 499 (1941); L. C. Allen, *Phys. Rev.*, **118,** 167 (1960); R. McWeeny, *Phys. Rev.*, **126,** 1028 (1962).

50 M. Mokarram and J. L. Ragle, *J. Chem. Phys.*, **59,** 2770 (1973).

51 C. T. O'Konski and Tae-Kyu Ha, *J. Chem. Phys.*, **49,** 5354 (1968).

52 D. B. Neumann and J. W. Moskowitz, *J. Chem. Phys.*, **50,** 2216 (1969).

53 R. Bonaccorsi, E. Scrocco, and J. Tomasi, *J. Chem. Phys.*, **50,** 2940 (1969).

54 T. H. Dunning, Jr., *J. Chem. Phys.*, **53,** 2823 (1970).

55 P. F. Franchini, R. Moccia, and M. Zandomeneghi, *Intern. J. Quantum Chem.*, **4,** 487 (1970).

56 G. P. Arrighini, C. Guidotti, and J. Tomasi, *Theoret. Chim. Acta*, **18,** 329 (1970).

57 S. D. Gornostansky and C. W. Kern, *J. Chem. Phys.*, **55,** 3253 (1971).

58 T. H. Dunning Jr., *J. Chem. Phys.*, **55,** 3958 (1971).

59 K. E. Kari and I. G. Czimadia, *Theoret. Chim. Acta*, **22,** 1 (1971).

60 W. C. Ermler and C. W. Kern, *J. Chem. Phys.*, **58,** 3458 (1973).

61 S. Green, *Phys. Rev.*, **A4,** 251 (1971).

62 P. Grigolini and R. Moccia, *J. Chem. Phys.*, **57,** 1369 (1972).

63 K. K. Docken and J. Hinze, *J. Chem. Phys.*, **57,** 4928 (1972).

64 D. H. Liskov, C. F. Bender, and H. F. Schaefer III, *J. Am. Chem. Soc.*, **94,** 5178 (1972).

65 P. S. Bagus, B. Liu, and H. F. Schaefer III, *J. Am. Chem. Soc.*, **94,** 6635 (1972).

66 D. H. Liskov and H. F. Schaefer III, *J. Am. Chem. Soc.*, **94,** 6641 (1972).

67 D. R. Yarkony and H. F. Schaefer III, *Chem. Phys. Letters*, **15,** 514 (1972).

68 G. L. Bendazzoli, M. Dixon, and P. Palmieri, *Intern. J. Quantum Chem.*, **7,** 223 (1973).

69 Tae-Kyu Ha and C. T. O'Konski, *Intern. J. Quantum Chem.*, **7,** 609 (1973).

70 E. C. Cook Jr. and D. Ebbing, *Intern. J. Quantum Chem.*, **7,** 707 (1973).

71 G. L. Bendazzoli, D. G. Lister, and P. Palmieri, *J. Chem. Soc.*, Faraday II, **69,** 791 (1973).

72 S. Rothemberg and H. F. Schaefer III, *J. Am. Chem. Soc.*, **95,** 2095 (1973).

73 P. S. Bagus and H. F. Schaefer III, *J. Chem. Phys.*, **58,** 1844 (1973).

74 R. W. Hand, W. J. Hunt, and H. F. Schaefer III, *J. Am. Chem. Soc.*, **95,** 4517 (1973).

75 J. J. Eberhardt, R. Moccia, and M. Zandomeneghi, *Chem. Phys. Letters*, **24,** 524 (1974).

76 B. P. Stoicheff, *Canad. J. Phys.* **32,** 339 (1954); B. Bak, L. Hansen-Nygaard and J. Rastrup-Andersen, *J. Mol. Spectry.* **2,** 361 (1958) and *Disc. Faraday Soc.* **19,** 30 (1955); T. C. Wang and L. S. Bartell, *J. Chem. Phys.* **61,** 2840 (1974); B. Bak, D. Christensen, L. Hansen and J. Rastrup-Andersen, *J. Chem. Phys.* **24,** 720 (1956).

6. STRUCTURAL PHASE TRANSITIONS IN RMX$_3$ (PEROVSKITE) AND R$_2$MX$_6$ (ANTIFLUORITE) COMPOUNDS

Robin L. Armstrong and Henry M. van Driel

Department of Physics, University of Toronto, Canada M5S 1A7

I. INTRODUCTION

The various techniques of magnetic resonance spectroscopy including nuclear magnetic resonance (NMR), nuclear quadrupole resonance (NQR) and electron paramagnetic resonance (EPR) have been applied to the study of a wide variety of phase transitions in aggregated systems. For example, superconducting,[1] ferro- and antiferromagnetic,[2,3] ferro- and antiferroelectric[4,5] and purely displacive[6] phase transitions have all been investigated using the methods of magnetic resonance. One of the principal reasons that these techniques are especially useful for the study of phase transitions is that they provide a microscopic probe and thereby make it possible to investigate local aspects associated with the transitions. The information obtained is therefore complementary to that provided by techniques such as light scattering and neutron scattering which are sensitive to the collective excitations in the crystal.

In this paper we will discuss only those purely displacive phase transitions which in some cases destroy the cubic symmetry characteristic of the high temperature phases of antifluorite (R$_2$MX$_6$) and perovskite (RMX$_3$) structures. Although the discussion will focus on the results of NQR and NMR experiments, occasionally data from other experiments will be introduced when they are necessary to provide complementary information. It is our firmly held view that the most successful way to investigate complex physical phenomena such as phase transitions is through a combination of diverse experimental methods.

The antifluorite and perovskite families have been selected for the present discussion because they constitute a complementary pair. The two crystal structures are similar. The R$_2$MX$_6$ lattice may be viewed as a 'half empty' RMX$_3$ lattice. In both structures the purely displacive phase transitions are

179

describable in terms of a reorientation of the equilibrium position of the MX_6 octahedra combined with a distortion of the R ion cages in which they reside. However, the coupling between octahedra is direct for the RMX_3 compounds but indirect for the R_2MX_6 compounds.

No attempt will be made in this review to summarize the results of each and every magnetic resonance investigation of a compound of the R_2MX_6 and RMX_3 types in which a purely displacive phase transition has been observed. But rather the emphasis will be directed towards the various physical aspects of the problem and the theoretical approaches to them. Experimental results will be chosen for those particular compounds which best illustrate the points that we wish to establish.

In Section II, several general theoretical models are discussed. First an empirical geometrical model which relates lattice dimensions to transition temperatures is introduced. Then Landau's famous treatment of second-order phase transitions based on symmetry considerations is reviewed. Finally dynamical considerations are introduced which relate displacive phase transitions to the occurrence of soft lattice modes.

In Section III, experiments to observe the static behaviour of the phase transitions are considered. NQR and NMR spectra are extremely sensitive to the changes in symmetry which often occur at the sites of the resonant nuclei as a result of a displacive phase transition. The temperature dependence of the NQR frequency may reflect the softening of the lattice mode which triggers the phase transition. Magnetic resonance techniques are well suited to the direct observation of the temperature dependence of the order parameter associated with a displacive phase transition. In this way, a critical exponent associated with the order parameter may be obtained.

In Section IV, experiments to observe the dynamic behaviour of the phase transitions are discussed. The temperature dependence of the nuclear spin–lattice relaxation time measurements may reflect the softening of the lattice mode which triggers the phase transition. Within the critical region, nuclear spin–lattice relaxation time data provide a monitor for dynamic critical phenomena.

II. GEOMETRICAL, SYMMETRY AND DYNAMICAL CONSIDERATIONS
A. High Temperature Structures

The cubic antifluorite structure[7] R_2MX_6 belongs to the space group O_h^5. The basis consists of an MX_6 octahedron with M atom coordinates $(0, 0, 0)$ X atom coordinates $(\pm ua, 0, 0), (0, \pm ua, 0), (0, 0, \pm ua)$ where $u \leq 0.25$ and two R ions with coordinates $\pm\frac{1}{4}(a, a, a)$. The parameter a is the lattice constant. This structure is illustrated in Fig. 1. The first Brillouin zone is a truncated octahedron. Figure 2 shows the locations of some of the special symmetry points $(\Gamma, X, K \ldots)$ in the zone.

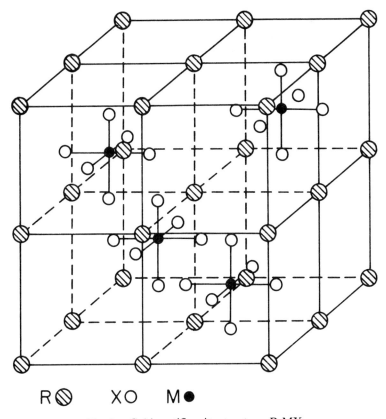

R⊚   XO   M●

Fig. 1.   Cubic antifluorite structure, R_2MX_6.

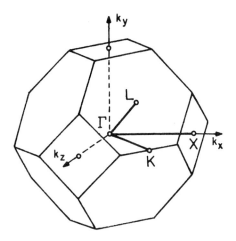

Fig. 2.   Face-centred cubic Brillouin zone with high-symmetry points Γ, X, L, K designated.

181

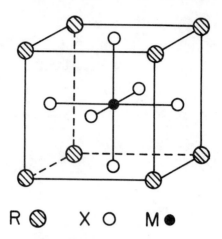

R ⊘   X ◯   M ●

Fig. 3.   Cubic perovskite structure, RMX₃.

The cubic perovskite structure[8] RMX₃ belongs to the space group O_h^1. The basis consists of an MX₃ half-octahedron with M atom coordinates $(0, 0, 0)$, X atom coordinates $\frac{1}{2}(a, 0, 0)$, $\frac{1}{2}(0, a, 0)$, $\frac{1}{2}(0, 0, a)$ and an R ion with coordinates $\frac{1}{2}(a, a, a)$ where a is the lattice constant. The structure is illustrated in Fig. 3. The first Brillouin zone is a simple cube. Figure 4 shows the locations of some of the special symmetry points in the zone.

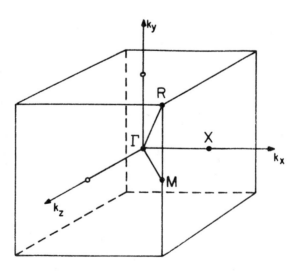

Fig. 4.   Simple cubic Brillouin zone with high-symmetry points Γ, R, X, M designated.

B. Nature of the Phase Transitions

Many of these structures undergo one or more structural phase transitions as the temperature is lowered. Some of these are purely displacive in nature while others are ferroelectric, antiferroelectric or antiferromagnetic. The present discussion will be confined to only those high temperature transitions which are of the purely displacive variety. For example, at 105.5 K, the cubic perovskite $SrTiO_3$ undergoes a transition in which the equilibrium positions of the TiO_6 octahedra begin to rotate away from their high temperature positions[9] (see Fig. 5) and the lattice undergoes a tetragonal distortion about a [100]

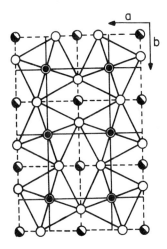

Fig. 5.   Orientation of TiO_6^{2-} octahedra in the *ab* plane in the tetragonal phase of the perovskite $SrTiO_3$. The tetragonal distortion is along the *c*-axis.

tetrad axis. A similar transition[10] occurs in $LaAlO_3$ at ~ 800 K except that in this case the cubic lattice undergoes a trigonal distortion about the [111] triad axis. An example from the cubic antifluorite family is provided[11,12] by K_2ReCl_6. At 110.9 K the equilibrium positions of the $ReCl_6^{2-}$ octahedra begin to rotate away from their high temperature positions (see Fig. 6) and the lattice undergoes a tetragonal distortion. The octahedra perform a ferro-rotation so that the size of the unit cell is the same in both phases. A similar transition occurs in K_2TeBr_6 except that in this case the rotations of adjacent octahedra are in the opposite sense and the dimension of the unit cell doubles in the low temperature phase.[13] These phase transitions are accompanied by the occurrence of soft low-lying lattice modes. For the perovskite compounds it is always the softening of a zone boundary phonon that triggers the transitions; for an antifluorite compound either a zone centre or a zone boundary phonon may be the important one.

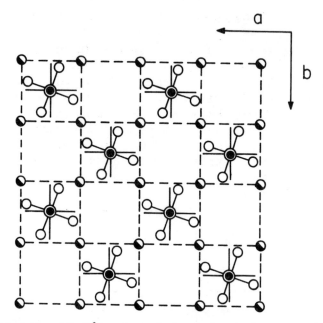

Fig. 6.   Orientation of $ReCl_6^{2-}$ octahedra in the *ab* plane in the tetragonal phase of the antifluorite K_2ReCl_6. The tetragonal distortion is along the *c*-axis.

The question is often asked whether a particular phase transition is first-order or second-order. It is our opinion that because of the inevitable lattice imperfections and the associated lattice strains, no displacive phase transition can ever provide an example of a perfect second-order phase transition (unlike the case of a superconducting phase transition). Nonetheless, many of the transitions are predominantly second-order in the sense that the second-order characteristics probably account for most of the energy change of the total transformation.[14] Therefore, the symmetry implications of Landau's theory of second-order phase transitions provide useful starting conditions for the discussion of displacive transitions.

C. Empirical Geometrical Model

In 1964 Brown[13] introduced an empirical geometrical model that permits one to predict which R_2MX_6 compounds will show a distortion at 293 K, which will distort at a lower temperature and which will be stable at 0 K. The criterion devised is based on the relative sizes of the cations and anions. At high temperatures the anions form a face-centred array and are oriented such that each halogen (X) atom is in contact with four other halogens on neighbouring anions. This leaves cavities in the structure into which the cations fit. Each

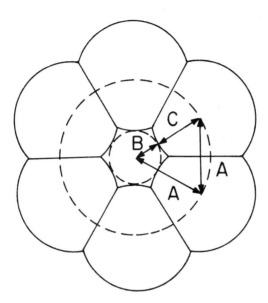

Fig. 7. Section through the environment of the cation (R) site in a crystal with the R_2MX_6 structure. The space available to the cation has radius B. This space is defined by the distribution of neighbouring anions (MX_6) of radius C on a sphere with radius A.

cavity is formed by twelve halogen atoms, three from each of the four anions that surround the cavity. The centres of the halogen atoms lie on a sphere with radius A equal to the halogen–halogen contact distance. Figure 7 shows a section of this sphere. Since the cations are usually much smaller than the halogens, the size of the cavity is determined by the halogen–halogen contacts between anions rather than halogen–cation contacts. If the cation is much smaller than the cavity into which it fits, it will be free to rattle around inside the cavity unless the anions reorient themselves in such a way as to reduce the effective size of the cavity and thereby lock the cations into place. An obvious quantity which should be important for determining whether or not such a distortion will occur is the ratio of the size of the cation to the size of the cavity into which it must fit. This parameter is called the radius ratio and designated as r. An analysis of available data showed that for $r < 0.89$ the lattice is distorted at 293 K, for $0.89 < r < 0.98$ the lattice is cubic at 293 K but distorts at some lower temperature, and for $r > 0.98$ the lattice is stable to 0 K. For example, in K_2PtI_6, $r = 0.87$ and the lattice is distorted at 293 K; in K_2PtBr_6, $r = 0.91$ and although the lattice is cubic at 293 K it distorts at 169 K; in Cs_2PtCl_6, $r = 1.16$ and the lattice is stable to 0 K.

In the perovskites it may also be noted that transitions involving the tilting of MX_6 octahedra occur when the R cation is small compared to the volume available for it. For example, in $CsPbCl_3$, $r = 0.82$ and the lattice is distorted at 293 K whereas in $BaTiO_3$, $r \approx 1$ and no tilting transition occurs.

D. Landau and Lifshitz Symmetry Conditions

As early as 1937, Landau's theory[15] was used to describe structural (as well as other) phase transitions of the second order. This mean field theory introduces the concept of an order parameter η which in some sense describes the difference in the configurations of the atoms in the two phases. The theory assumes that the crystal is spatially homogeneous.

The density function $\rho_0(\mathbf{r})$ gives the probability distribution of atoms in the high temperature phase. The change $\Delta\rho(\mathbf{r})$ from $\rho_0(\mathbf{r})$ caused by the phase transition can be expressed as an expansion in terms of the basis functions for the irreducible representations of the high temperature space group $\mathscr{G}_0$. Quite generally one can write

$$\Delta\rho(\mathbf{r}) = \sum_n{}' \sum_i c_i^{(n)} \phi_i^{(n)}$$

where the functions $\phi_i^{(n)}$, with n labelling the representation and i labelling the 'row' of that representation, transform among themselves under all operations of $\mathscr{G}_0$. The prime indicates that the identity representation is omitted since $\Delta\rho(\mathbf{r})$ results from a change in symmetry. The complete density function $\rho(\mathbf{r})$ given by

$$\rho(\mathbf{r}) = \rho_0(\mathbf{r}) + \Delta\rho(\mathbf{r})$$

is a function of pressure P, temperature T and the coefficients $c_i^{(n)}$. Since the crystal has symmetry $\mathscr{G}_0$ at the transition point and above, all $c_i^{(n)}$ must be zero and $\Delta\rho(\mathbf{r})$ is zero in this regime. Since the change in state of the crystal in a second-order phase transition is continuous, the $c_i^{(n)}$ must tend to zero through arbitrarily small values near the transition point. Therefore, any thermodynamic potential (such as the Gibbs free energy G) used to describe the crystal can be written as an expansion in powers of the $c_i^{(n)}$. Since the Gibbs free energy must be independent of the coordinate system used to describe it, the expansion of G can contain only invariant combinations of the $c_i^{(n)}$ to the appropriate power. No linear invariant can exist, however, if a change in symmetry occurs at the transition. Therefore, since one and only one second-order invariant exists for any given irreducible representation, the leading terms in the expansion of G are

$$G = G_0 + \sum_n{}' A^{(n)}(P, T) \sum_i [c_i^{(n)}]^2$$

G will have a minimum value as required for stability of the lattice if $A^{(n)} > 0$ for all n when every $c_i^{(n)} = 0$. This is the case above the transition point. For some $c_i^{(n)}$ to be non-zero below the transition point, the associated $A^{(n)}$ must be less than zero so that $A^{(n)}(T_c) = 0$. Since the $A^{(n)}$ are the functions of P and T, two distinct $A^{(n)}$ will in general not vanish for the same value of T_c. Therefore, it is

186

sufficient to consider

$$\Delta\rho(\mathbf{r}) = \sum_i c_i^{(n_0)}\phi_i^{(n_0)}$$

where $n = n_0$ designates the particular irreducible representation associated with the transition. The order parameter η is then defined by

$$\eta^2 = \sum_i (c_i^{(n_0)})^2$$

so that

$$G = G_0 + \eta^2 A^{(n_0)}(P, T)$$

Continuing the expansion

$$G = G_0 + \eta^2 A^{(n_0)}(P, T) + \eta^3 \sum_\alpha B_\alpha^{(n_0)}(P, T) f_\alpha^{(3)}(\gamma_i^{(n_0)})$$

$$+ \eta^4 \sum_\alpha C_\alpha^{(n_0)}(P, T) f_\alpha^{(4)}(\gamma_i^{(n_0)}) \dots$$

where the $f_\alpha^{(i)}$ are invariants of order i constructed from coefficients $\gamma_i^{(n_0)}$ defined such that

$$c_i^{(n_0)} = \eta\gamma_i^{(n_0)}$$

Since for a second-order phase transition η must be a continuous function of P, T near T_c, it is required that there be no third-order invariant. In the formalism of group theory

$$[\Gamma^3] \cap \Gamma^{1+} = 0$$

where Γ is the irreducible representation of $\mathscr{G}_0$ which induces the transition and Γ^{1+} is the identity representation. This is the Landau condition. The equation states that the symmetrized cube of a representation which is capable of inducing the transition does not contain the identity representation. In addition, lattice stability requires that all coefficients $C_\alpha^{(n_0)}(P, T) > 0$.

Lifshitz further postulated[15] that the unit-cell size in the distorted phase should be a simple multiple of the high temperature unit-cell size. Since the basis functions $\phi_i^{(n)}(\mathbf{r})$ can be written in the form

$$\phi_i^{(n)}(\mathbf{r}) = u_{i\mathbf{k}}^{(n)}(\mathbf{r}) \exp(i\mathbf{k} \cdot \mathbf{r})$$

where the functions $u_{i\mathbf{k}}^{(n)}(\mathbf{r})$ reflect the periodicity of the lattice, it follows that the irreducible representations are characterized by wave vectors $\mathbf{k}$ in the reciprocal lattice. This latter postulate then requires that only those values of $\mathbf{k}$ are allowed which can be written as simple fractions of a (small) reciprocal lattice vector. In the terminology of group theory the Lifshitz condition becomes

$$\{\Gamma^2\} \cap \mathbf{V} = 0$$

This equation states that the antisymmetrized square of a representation which can induce the phase transition does not include the vector representation **V**.

E. Dynamical Considerations and Soft Modes

The Landau theory was cast into the formalism of lattice dynamics beginning with Ginzburg[17] in 1960. The density function $\rho_0(\mathbf{r})$ can be written as

$$\rho_0(\mathbf{r}) = \sum_{l\kappa} \sigma_\kappa (\mathbf{r} - \mathbf{r}_{l\kappa})$$

where $\sigma_\kappa(\mathbf{r})$ is the density of atoms of type κ situated in the *l*th unit cell at $\mathbf{r}_{l\kappa}$. It follows that

$$\Delta\rho(\mathbf{r}) = \sum_{l\kappa} \mathbf{u}_{l\kappa} \cdot \operatorname{grad} \sigma_\kappa (\mathbf{r} - \mathbf{r}_{l\kappa})$$

The atomic displacements $\mathbf{u}_{l\kappa}$ may be expressed in terms of phonon coordinates as

$$\mathbf{u}_{l\kappa} = (Nm_\kappa)^{-\frac{1}{2}} \sum_{kj} Q_j(\mathbf{k}) \mathbf{e}_{\kappa j}(\mathbf{k}) \exp(i\mathbf{k} \cdot \mathbf{r}_{l\kappa})$$

so that
$$\Delta\rho(\mathbf{r}) = N^{-1/2} \sum_{kj} Q_j(\mathbf{k}) \left[\sum_{l\kappa} m_\kappa^{-1/2} \mathbf{e}_{\kappa j}(\mathbf{k}) \cdot \operatorname{grad} \sigma_\kappa (\mathbf{r} - \mathbf{r}_{l\kappa}) \right.$$
$$\left. \times \exp i\mathbf{k} \cdot (\mathbf{r}_{l\kappa} - \mathbf{r}) \right] \exp (i\mathbf{k} \cdot \mathbf{r})$$

A comparison of this equation with the Landau theory shows that each $Q_j(\mathbf{k})$ is an order parameter and that the term in brackets [ ] is a function with the lattice periodicity. With the assumption that for a second-order phase transition, a term in the free energy which is quadratic in $Q_j(\mathbf{k})$ must vanish, and conjecturing that it is very improbable that at a particular temperature this should occur for more than one $(\mathbf{k}j)$, the two approaches become identical.

The corresponding form for the free energy as given by Cowley[18] and by Miller and Kwok[19] is

$$G = \tfrac{1}{2} \sum_{kj} \omega_j^2(\mathbf{k}) |Q_j(\mathbf{k})|^2$$
$$+ \frac{1}{n!} \sum_{n>2} \sum_{k_1 \ldots k_n j_1 \ldots j_n} V_{j_1 \ldots j_n}^{(n)}(\mathbf{k}_1 \ldots \mathbf{k}_n) Q_{j_1}(\mathbf{k}_1) \ldots Q_{j_n}(\mathbf{k}_n)$$

A comparison of this equation for the single relevant phonon $(\mathbf{k}j)^0$ with the Landau theory reveals that the phonon frequency $[\omega_j^2(\mathbf{k})]^0$ corresponds to the quantity $A^{(n_0)}$ and that the phonon amplitude $[Q_j(\mathbf{k})]^0$ corresponds to the order parameter η. Therefore, the lattice dynamical analysis predicts that

(1) the frequency of the soft-mode phonon approaches zero as T approaches T_c;
(2) the distortion of the lattice below T_c is brought about by the freezing out of the eigenvector of the same soft-mode phonon.

The Landau and Lifshitz conditions place severe restrictions on the possible soft-mode candidates, particularly if the low-temperature space group is known. They restrict the points in $\mathbf{k}$-space that need be considered to those of significant symmetry such as the Γ, X, R, M ... points in the Brillouin zone. In particular, from the Lifshitz condition it follows that only those second-order transitions are possible for which the group of the wave vector under consideration contains the inversion operation.

For the R_2MX_6 structure one is thereby limited to representations at the Γ, X and L points of the Brillouin zone. The L point can be eliminated as a result of experiments[11] which have shown that none of the distorted phases of R_2MX_6 structures investigated to date possess more than four molecules per primitive unit cell. The Landau condition shows that only the X^{2+}, X^{4+} and Γ^{15+} modes do not possess third-order invariants. Of these remaining candidates the X^{2+} mode predominantly involves internal motions of the MX_6 octahedra. This mode is expected to be relatively high in energy and therefore an extremely unlikely candidate for a soft mode. The other two possibilities correspond to zone-centre and zone-boundary longitudinal rotary-mode phonons.

Several authors[20,21] have tabulated the possible soft-mode candidates for the RMX_3 structure. If one limits one's consideration to the tilting type of displacive phase transition the only possible candidates involve one or more of the branches of the triply degenerate R^{25} phonon or the M^3 phonon, both of which are describable as pseudo-rotary modes.

Finally, a comment about the temperature dependence of the order parameter is appropriate. According to the previous discussion $A^{(n_0)}(T_c) = 0$, $A^{(n_0)}(T) < T$ for $T < T_c$ and $A^{(n_0)}(T) > 0$ for $T > T_c$. It is therefore reasonable to assume that sufficiently near to T_c the temperature dependence $A^{(n_0)}(T)$ is of the form

$$A^{(n_0)}(T) = A_0(T - T_c)$$

Since minimization of the free energy with respect to η requires that $\eta^2 = 0$ for $T > T_c$ and $\eta^2 = -A/2C$ for $T < T_c$, the Landau theory predicts that the temperature dependence of the order parameter be given by

$$\eta^2 = C_0(T - T_c)$$

where C_0 is a constant. In other words, the critical exponent β associated with the order parameter in the Landau theory is $1/2$. Further considerations[22] have shown that the critical exponents associated with the compressibility are $\gamma = \gamma' = 1$ and with the specific heat at constant volume are $\alpha = \alpha' = 0$ for a mean field theory.

F. Limitations of the Landau Theory

Experiments have shown that the Landau theory breaks down for temperatures very close to T_c. This is to be expected since the Landau theory is a mean field theory in which fluctuations are neglected. Once the fluctuations in η become comparable to the average value of η, deviations from Landau behaviour should be observable. However, it is still possible to use the Landau theory outside of the region of critical fluctuations by assuming that, although the fluctuations cause a renormalization of the transition temperature, the expansion of the free energy can be regarded as taking place about an apparent transition temperature T_c'. Also near the transition temperature the critical exponent β associated with the order parameter deviates from the Landau theory value of $1/2$. In $SrTiO_3$, T_c' is displaced by $\sim 5\%$ from T_c as a result of critical fluctuations and $\beta = 1/3$ in the critical region.[23]

The above provides one example of what is meant by the statement that Landau's theory is incomplete rather than incorrect. A second example which involves the coupling of phonon modes has been discussed by Axe *et al.*[24] A second-order phase transition induced by a single soft mode and describable by a well-defined order parameter, can have other order parameters associated with modes that are coupled to the driving mode. Consider for example a crystal in which the essential terms in the free energy are

$$G = \tfrac{1}{2}A_1\eta_1^2 + \tfrac{1}{4}C_1\eta_1^4 + \tfrac{1}{6}D_1\eta_1^6 + \tfrac{1}{2}A_2\eta_2^2 + H_{12}\eta_1^2\eta_2$$

If A_1 is taken to be proportional to $(T - T_c)$ and all other coefficients are assumed to be temperature independent, then lattice stability implies that

$$G = \tfrac{1}{2}A_1\eta_1^2 + \left(\tfrac{1}{4}C_1 - \tfrac{1}{2}\frac{H_{12}^2}{A_2}\right)\eta_1^4 + \tfrac{1}{6}D_1\eta_1^6$$

Provided that $C_1 > 2H_{12}^2/A_2$, the crystal will undergo a second-order phase transition at $T = T_c$ with $\eta_1 \propto (T_c - T)^{1/2}$ and $\eta_2 \propto (T_c - T)$ in the low-temperature phase. The parameter η_1 is the primary-order parameter and η_2 the secondary-order parameter. At the phase transition, the frequency corresponding to the primary-order parameter becomes zero. The cubic to tetragonal ferroelectric transition in $BaTiO_3$ has been discussed in this manner, η_1 being taken as the spontaneous polarization and η_2 the strain. In both the perovskite and antifluorite compounds, the tilting transitions are accompanied by a small tetragonal distortion of the lattice leading one to imagine that a secondary-order parameter may be involved in these transitions as well.

G. Beyond the Landau Theory

Since the appearance of the lattice dynamical interpretation of Landau's theory, more sophisticated models for displacive phase transitions have appeared. These new approaches go beyond the pseudoharmonic lattice. For example, anharmonic corrections to the theory and phonon renormalization by anharmonic interactions have been introduced. Among the more extensive developments along these lines is that by Pytte and Feder.[25] They have put forward a microscopic theory based on a model Hamiltonian and including correlations. They were able *ab initio* to determine the temperature dependence of the soft modes and the order parameter through an evaluation of correlation functions. Still, the results are mostly within the mean field category and as such are in substantial agreement with the consequences of the Landau expansion.

More recently Cowley and Bruce[26] have proposed a theory making use of the techniques of the renormalization group to account for the critical phenomena in RMX_3 compounds.

III. STATIC INVESTIGATIONS

A. Nuclear Quadrupole Resonance Line Splittings

The Hamiltonian for the interaction of the quadrupole moment Q of a nucleus with its local electric field gradient ∇E is given by the tensor–scalar product

$$\mathcal{H}_Q = -Q : (\nabla E)$$

In a pure nuclear quadrupole resonance experiment, the frequencies ν of the transitions are determined by the secular part of this Hamiltonian. In terms of a set of principal axes, only two parameters are required to characterize the electric field gradient. These are the field gradient parameter eq and the asymmetry parameter η. For a spin $3/2$ nucleus with scalar quadrupole moment Q

$$\nu = \frac{1}{2} \cdot \frac{e^2 qQ}{h} (1 + \eta^2/3)^{1/2}$$

and each resonance line in the spectrum corresponds to a set of nuclei in a unique physical environment. Therefore, when a structural change is accompanied by alterations in the local electric field gradients, resonance lines are displaced or they split. The earliest applications of NQR to the study of phase transitions involved the observation of line splittings. This is a very sensitive technique for the detection of changes in atomic positions—more sensitive than X-ray or neutron scattering. To illustrate this fact, consider a point charge *e* at a distance *r* from a resonant nucleus. It follows that

$$\left|\frac{\Delta r}{r}\right| = \left|\frac{\Delta \nu}{3\nu}\right|$$

For $\Delta\nu \sim 6\,\text{kHz}$ and $\nu \sim 20\,\text{MHz}$, $|\Delta r/r| \sim 10^{-4}$. That is, a change in atomic displacements of $\sim 0.01\%$ is easily detected.

Magnetic resonance techniques provide a microscopic probe for the study of phase transitions. Therefore, they yield more detailed information than specific heat, thermal expansion and other measurements of thermodynamic quantities. Although the observation of line splittings provides information about which atoms move, such observations cannot in general provide the exact symmetry of the new phase.

In the past, experimenters have often commented on the order of a transition from the continuity or otherwise of the NQR frequencies at the transition temperature. Although the observation of hysteresis can confirm the first-order character of a transition, the absence of hysteresis does not necessarily imply a second-order nature.

For 'nearly' continuous transitions, characterized by small rates of change of frequency with temperature ($\partial\nu/\partial T$), the accurate location of the transition temperature is limited by the resonance line width ($1/T_2^*$). For example, for the chlorine resonance in K_2ReCl_6 at 110.9 K, $\partial\nu/\partial T \sim 0.15\,\text{kHz K}^{-1}$ on the high-temperature side and $\sim 5\,\text{kHz K}^{-1}$ on the low-temperature side, and $(1/T_2^*) \sim 10\,\text{kHz}$. In this instance, T_c can be measured only to within a few tenths of a degree. In favourable cases, however, T_c can be located to within 0.001 K.

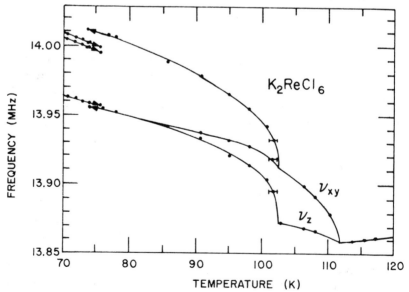

Fig. 8. Temperature dependence of the ^{35}Cl NQR frequency in K_2ReCl_6.

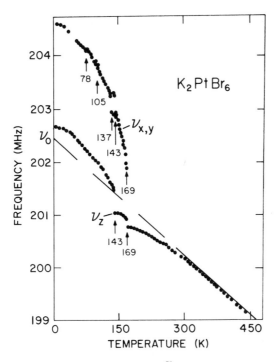

Fig. 9.   Temperature dependence of the ^{79}Br NQR frequency in K_2PtBr_6.

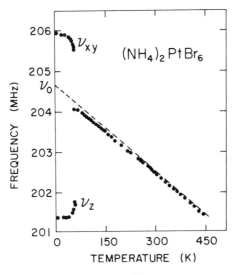

Fig. 10.   Temperature dependence of the ^{79}Br NQR frequency in $(NH_4)_2PtBr_6$.

Kubo, Nakamura, Ikeda and co-workers[27,28] have identified many structural phase transitions in R_2MX_6 compounds from measurements at liquid nitrogen, dry ice and room temperature. Detailed measurements of the line splittings as a function of temperature are available for a number of these compounds. Examples of the ^{35}Cl resonance[11] in K_2ReCl_6 and the ^{79}Br resonances[29,30] in K_2PtBr_6 and $(NH_4)_2PtBr_6$ are shown in Figs. 8, 9, and 10. Figure 8 shows that K_2ReCl_6 undergoes several phase transitions. The high-temperature transition at 110.9 K is an example of a 'nearly' continuous transition. The two lines below 110.9 K have intensities in the ratio 2 : 1 which is consistent with a lowering of the symmetry from cubic to tetragonal. The transition at 76 K shows hysteresis and is therefore a first-order transition. Figure 9 indicates that K_2PtBr_6 also undergoes several phase transitions. The high-temperature transition at 169 K shows a slight discontinuity but also exhibits second-order behaviour as evidenced by the temperature variation of the line splitting below 169 K. Figure 10 shows a single-phase transition in $(NH_4)_2PtBr_6$ at 58 K. The discontinuity at the transition temperature is clear.

The O'Leary and Wheeler[11] model to describe the high-temperature phase transition in R_2MX_6 compounds has already been mentioned in Section II, Part B. The angle of rotation ϕ of the equilibrium positions of the MX_6 octahedra away from their equilibrium positions in the high-temperature phase is taken as the primary-order parameter. The relative change in the lattice constant $\Delta a/a$ in the direction of the axis of distortion is a secondary-order parameter. The observed splitting of the resonance lines below T_c reflects the change in the order parameters with temperature. A point-charge calculation predicts that

$$\Delta \nu = A\phi^2 + B(\Delta a/a)$$

where $\Delta \nu = |\nu_{xy} - \nu_z|$ is the line splitting below T_c and A, B are constants. If one makes the reasonable assumption that the anion contacts determine the cage size then $\Delta a/a \propto \phi^2$ and

$$\Delta \nu = A'\phi^2$$

An analysis of the NQR data for K_2ReCl_6 in the range $103 < T < 110.9$ K shows that $|\nu_{xy} - \nu_z|$ varies linearly with temperature as shown in Fig. 11. This indicates that the critical exponent associated with the order parameter ϕ is $\beta = 1/2$ in agreement with the Landau theory. The extrapolated temperature of the phase transition is ~ 115 K so that $\phi \propto (T - T_0)^{1/2}$ with $T_0 \sim 115$ K. Insufficient data are available to say anything about the temperature dependence of ϕ close to T_c.

Substitution of the appropriate parameters in the detailed expression[30] for $\Delta \nu$ yields the values $\phi \sim 4°$ in K_2ReCl_6, $\phi \sim 7°$ in K_2PtBr_6 and $\phi \sim 11°$ in $(NH_4)_2PtBr_6$, a few degrees below T_c in each case. These values for K_2ReCl_6 and $(NH_4)_2PtBr_6$ are in remarkable agreement with independent determinations using other experimental techniques.[11,31]

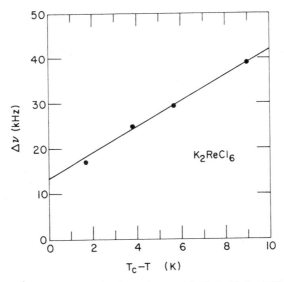

Fig. 11. Temperature dependence of $\Delta\nu = (\nu_{xy} - \nu_z)$ in K_2ReCl_6 for $103 < T < 110.9$ K.

In the foregoing discussion it has been implicitly assumed that the asymmetry parameter η is zero in the cubic phase of R_2MX_6 compounds at the X nuclear sites and is sufficiently small in the distorted phase to be safely neglected. Model calculations (see Section III, Part E) indicate that $\eta \leq 0.003$ in the distorted phase.

For a spin $5/2$ nucleus, two resonance lines occur for each value of the local electric field gradient. Therefore it is possible to determine e^2qQ/h and η separately at each temperature. A systematic investigation of ^{127}I resonances in K_2MI_6 compounds is presently under way in our laboratory.

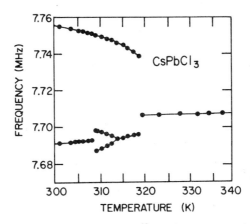

Fig. 12. Temperature dependence of ^{35}Cl NQR frequency in $CsPbCl_3$.

There have been fewer pure NQR studies of RMX_3 compounds. The most frequently studied perovskites are those for which the X nuclei are either ^{16}O or ^{19}F. These nuclei do not possess quadrupole moments. One exception, however, is the ^{35}Cl resonance[32] in $CsPbCl_3$. Figure 12 shows the temperature variation of the resonance frequency over the range $300 < T < 340$ K. Phase transitions are indicated at three temperatures. The nature of the high-temperature transition has been the subject of a considerable controversy.[33-36] Recent neutron scattering[37] and X-ray data[38] have, however, clearly shown that this transition is of the pure tilting variety with an associated tetragonal distortion and is driven by the softening of the pseudo-rotary mode at the M point in the Brillouin zone. The two lower-temperature phase transitions result from the freezing out of two branches of the mode that corresponds to the R^{25} mode in the cubic phase.

The NQR frequency data are consistent with the model proposed to account for the neutron scattering and X-ray results.

B. Quadrupole Coupling Constants for Samples in Applied Static Magnetic Fields

The Hamiltonian for a nucleus of spin $I\hbar$ and quadrupole moment Q in an applied static magnetic field H is given by

$$\mathcal{H} = -\gamma\hbar\mathbf{H}.\mathbf{I} - Q:(\nabla E)$$

If the quadrupole coupling is weak compared to the magnetic interaction (Case I), the effect of the quadrupole interaction is to shift the Zeeman levels by an amount proportional to the quadrupole coupling constant (e^2qQ/h). In the high-temperature cubic phase, the electric field gradients at both the R and M nuclear sites in R_2MX_6 and RMX_3 compounds are rigorously zero since the site symmetries are cubic. In the distorted phase, small non-zero quadrupole coupling constants occur which may be related to an order parameter associated with the phase transition. The information obtained from studies of the R and M nuclei is complementary to that obtained from studies of the X nuclei.

To date no such NMR investigations of R or M nuclei in R_2MX_6 compounds have been reported. Several, however, have been carried out in RMX_3 substances. One such example[39] is in $KMnF_3$. Measurements of the ^{39}K resonance below T_c give the temperature dependence of the quadrupole coupling constant. These results are shown in Fig. 13. The c/a ratio has been measured as 1.005. This distortion is not nearly enough by itself to account for the observed coupling constant. The rotation of the MnF_6^{4-} octahedra, however, results in a contribution

$$|eq| \sim 30|1 - \gamma_\infty|a^{-3}\phi^2$$

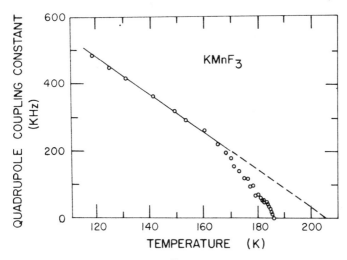

Fig. 13. Temperature dependence of the ^{39}K quadrupole coupling constant in KMnF$_3$.

to the electric field gradient. In this expression, γ_∞ is the Sternheimer antishielding factor, a the lattice constant and ϕ the rotation angle of the octahedron. This contribution is almost certainly the dominant one and therefore the temperature dependence of the quadrupole coupling constant reflects the temperature dependence of ϕ^2 where ϕ is the order parameter for the transition. From Fig. 13, it follows that in the range $120 < T < 170$ K, $\phi \propto (T_0 - T)^{1/2}$ with $T_0 \simeq 205$ K, in agreement with the Landau theory. Near to $T_c = 186$ K, $\phi \propto (T_c - T)^{1/3}$, as expected for a transition which is accompanied by large critical fluctuations. It might be noted that for this example the transition temperature is renormalized by $\sim 10\%$ (as compared to $\sim 5\%$ in SrTiO$_3$). This result is interpreted as evidence that the critical fluctuations in KMnF$_3$ tend to be two-dimensional (perpendicular to the tetragonal axis) in character.

Measurements of quadrupole coupling constants are also able to provide an indication of the relative phases of rotation, of the MX$_6$ octahedra between planes in the perovskite structures.[40] If the rotations of octahedra are in the opposite sense in adjacent planes, then it can be concluded that the structure must have resulted from the freezing in of the R^{25} $[k = (2\pi/a)(\frac{1}{2}, \frac{1}{2}, \frac{1}{2})]$ pseudo-rotary mode so that at an R site

$$\frac{e^2qQ}{h} \propto \phi^2, \qquad \eta = 0, \qquad z \| c$$

where η is the asymmetry parameter and z the direction of that principal axis of the electric field gradient along which the magnitude of the field gradient is a maximum. On the other hand, if the rotations of octahedra are in the same sense in adjacent planes, then it can be concluded that the structure results from the freezing in of an $M[(k=(2\pi/a)(\frac{1}{2},\frac{1}{2},0)]$ pseudo-rotary mode so that

$$\frac{e^2qQ}{h}\propto\phi,\qquad \eta=1,\qquad z\perp c$$

In LaAlO$_3$ the quadrupole coupling constants[41] at both La and Al sites are such that $(e^2qQ/h)\propto\phi^2$ indicating that the softening of an R mode is responsible for initiating the transition. In the non-stoichiometric perovskites, Na$_x$WO$_3$, the quadrupole coupling constant at the Na sites is such that $(e^2qQ/h)\propto\phi$ indicating that the softening of an M mode is in this case responsible for the transition.

If the quadrupole coupling is strong compared to the magnetic interaction (Case II), the effect of the magnetic interaction is to split the pure quadrupole levels. For a spin 3/2 nucleus, the transition frequencies between the Zeeman split states are given by

$$\nu=\frac{1}{2}\frac{e^2qQ}{h}\pm\left(\frac{\gamma H}{2\pi}\right)(3-f)\cos\theta$$

where $f=[1+4\tan^2\theta]^{1/2}$ with θ the angle between H and the high-symmetry axis of the electric field gradient. The observation of these transitions for the chlorine nuclei in a single crystal of K$_2$ReCl$_6$ has been used to measure the

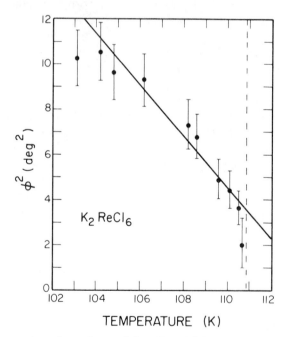

Fig. 14. Temperature dependence of the square of the rotation angle ϕ of the ReCl$_6^{2-}$ octahedra in K$_2$ReCl$_6$ below the 110.9 K phase transition.

rotation angle ϕ of the $ReCl_6^{2-}$ octahedra below the 110.9 K phase transition. In the cubic phase, the crystal is aligned to make θ equal to zero. In the tetragonal phase, the equilibrium positions of the $ReCl_6^{2-}$ octahedra rotate about the axis of distortion. The amount of rotation at any temperature can be measured[31] by rotating the crystal about the distortion axis so as to obtain again the $\theta = 0$ condition. Figure 14 shows a plot of the square of the rotation angle ϕ versus temperature. The solid line is consistent with the expression $\phi \propto (T - T_0)^{1/2}$ with $T_0 = 114$ K. That is, the temperature dependence of the order parameter ϕ may be described by the Landau theory to within 0.4 K of the actual transition temperature.

C. Nuclear Quadrupole Resonance Detection of Soft Modes

The NQR frequency changes with temperature even within a temperature range in which no phase transitions occur. The observed variation can result from various effects including: (1) the change in the amplitudes of the lattice vibrations responsible for the time averaging of ∇E; (2) the softening of a low-lying lattice mode indicating the onset of a phase transition; (3) the change in volume due to thermal expansion; (4) the destruction of π-bonding. In this section the first two of these effects will be considered. That is, a constant volume theory applicable to compounds in which π-bonding effects are negligible will be discussed.

In the first model proposed to account for the temperature dependence of the NQR frequency, Bayer[42] assumed that the time averaging of the electric field gradient arose from a single torsional mode bending the bond between the particular atom of interest and a neighbouring atom. The resonant nucleus was assumed to experience an axially symmetric electric field gradient in both the presence and absence of lattice vibrations. Treating the torsional mode as an Einstein oscillator, Bayer showed that the temperature variation of the NQR frequency could be described by the equation

$$\nu = \nu_0 \left[1 - \frac{3\hbar}{4I\omega_t} \coth \left(\frac{\hbar\omega_t}{2k_B T} \right) \right]$$

The torsional oscillation frequency ω_t may often be identified with one of the low-frequency lattice modes observed by conventional spectroscopic methods. The moment of inertia I for the motion is taken as the appropriate geometrical value and ν_0 as the static lattice resonance frequency. Although this single-mode model is rather unsatisfactory on physical grounds, its simplicity is appealing and it has been used more or less successfully as the basis for the analysis of a large amount of NQR data.

Wang[43] extended the Bayer theory to provide a more realistic treatment of torsional oscillation modes. In the simplest case, the atom containing the resonant nucleus is assumed to execute torsional oscillations about two principal axes of the electric field gradient tensor. Assuming axial symmetry with respect to an axis labelled z, this model leads to the equation

$$\nu = \nu_0\{1 - \tfrac{3}{2}[\langle\theta_x^2\rangle + \langle\theta_y^2\rangle]\}$$

The mean square angular displacements $\langle\theta_x^2\rangle$ and $\langle\theta_y^2\rangle$ of appropriate Einstein oscillators are given by

$$\langle\theta_i^2\rangle = \frac{\hbar}{2I_\alpha\omega_\alpha}\coth\left(\frac{\hbar\omega_\alpha}{2k_BT}\right)$$

The most general constant volume theory is due to Kushida.[44] In his theory, the angular displacements of the principal axes of the field gradient are expanded in normal coordinates of the lattice modes. The resultant expression for the NQR frequency for an axially symmetric field gradient is

$$\nu = \nu_0\left[1 - \frac{3\hbar}{4}\sum_i \frac{A_i}{\omega_i}\coth\left(\frac{\hbar\omega}{2k_BT}\right)\right]$$

where i runs over the normal modes. The quantities A_i contain the expansion coefficients; in general, they have no simple physical interpretation. For a torsional oscillation mode in which the motion is perpendicular to the axis of symmetry of the field gradient, A_i can be identified as the reciprocal moment of inertia for the motion and the effect of such a mode on the NQR frequency is the same as predicted by Bayer.

In the application of the Kushida expression, it is generally assumed that bond stretching motions are of negligible importance relative to bond bending motions in averaging out the electric field gradient. This is a reasonable statement for several reasons. Bond stretching modes cause only slight alterations in the electron distribution around the nucleus on the time scale of an NQR experiment. The amplitudes of stretching motions are so small as to be relatively insignificant in modifying the electric field gradient tensor. This is reflected by the high-temperature contribution of such modes which is proportional to (T/ω_i^2). Bond bending modes time-average the electric field gradient by means of an orientational averaging of the principal axes of the field gradient tensor relative to a set of crystal-fixed axes. To lowest order, the largest component q_{zz} of the diagonalized field gradient tensor appears as

$$q_{zz} = q_0[1 - 3\langle\theta^2\rangle]$$

where $\langle\theta^2\rangle$ is the mean square angular displacement of the bond axis in a direction perpendicular to the bond.

NQR frequency measurements therefore should provide a useful technique for the study of low-lying librational modes. The model proposed for the high-temperature phase transitions in the antifluorite structures suggests that these transitions are triggered by the softening of a rotary lattice mode. In the cubic phase of R_2MX_6 substances the rotary mode is both infrared and Raman inactive. O'Leary[12] suggested that NQR offered a practical means to observe the softening of the rotary mode. The field gradient at an X nucleus may be considered as the sum of two parts, a contribution q_{ci} from the MX_6 complex ion to which the X nucleus belongs and a contribution q_{ni} from the neighbouring ions, in particular the R ions which define the cage in which the MX_6 ion resides. The two contributions are of opposite sign. A computer calculation has shown that the latter contribution at an X site in R_2MX_6 is $\sim 5\%$ of the former. It is assumed that the lattice mode contribution to the electric field gradient in a substance such as K_2PtBr_6 has an even smaller effect in determining the temperature dependence of the NQR frequency since the translatory lattice modes, which involve motions of the K^+ ion cage relative to the Br nuclei, have much higher frequencies than the rotary modes. This may not be a very good assumption; it is certainly a bad assumption for R_2MCl_6 compounds (see Section III. E). Nonetheless, in what follows only the bending components of the internal modes and the rotary lattice modes will be considered.[45]

From the theory of lattice dynamics in the harmonic approximation,[46] the mean-square angular displacement $\langle \theta_\alpha^2 \rangle$ of the M—X bond is

$$\langle \theta_\alpha^2 \rangle = \frac{\hbar}{2NR^2} \sum_{kj} \left| \frac{e_\alpha(X|\mathbf{k}j)}{m_X^{1/2}} - \frac{e_\alpha(M|\mathbf{k}j)}{m_M^{1/2}} \right|^2 \left[\coth\left(\frac{\hbar\omega_j(\mathbf{k})}{2k_BT} \right) \right] \frac{1}{\omega_j(\mathbf{k})}$$

N is the total number of unit cells in the lattice, R is the M—X bond length, m_κ is the mass of nucleus κ; $e_\alpha(\kappa|\mathbf{k}j)$ is the αth component of the polarization vector associated with the displacement of nucleus κ in the normal mode of branch j and frequency $\omega_j(\mathbf{k})$ at point $\mathbf{k}$ in the Brillouin zone. The summation is taken over the fifteen internal modes of an MX_6 ion and the three rotary lattice modes. The frequency shift $\Delta\nu_{exp}(T)$ from the static lattice resonance frequency ν_0 at temperature T can be written as

$$\Delta\nu_{exp}(T) = \Delta\nu_{int}(T) + \Delta\nu_{rot}(T)$$

where

$$\Delta\nu_{int}(T) = \frac{3\hbar\nu_0}{2NR^2} \sum_{kj}^{\text{int modes}} \left| \frac{e_\alpha(X|\mathbf{k}j)}{m_X^{1/2}} - \frac{e_\alpha(M|\mathbf{k}j)}{m_M^{1/2}} \right|^2 \left[\coth\left(\frac{\hbar\omega_j(\mathbf{k})}{2k_BT} \right) \right] \frac{1}{\omega_j(\mathbf{k})}$$

$$\Delta\nu_{rot}(T) = \frac{3\hbar\nu_0}{2NR^2} \sum_{kj}^{\text{rot modes}} \frac{e_X^2(X|\mathbf{k}j)}{m_X} \left[\coth\left(\frac{\hbar\omega_j(\mathbf{k})}{2k_BT} \right) \right] \frac{1}{\omega_j(\mathbf{k})}$$

The contribution $\Delta\nu_{int}(T)$ can be evaluated assuming the normal modes of an isolated MX_6 molecule (see Fig. 15), neglecting dispersion, which is expected to be small, and substituting the vibrational frequencies which are in many instances available from infrared and Raman experiments.[47] The difference between $\Delta\nu_{exp}(T)$ and $\Delta\nu_{int}(T)$ is taken to be the contribution $\Delta\nu_{rot}(T)$. In the high-temperature limit and in the approximation that the longitudinal and

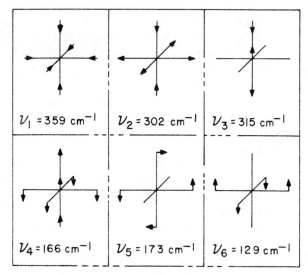

$V_1 = 359$ cm^{-1} $V_2 = 302$ cm^{-1} $V_3 = 315$ cm^{-1}

$V_4 = 166$ cm^{-1} $V_5 = 173$ cm^{-1} $V_6 = 129$ cm^{-1}

Fig. 15.   Schematic representation of the $\mathbf{k} = 0$ internal modes of an R_2MX_6 crystal. The frequencies quoted are those for K_2ReCl_6.

transverse rotary modes are degenerate,

$$\Delta\nu_{\text{rot}}(T) = \frac{3\hbar\nu_0}{2NI}\sum_{\mathbf{k}}\frac{2k_BT}{\hbar\omega_{\text{rot}}^2(\mathbf{k})} = \frac{3\nu_0 k_B T}{I}4\pi V\int_{\mathbf{k}}\frac{k^2\,dk}{\omega_{\text{rot}}^2(\mathbf{k})}$$

where I is the moment of inertia of the MX_6 octahedron about a principal axis and V is the volume of the unit cell in the actual lattice. The integration is performed through one unit cell in the reciprocal lattice. The neglect of the $\mathbf{k}$ dependence of the polarization vector is equivalent to the assumption that the rotary mode undergoes negligible mixing with other lattice modes away from $\mathbf{k} = 0$. This is reasonable since the only modes that can mix with the rotary mode are separated from it by a large energy gap. Neglect of the dependence of $\omega(\mathbf{k})$ on the direction of $\mathbf{k}$ is justified on the basis of a model calculation of the dispersion curves (see Section III, Part E). Finally, we define a zone-averaged rotary mode frequency

$$\bar{\omega}_\nu = \left[4\pi V\int_{\mathbf{k}}\frac{k^2\,dk}{\omega_{\text{rot}}^2(\mathbf{k})}\right]^{-1/2}$$

and write the expression for $\Delta\nu_{\text{rot}}(T)$ in the form

$$\Delta\nu_{\text{rot}}(T) = \frac{3\nu_0 k_B T}{I\bar{\omega}_\nu^2}$$

If the rotary mode softens at some point in the Brillouin zone as the transition temperature is approached from above, then $\bar{\omega}_\nu$ should show a corresponding temperature variation.

202

The ^{79}Br NQR frequency data (see Fig. 9) obtained in the high-temperature phase for K$_2$PtBr$_6$ have been analysed in this manner. A value for ν_0 of 202.47 ± 0.03 MHz was obtained from an extrapolation to $T = 0$ K of the linear fit to the data for $350 < T < 450$ K. Since no infrared or Raman data are available for K$_2$PtBr$_6$, the internal mode frequencies[47] for (NH$_4$)$_2$PtBr$_6$ were substituted. It is well established that the internal mode frequencies in R$_2$MX$_6$ compounds are quite insensitive to the crystalline environment. The contribution of $\Delta\nu_{int}(T)$ accounts for only ~ 10% of $\Delta\nu_{exp}$. The deduced temperature dependence of $\bar{\omega}_\nu$ is illustrated in Fig. 16 As the temperature decreases towards

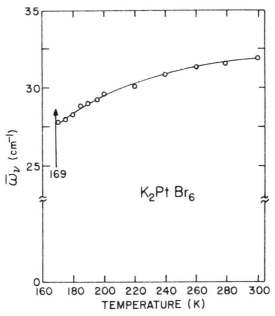

Fig. 16.   Temperature dependence of the average rotary mode frequency $\bar{\omega}_\nu$ in K$_2$PtBr$_6$ as deduced from the NQR frequency data in the high-temperature cubic phase.

the transition, $\bar{\omega}_\nu$ decreases from 32 cm^{-1} at 300 K to 28 cm^{-1} at 170 K. That is, a 12% softening of the average frequency of the rotary lattice mode is indicated. This amount of softening is consistent with a much larger softening of the rotary lattice mode at $k = 0$ and therefore in agreement with the model proposed to account for the structural phase transition in K$_2$PtBr$_6$ at 169 K.

The ^{35}Cl NQR frequency data[48] obtained in a series of R$_2$PtCl$_6$ compounds have also been analysed in this fashion.[49] The frequency shift data $\Delta\nu(T) = \nu_0 - \nu(T)$ for Rb$_2$PtCl$_6$ and K$_2$PtCl$_6$ are shown in Fig. 17. No phase transition is observed in either compound. The solid lines are calculated empirical fits to the data obtained using an iterative procedure with ν_0 and $\bar{\omega}_\nu(T)$ selected as the

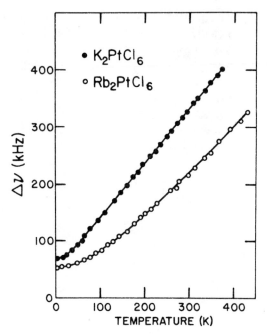

Fig. 17. Temperature dependence of the ^{35}Cl NQR frequency shift data in K$_2$PtCl$_6$ and Rb$_2$PtCl$_6$. The solid lines are calculated empirical fits.

variable parameters. The deduced temperature dependences of $\bar{\omega}_\nu$ are plotted in Fig. 18. For Rb$_2$PtCl$_6$, $\bar{\omega}_\nu$ decreases linearly as the temperature increases. This is just the behaviour to be expected in compounds which are structurally stable. The decrease reflects the reduction in the restoring force experienced by the PtCl$_6^{2-}$ octahedra as the solid expands. For K$_2$PtCl$_6$ the behaviour of $\bar{\omega}_\nu$ is remarkably different from that in Rb$_2$PtCl$_6$ and strongly resembles that deduced for K$_2$PtBr$_6$ (which is also shown in Fig. 18). The obvious conclusion is that the ^{35}Cl NQR frequency data in K$_2$PtCl$_6$ indicate an incipient phase transition.

The basic result of the empirical geometrical model presented in Section II, Part C is that if the radius of the cation is less than a certain fraction of the size of the cavity that it occupies in the prototype phase, then at some lower temperature a reorientation of the anions will occur so as to reduce the size of the cavity. The higher the temperature the larger the effective size of the cation as a result of the increased amplitudes of the lattice vibrations. On the assumption that the mean squared amplitude of the K$^+$ ion motion is proportional to the temperature, it was predicted that the K$_2$PtCl$_6$ structure would experience a structural phase transition near 0 K. The NQR frequency data are in agreement with this prediction.

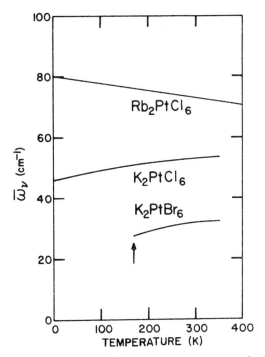

Fig. 18. Temperature dependence of the average rotary mode frequency $\bar{\omega}_\nu$ in Rb$_2$PtCl$_6$, K$_2$PtCl$_6$ and K$_2$PtBr$_6$.

Incipient behaviour has also been observed in the perovskite family in connection with an anticipated ferroelectric transition in KTaO$_3$. The crystal is barely stable against the ferroelectric mode at 4 K.[50,51]

D. Effects of Volume Changes and π-bonding on Nuclear Quadrupole Resonance Frequencies

The contribution of volume changes to the temperature dependence of nuclear quadrupole resonance frequencies was first considered by Kushida, Benedek and Bloembergen.[52] Based on the assumption that $\nu \equiv \nu(V, T)$ they showed that

$$\left(\frac{\partial \nu}{\partial T}\right)_V = \left(\frac{\partial \nu}{\partial T}\right)_P - \left(\frac{\partial V}{\partial T}\right)_P \left(\frac{\partial \nu}{\partial V}\right)_T = \left(\frac{\partial \nu}{\partial T}\right)_P + \frac{\alpha}{\beta}\left(\frac{\partial \nu}{\partial P}\right)_T$$

where α is the isobaric volume thermal expansion coefficient and β is the isothermal compressibility of the solid. In general, measurements of both the temperature and pressure dependence of the NQR frequency and the ratio α/β are required before meaningful comparisons with constant volume theoretical

205

expressions, such as that provided by the Kushida theory, can be made. In the absence of the necessary data, it is often assumed that the lattice frequencies decrease linearly as the temperature increases, in an attempt to take account of volume effects in an empirical fashion. In the previous section we saw that an analysis of the ^{35}Cl NQR constant pressure data in Rb_2PtCl_6 yielded an averaged rotary-mode frequency which indeed decreased linearly with increasing temperature.

To date it has generally been assumed for the perovskites that volume changes have a negligible effect on the behaviour of those quantities related to structural phase transitions, such as transition temperatures, soft mode frequencies and order parameters. However, in several perovskites that exhibit ferroelectric behaviour, it is known that volume effects can be quite large.[53]

For certain of the antifluorites there is strong evidence to suggest that volume effects completely dominate the observed temperature variation of the NQR frequency. The derivatives $(\partial v/\partial P)_T$ have been measured[54-57] for the ^{35}Cl nuclei in the series K_2MCl_6 with $M \equiv Pt$, Ir, Os, Re for a range of temperatures near 300 K and for hydrostatic pressures up to 5 000 kg. cm^{-2}.

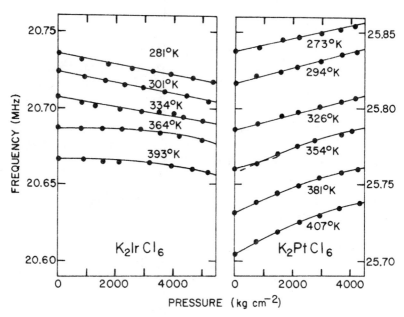

Fig. 19.   Pressure variation of the ^{35}Cl NQR frequencies in K_2IrCl_6 and K_2PtCl_6 at several temperatures.

Some typical data are shown in Fig. 19. From these measurements it was noted that the derivatives were temperature and pressure independent but substance dependent. The $(\partial v/\partial P)_T$ values are listed in Table 1. It should be noted that no

Table 1. Summary of $(\partial\nu/\partial T)_P$ and $(\partial\nu/\partial P)_T$ values in K_2MCl_6 compounds

Substance	$(\partial\nu/\partial T)_P$ (kHz K^{-1})	$(\partial\nu/\partial P)_T$ (kHz kg^{-1} cm^2)
K_2PtCl_6	-0.933 ± 0.014	0.0050 ± 0.004
K_2IrCl_6	-0.570 ± 0.008	-0.0037 ± 0.003
K_2OsCl_6	-0.179 ± 0.008	-0.0071 ± 0.003
K_2ReCl_6	$0.06\ \ \pm0.01$	-0.0112 ± 0.003

measurements were taken near to a phase transition. Also, at temperatures in the vicinity of 400 K, the NQR frequency no longer varies linearly with pressure. This change in behaviour has been shown to be related to the onset of hindered rotations of MCl_6 groups at high temperatures and is of no significance for the present discussion. In the same table $(\partial\nu/\partial T)_P$ values are given as obtained from experimental measurements of the temperature dependence of the resonance frequencies in a range of temperatures about 300 K.

To assess the importance of volume effects for the various members of the K_2MCl_6 series, it is necessary to know the quantity α/β in each case. Unfortunately, very few measurements of either α or β have been reported. However, O'Leary[58] has estimated that $\alpha/\beta \sim 40$ kg cm^{-2} K^{-1} for K_2ReCl_6. Therefore, in K_2ReCl_6 (α/β) $(\partial\nu/\partial P)V \sim -0.49$ kHz K^{-1}. This compares with the value $(\partial\nu/\partial T)_P = 0.06$ kHz K^{-1} near 300 K, obtained from the ^{35}Cl NQR frequency

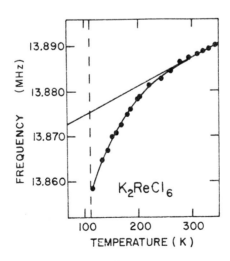

Fig. 20. Temperature dependence of the ^{35}Cl NQR frequency in the high-temperature phase of K_2ReCl_6.

data shown in Fig. 20. That is,

$$\left(\frac{\partial \nu}{\partial T}\right)_V \simeq \left(\frac{\alpha}{\beta}\right)\left(\frac{\partial \nu}{\partial P}\right)_T$$

and the usual approximation $(\partial \nu / \partial T)_V \simeq (\partial \nu / \partial T)_P$ is invalid in this case.

Even if α/β is known for temperatures far from a phase transition, this is not sufficient, since α and β may themselves exhibit a considerable variation in the region of a phase transition. For example, α has recently been measured[59] for single crystals of K_2ReCl_6 and K_2OsCl_6. In K_2OsCl_6, α undergoes an anomalous increase by a factor of three near the structural phase transition at 45 K. In K_2ReCl_6, the behaviour of α near the 110.9 K phase transition is even more dramatic. An increase by a factor of 20 is noted. On the basis of this type of evidence, we conclude that in order to achieve a complete and correct understanding of the phase transitions and the associated lattice dynamics of these compounds, experiments should be carried out to determine the importance of explicit volume effects.

Returning to Fig. 20, it can be noted that the deviation of the NQR frequency for temperatures near to the phase transition at 110.9 from the extrapolated straight line behaviour, is similar to that observed in the prototype phase of K_2PtBr_6 as depicted in Fig. 9. Qualitatively, a softening of the rotary lattice mode frequency in K_2ReCl_6 is indicated. However, in view of the complexity indicated above, quantitative conclusions concerning this phase transition in K_2ReCl_6 will be difficult to obtain.

The Kushida term $(\partial \nu / \partial T)_V$ in the Kushida, Benedek and Bloembergen relation may be altered from the value calculated in the previous section as the result of a second electric field gradient averaging mechanism[60,61] postulated to account for the differences in the behaviour of the NQR data within the K_2MCl_6 series of compounds. Although their electronic structures are very similar, the number of electrons in the antibonding π^* orbital decreases continually as one progresses through the series. In K_2PtCl_6 the π^* orbital is just filled and, from a consideration of the form of the molecular orbitals, it can be concluded that no net π-delocalization of the halogen lone-pair electrons can occur. It is expected, however, that π-bonding should occur in K_2IrCl_6 and increase in a regular manner through K_2OsCl_6 and K_2ReCl_6. Kubo and Nakamura[62] introduced a parameter Π as a quantitative measure of the amount of π-bonding. They assumed that Π is proportional to n, the number of holes in the t_{2g}^* orbital. A reformulation of their theory in terms of the molecular orbital approach[63] shows that, in the absence of sp hybridization, the parameter Π is proportional to nf_π, where f_π is the fractional population in the p_π orbitals of the ligands, which is transferred to the t_{2g}^* orbitals of the metal ion. The Kubo and Nakamura assumption is therefore equivalent to the assumption that f_π is a constant for $M \equiv Ir, Os, Re$. The parameter f_π for the Ir—Cl bonds has been deduced from EPR measurements in $(NH_4)_2IrCl_6$ to be $\sim 8\%$ by Thornley.[64]

Recently it has been shown[65] from measurements in K_2ReCl_6 that $f_\pi \sim 8\%$ for the Re—Cl bonds as well. These results provide experimental verification of the Kubo and Nakamura assumption.

Ikeda, Nakamura and Kubo[60] suggested that the thermal vibrations of the lattice partially destroy the relatively weak π-bonding. As a result the electron charge distribution at the chlorine sites is modified in such a manner as to cause an increase in the time averaged electric field gradient. Since the destruction of π-bonding increases with the temperature, this mechanism predicts an increase in the NQR frequency with increasing temperature. That is, this effect tends to cancel the usual Bayer–Kushida effect.

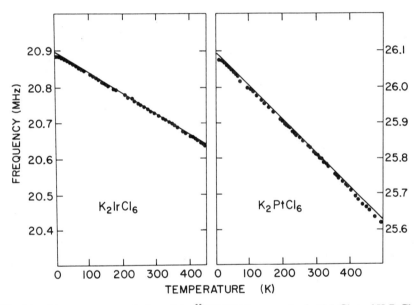

Fig. 21.   Temperature variation of the ^{35}Cl NQR frequencies in K_2IrCl_6 and K_2PtCl_6.

Figure 21 shows the ^{35}Cl NQR frequency data[55] for K_2PtCl_6 and K_2IrCl_6. The slopes of the straight lines give the corresponding $(\partial \nu/\partial T)_P$ values. For a value of $\alpha/\beta \sim 10 \text{ kg cm}^{-2} \text{ K}^{-1}$ it follows that $(\partial \nu/\partial T)_P \simeq (\partial \nu/\partial T)_V$ for these compounds. The influence of π-bonding is clearly evident. The value of $(\partial \nu/\partial T)_P$ is less negative in K_2IrCl_6 than in K_2PtCl_6 as expected on the basis of the π-bonding destruction model.

In Fig. 22 the $(\partial \nu/\partial T)_P$ values from Table 1 are plotted as a function of the number of t_{2g}^* holes in the antibonding π^* molecular orbital of the MCl_6^{2-}

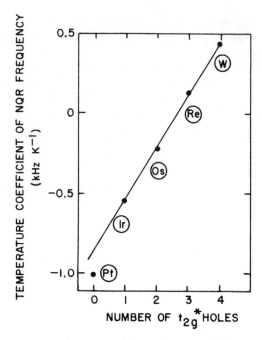

Fig. 22.   Temperature coefficient of ^{35}Cl NQR frequency versus number of π-antibonding orbital vacancies for the K$_2$MCl$_6$ series.

complex ion. This plot suggests that a simple monotonic relation exists between the amount of field gradient averaging by the π-bonding destruction model and the number of t^*_{2g} holes.

The π-bonding destruction mechanism also provides a simple qualitative explanation for the NQR pressure data mentioned above. As the pressure is increased, the amplitudes of the vibrational motions are reduced; therefore, the amount of destruction of π-bonding is reduced. That is, this mechanism makes a negative contribution to $(\partial \nu / \partial P)_T$ in agreement with the observation that $(\partial \nu / \partial P)_T$ is increasingly negative as M $\equiv$ Ir, Os, Re but positive for K$_2$PtCl$_6$. (See Table 1.)

A quantitative formulation of the destruction of π-bonding has been given by Haas and Marram.[61] The bending vibrations tend to alter the overlap between the p-orbital of the halogen and the d-orbital of the metal ion which form the π-bond. Increasing amplitudes of these modes therefore result in a change of the time-averaged population in the chlorine orbitals which is proportional to the mean squared angular amplitude of the bending motion. It follows that

$$\Delta N_\pi(\theta) = 5\xi n \langle \theta^2 \rangle$$

210

where ξ is a constant that depends on the nature of the molecular orbitals of the MCl_6^{2-} complex and the strength of the π-bond. The expression for $(\partial v/\partial T)_V$ can be written in the form

$$\left(\frac{\partial v}{\partial T}\right)_v = v_0\left(\frac{5v_{at}\xi n}{v_0} - 3\right)\frac{\partial\langle\theta^2\rangle}{\partial T}\bigg|_v$$

where v_{at} is the NQR frequency for a free chlorine atom in its ground state.

A comparison of the NQR data for K_2PtCl_6 and K_2IrCl_6, assuming that $(\partial v/\partial T)_P \simeq (\partial v/\partial T)_V$ in these compounds, and taking account of the π-bonding destruction in K_2IrCl_6 by the lattice vibrations, reveals that the rotary lattice mode in K_2IrCl_6 also softens as the temperature approaches 0 K. Conventional spectroscopic measurements in the antiferromagnetic phase of K_2IrCl_6 ($T <$ 3.08 K) give further evidence of the incipient phase transition in this substance (R. G. Wheeler, personal communication). This conclusion is in agreement with the empirical geometrical model of Section II, Part C.

E. Lattice Dynamical Model Calculations: R_2MX_6

The most thoroughly studied of the R_2MX_6 compounds is K_2ReCl_6. Specific heat,[66] thermal expansion,[58,59] X-ray scattering,[11] conventional spectroscopic[11,47,67] and magnetic resonance[11,31,57] data are all available. O'Leary and Wheeler[11] (henceforth referred to as OLW) have constructed a rigid-ion model[68] for the K_2ReCl_6 lattice. In an attempt to verify that the 'tilting octahedra' model provides a correct description of the 110.9 K phase transition and as a means to test some of the lattice dynamical assumptions previously used in the analysis of NQR data, we have used essentially the same model.[69]

In a non-rigid model, it is assumed that the ions are non-polarizable and that they interact via central forces which can be separated into a short range repulsive interaction between nearest neighbours and a long range attractive contribution due to ionic charges. The failure to include ionic polarizabilities in this model means that it cannot hope to provide an adequate description[70] of the high-frequency dielectric constant. Although this represents a serious deficiency in a model to describe a ferroelectric substance, it is not nearly so serious in a model to describe a substance in which a purely displacive phase transition occurs.

An obvious objection to the use of a rigid-ion model to describe an R_2MX_6 compound is the presence of the strongly covalent bonded MX_6^{2-} complexes. Fortunately the influence of the covalent bonding will mainly effect the intra-octahedron force constants and only to a lesser extent the inter-octahedron and octahedron–R ion interactions which are of most interest to our studies. Furthermore the rigid-ion model is known to provide a good reproduction of the dispersion curves[71,72] in CaF_2 and UO_2 which have the same crystal structure as the R_2MX_6 salts. In addition, Dorain[73] has shown that the rigid-ion model applied to K_2ReCl_6 provides a density of states which agrees quite well with the density of states as deduced from optical sideband data.

Using the notation of Woods et al.,[74] the equations of motion for the rigid ion model can be written as

$$M_K\omega^2\mathbf{U}(K) = \sum_{\kappa'}[R(\kappa, \kappa') + Z(\kappa)C(\kappa, \kappa')Z(\kappa')]\mathbf{U}(\kappa')$$

where M_κ is the mass of the κth ion, $\mathbf{U}(\kappa)$ is its displacement associated with a particular normal mode of frequency ω, and $Z(\kappa)$ its charge. The matrices $R(\kappa, \kappa')$ and $C(\kappa, \kappa')$ specify the short-range repulsive and long-range attractive interactions, respectively.

To apply the rigid-ion model to a particular structure, it is necessary to choose which attoms are to interact via the short range force. The rough criterion used is that all atoms within a distance equal to one half the lattice parameter are assumed to interact with one another (see Fig. 23). With the

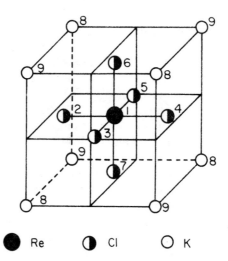

Fig. 23.   Local environment of a $ReCl_6^{2-}$ octahedron in the fcc crystal K_2ReCl_6.

exception of the Re–K pair, all atoms within the unit cell in K_2ReCl_6 are assumed to interact with one another and with symmetry related atoms. In addition octahedron–octahedron interactions are included by allowing a chlorine atom of a given octahedron to interact with the two nearest chlorine atoms on adjacent octahedra (see Fig. 24). These inter-unit cell pairs are not related by symmetry to any intra-unit cell pairs because the positions of the chlorine atoms are not determined by symmetry. The parameters used to specify a particular short-range interaction consist of the first and second derivatives of the central potential—these are referred to as the perpendicular

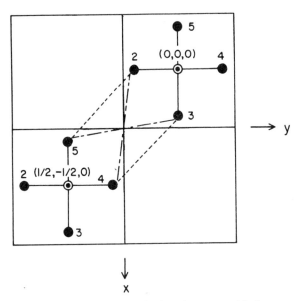

Fig. 24.   A chlorine atom of a given octahedron interacts with the two nearest chlorine atoms on adjacent octahedra.

and parallel force constants respectively. In this manner, fourteen parameters are generated. The inclusion of the Coulomb interaction adds three more parameters, namely the charges on the rhenium, chlorine and potassium atoms. The requirement of charge neutrality reduces the number of independent charges to two. The Coulomb part of the dynamical matrix requires careful consideration since the sums involved converge very slowly and exhibit a discontinuous behaviour as a function of wave vector at zero wave vector. The handling of these sums is achieved by means of the Ewald theta transformation[75] which divides the sum over real space into a sum over real space and a sum over reciprocal space, both of which converge rapidly.

There are two additional constraints which further reduce the number of independent parameters. In order that a chlorine atom be in a position of equilibrium, it is necessary that the total force on the atom be zero (the forces on the other two types of atoms vanish by symmetry). Since a stable crystal must be in a state of zero stress, a constraint is imposed on the perpendicular force constants and charges. Finally, the fitting procedure revealed that the perpendicular force constant between adjacent potassium atoms was too small to be fitted too; it was set equal to zero. Therefore, there are thirteen independent parameters to be fitted. Twenty-two phonon energies at the centre and edge of the Brillouin zone have been measured for the prototype phase of K_2ReCl_6. The corresponding mode eigenvectors are known for fourteen of these. The identification of certain zone-boundary phonon energies

Table 2. Observed lattice modes and their energies for the prototype phase of K_2ReCl_6

Mode	How observed	Observed energy[a]	Calculated energy (OLW Model I)
Γ^{15+} rotary	I.R., Raman	26 cm^{-1}	45.2 cm^{-1}
Tran optic	I.R. (trans., refl.)	75^d	77.0
Γ^{25+}	Raman	84^d	84.9
Long optic	I.R. (trans. + refl.)	106^d	104.9
Tran ν_4	I.R. (trans. + refl.)	166^d	168.4
ν_5	Raman	172^d	171.6
Long ν_4	I.R. (refl.)	180^d	179.4
ν_2	Raman	302^d	300.8
Tran ν_3	I.R. (trans. + refl.)	320	316.7
ν_1	Raman	359^d	359.6
X^{4-} TA	I.R.	43^d	36.7
X^{5+}	Raman[c]	62	63.5
X^{2-}	I.R. (2 phonon)	67	70.9
X^{4+} (rotary)	Raman	68	70.2
X^{5-}	I.R. (refl.)	90	92.8
X^{5+}	Raman[c]	95	96.3
ν_6 zone edge	Sidebands[b]	122^d	122.5
ν_6 zone edge	Sidebands[b]	129^d	129.6
ν_4 zone edge	Sidebands[b]	164^d	166.8
ν_4 zone edge	Sidebands[b]	166^d	168.4
ν_3 zone edge	Sidebands[b]	307^d	307.6
ν_3 zone edge	Sidebands[b]	313^d	312.0
X^{4-} LA	Sidebands	$\leq 65^d$	66.4

[a] measured in the vicinity of 4 K.
[b] indicates the existence of an ambiguity in the assignment of transverse and longitudinal components at the zone boundary.
[c] indicates the existence of an ambiguity in the mode assignments of the two X^{5+} modes.
[d] modes that were used by OLW in the fitting procedure.

in the prototype phase was made possible by the occurrence of the phase transitions. The phase transition at 103 K quadruples the size of the unit cell so that additional modes of infinite wavelength appear in the distorted phase which were located at the Brillouin zone boundary in the prototype phase. A small distortion of the crystal can provide information about the dispersion of various branches of the phonon spectrum. Table 2 summarizes the observed modes, their energies and method of detection. The essential symmetry character of some of these modes is shown in Fig. 15.

Several of the mode energies were shown to be relatively insensitive to changes in the temperature for temperatures from 4 to 100 K even near the 76 K phase transition. On the other hand, the energy of the rotary lattice mode decreases as the temperature increases from 4 K.

Spectroscopic measurements for K_2ReCl_6 have also been reported in the cubic phase by Debeau and Poulet[47] near 300 K and by Woodward and Ware[67] in solution. The internal mode energies reported agree substantially with one another and with the corresponding energies obtained by OLW at low temperatures. These results tend to justify the use of the low-temperature spectroscopic data in the construction of a model for the high-temperature cubic phase of K_2ReCl_6.

Table 2 also contains the energies as deduced by the OLW fitting procedure. The rms deviation for all of the observed modes, exclusive of the Γ^{15+} rotary lattice mode, is 2.4 cm^{-1}. This represents a remarkably good fit for a rigid-ion model calculation.

The dispersion curves deduced for each of the modes are shown in Figs. 25 and 26; the force constants and ionic charges derived from the calculation are listed in Table 3.

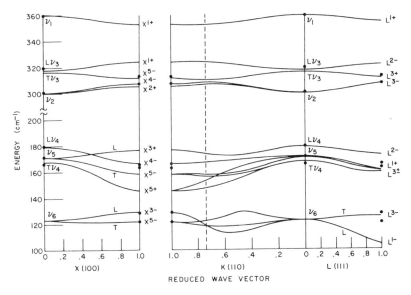

Fig. 25. Dispersion curves for the internal modes calculated from Model I. Symmetry labels and all mode energies deduced from the spectral data are indicated.

To this point, the model has been used in an effort to reproduce the behaviour of the crystal in the cubic phase far away from the phase transition. In an attempt to study the relationship between the rotary-mode frequency and the force constants, a second fitting procedure was considered in which the low-temperature rotary lattice mode frequency of 26 cm^{-1} was used in obtaining the fit. The force constants and ionic charges derived from this model are

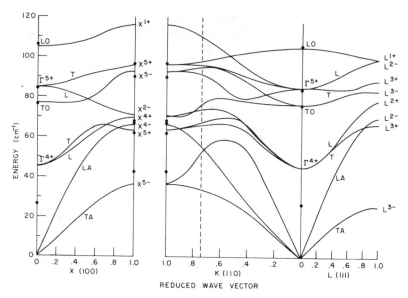

Fig. 26. Dispersion curves for the lattice modes calculated from Model I. Symmetry labels and all mode energies deduced from the spectral data are indicated.

Table 3. Lattice dynamical model parameters for K_2ReCl_6 (OLW Model I). Charges: Re 0.40 e; Cl -0.32 e; K 0.76 e. Force constants: (dyne cm^{-1}).

	parallel	perpendicular
Cl3(0, 0, 0)– Cl4(0, 0, 0)[a]	15 000	−2 700
Cl3(0, 0, 0)– Cl4($\frac{1}{2}$, −$\frac{1}{2}$, 0)	6 300	270
Cl2(0, 0, 0)– Cl4(0, 0, 0)	31 000	−2 000
Cl2(0, 0, 0)– Cl4($\frac{1}{2}$, −$\frac{1}{2}$, 0)	670	−1 500
Cl3(0, 0, 0)– K8(0, 0, 0)	6 700	−1 100
K8(0, 0, 0)– K9(0, 0, 0)	660	0
Re1(0, 0, 0)– Cl4(0, 0, 0)	131 000	14 000

[a] Atomic designations refer to Figs. 23 and 24.

given in Table 4. The rms deviation for all of the observed modes is in this case 2.9 cm^{-1}. OLW concluded from a comparison of Tables 3 and 4 that the force constants most affected (those which undergo the largest fractional changes) by variations in the rotary-mode frequency (as indicated by the boxes) are inter-octahedral in nature. The slight variations in the other force constants reflect the small changes necessary to keep the other lattice mode frequencies near to their original values and to maintain the lattice stability conditions.

216

Table 4. Lattice dynamical model parameters for K_2ReCl_6 (OLW Model II). Charges: Re 0.38 e; Cl -0.32 e; K 0.78 e. Force constants: (dyn cm^{-1}).

	parallel	perpendicular
Cl3(0, 0, 0)– Cl4(0, 0, 0)^a	16 000	$-2\,500$
Cl3(0, 0, 0)–Cl4($\frac{1}{2}$, $-\frac{1}{2}$, 0)	6 500	$\boxed{160}$
Cl2(0, 0, 0)–Cl3(0, 0, 0)	30 000	$-2\,600$
Cl2(0, 0, 0)– Cl4($\frac{1}{2}$, $-\frac{1}{2}$, 0)	$\boxed{290}$	$-1\,500$
Cl3(0, 0, 0)– K8(0, 0, 0)	6 800	$-1\,200$
K8(0, 0, 0)– K9(0, 0, 0)	750	0
Re1(0, 0, 0)– Cl4(0, 0, 0)	130 000	14 000

^{a} Atomic designations refer to Figs. 23 and 24.

Figure 27 shows the dispersion curves for the rotary lattice mode according to both Model I and Model II. It can be noted that as well as the zone centre softening (which is built into Model II), a substantial softening of the longitudinal rotary lattice mode at the zone boundary X point is achieved. OLW concluded that this result may indicate that the phase transition at 103 K is driven by this mode.

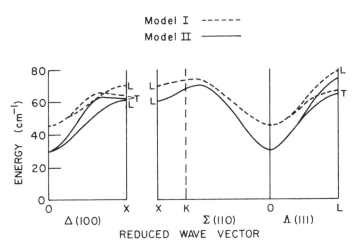

Fig. 27. Dispersion curves for the rotary mode calculated from both Model I and Model II. The labels L and T refer to the longitudinal and transverse branches, respectively.

The model that has been adopted throughout this article to account for the tilting phase transitions in the antifluorite structures (as well as in the perovskites) suggests that a weakening of the interaction between the R and X ions

217

(K^+ and Cl^- in K_2ReCl_6) should drive the transitions. This result is not consistent with the conclusions of OLW Model II. Therefore, the effect of varying only the two force constants associated with the K–Cl interaction was considered.[69] It is indeed possible to drive the zone centre rotary lattice mode from 46 cm^{-1} to zero while not severely affecting the energies of the other modes. At the stability limit, the new values for the parallel and perpendicular K–Cl force constants are 2 600 and 760 dyn cm^{-1}, respectively. The change of these force constants from their high-temperature values (6 700 and −1 100, respectively) is indicative of a weakening and increasingly attractive interaction.

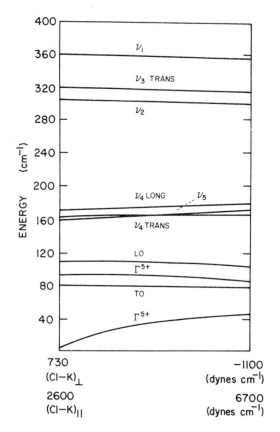

Fig. 28.   The dependence of certain zone-centre phonons on the two critical force constants $(Cl–K)_\parallel$ and $(Cl–K)_\perp$.

The dependence of the rotary-mode frequency and several other normal-mode frequencies on the two critical force constants is shown in Figs. 28 and 29. For other than the rotary lattice mode, the variation of the mode frequency

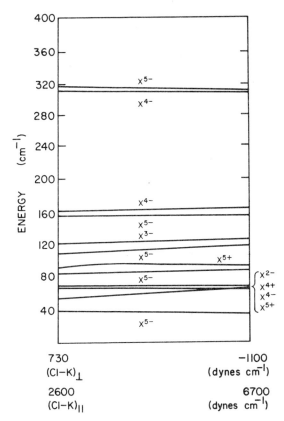

Fig. 29.   The dependence of certain zone-boundary phonons on the two critical force constants $(Cl-K)_{\parallel}$ and $(Cl-K)_{\perp}$.

is seen to be small. By making use of the many degrees of freedom still available in the model, one could presumably maintain the other mode frequencies at relatively fixed values. In our opinion this further degree of sophistication is not warranted in the absence of exact information concerning the temperature dependence of these modes near the phase transition. The simple approach taken is sufficient to indicate that the rigid-ion model presented is consistent with current thinking concerning the nature of the purely displacive phase transitions in R_2MX_6 compounds.

The association of a temperature dependence to harmonic force constants amounts to working in the pseudo-harmonic approximation. The origin of the temperature variation of the force constants lies in the anharmonic interactions.

The rigid-ion model has also been adapted[69] to permit calculation of the mean squared angular amplitude $\langle \theta^2 \rangle$ of the M—X bond in an R_2MX_6 salt. The prime motivation for this calculation was to check the validity of the multiple mode analysis of the NQR frequency data. The extension of the model derived for K_2ReCl_6 to other members of the K_2MCl_6 series of compounds is justified on the basis of the similarity of the observed energies of the optic modes as reported by Debeau and Poulet.[47]

From Section III.C the quantity $\langle \theta^2 \rangle$ is expressed in terms of the theory of lattice dynamics in the harmonic approximation as

$$\langle \theta^2 \rangle = \frac{\hbar}{NR^2} \sum_{\mathbf{k}j} \left| \frac{e_\alpha(X|\mathbf{k}j)}{m_X^{1/2}} - \frac{e_\alpha(M|\mathbf{k}j)}{m_M^{1/2}} \right|^2 \left[\coth \left(\frac{\hbar\omega_j(\mathbf{k})}{2k_BT} \right) \right] \frac{1}{\omega_j(\mathbf{k})}$$

This expression can be evaluated through the use of the dispersion curves presented in Figs. 25 and 26. The principal difficulty in obtaining accurate values for $\langle \theta^2 \rangle$ in such a calculation is the summation over $\mathbf{k}$ which would seem to imply the need for a large grid of points in $\mathbf{k}$-space. Such a procedure is extremely costly in terms of the computer time required. However, a technique has appeared in the literature[76–78] which greatly facilitates the computation of such expressions through the identification of a small number of 'special points' in the Brillouin zone. The locations of these 'special points' are independent of the particular function to be evaluated; they depend only on the symmetry of the lattice concerned. The method can be applied to any totally symmetric function $f(\mathbf{k})$ of wave vector $\mathbf{k}$, which is smoothly varying and which has periodicity $\mathbf{G}$ with $\mathbf{G}$ a reciprocal lattice vector. Such a function can be written as a Fourier expansion in the form

$$f(\mathbf{k}) = f_0 + \sum_{n=1}^{\infty} f_n \left[\sum_{|\mathbf{R}|=c_n} \exp(i\mathbf{k} \cdot \mathbf{R}) \right]$$

where f_0 is the average value of the function $f(k)$, $\mathbf{R}$ is a lattice vector in real space, and the summation indicated by $|\mathbf{R}| = c_n$ is taken over a shell of constant radius $|\mathbf{R}|$. Therefore, if a value $\mathbf{k} = \mathbf{k}^*$ could be found for which the second term on the right-hand side vanishes, then $f(\mathbf{k}^*)$ would simply be equal to f_0. Although this is in general not possible, it is possible to identify a small number of points in $\mathbf{k}$-space such that an appropriately weighted combination of the values of the function at these points will render the second term on the right entirely negligible. For example, for an fcc lattice the use of the two points $k_1 = (\frac{1}{4}, \frac{1}{4}, \frac{1}{4})$ and $k_2 = (\frac{3}{4}, \frac{1}{4}, \frac{1}{4})$ is sufficient to cause the first eight terms to vanish identically. Since $\langle \theta^2 \rangle$ is a smoothly varying periodic function of $\mathbf{k}$, it follows that to an excellent approximation

$$\langle \theta^2 \rangle = \tfrac{1}{4} \langle \theta^2(k_1) \rangle + \tfrac{3}{4} \langle \theta^2(k_2) \rangle$$

For OLW Model I parameters, this expression yields for $T = 300$ K

$$\langle \theta^2 \rangle = 3.47 \times 10^{-3} \text{ rad}^2$$

It is interesting to note that this value receives approximately equal contributions from the $ReCl_6^{2-}$ internal modes, from the lattice modes other than the rotary modes, and from the rotary modes. In the multiple mode analysis, the contribution from the internal modes is taken to be about 1/3 of the total and the contribution from the rotary lattice modes about 2/3 of the total. The contribution from the lattice modes other than the rotary modes is neglected. Clearly this is a bad assumption.

If the calculated value for $\langle \theta^2 \rangle$ is appropriate for K_2PtCl_6, the predicted value of the NQR frequency shift at 300 K is

$$|\Delta \nu| = 3\nu_0 \langle \theta^2 \rangle$$

$$= 273 \text{ kHz}$$

The experimentally deduced value is 305 kHz. Recent calculations of the electric field gradient at a chlorine site suggest that it may be more correct to substitute for ν_0 only that portion of the static lattice resonance frequency attributable to the host $PtCl_6^{2-}$ complex.[69] The effect of such a change is to increase the calculated value of $|\Delta \nu|$ to ~ 290 kHz.

The above calculation is certainly not appropriate to K_2ReCl_6 even far from a phase transition because of the presence of large specific volume effects and because of the π-bonding character of the Re—Cl bonds (see Section III, Part D).

If one associates a temperature of 110.9 K with our soft mode model parameters, the corresponding value for $\langle \theta^2 \rangle$ is 2.0×10^{-3} rad^2 as compared to

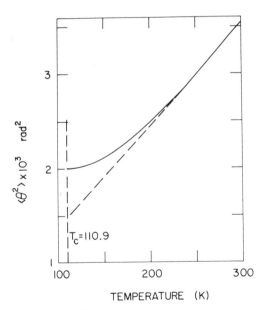

Fig. 30. The theoretically predicted variation of the mean squared angular displacement with temperature near to the structural phase transition in K_2ReCl_6.

1.49×10^{-3} for a stable lattice (OLW Model I). To produce an exact temperature dependence for $\langle \theta^2 \rangle$ between 110.9 K and 300 K would require detailed information on the temperature dependence of the force constants. In the absence of such information, it is reasonable to assume that the deviation of the critical force constants from their values at 110.9 K is proportional to $(T - T_c)$ near to T_c (according to mean field theory predictions) and becomes independent of T for T far from T_c. On this basis, the predicted variation of $\langle \theta^2 \rangle$ (which is proportional to the NQR frequency shift) with temperature is as shown in Fig. 30. This prediction might be compared in a qualitative way with the data on K_2PtBr_6 (see Fig. 9) in which the Pt—Br bonds have no π character and in which specific volume effects are probably not important. The agreement is excellent, thereby suggesting that the model calculations provide an adequate description of the qualitative behaviour of the NQR frequency in the vicinity of a structural phase transition of the tilting variety.

Numerous methods of calculating electric field gradients have been cited in the literature.[79-82] Most of these techniques take special care to define the boundaries of the lattice within which the lattice sums will be performed or the order in which the sub-sums will be carried out so that convergence will be as rapid as possible with a minimum of oscillatory behaviour. Typically, a summation region of diameter 80–100 Å is required to provide convergence to within 1%. The use of the Ewald theta transformation[75] to achieve rapidly convergent expressions for the electric field gradient was noted by de Wette[83] in 1961 but not employed since he concluded that the method was too cumbersome for easy computation. With advances in computer technology, such is no longer the case. We have found[70] that with this method, convergence to within 0.1% is obtained by summing over spheres of diameter ~ 3 times the nearest-neighbour distance in both direct and reciprocal space. Boundary effects are unimportant for such a calculation.

As an example of the use of the theta transformation to calculate electric field gradients, we have made use of the rigid-ion model for the K_2ReCl_6 lattice. First, the lattice contribution to the electric field gradient at a ^{35}Cl site was computed. Employing the Sternheimer antishielding factor $\gamma = -23$, appropriate for a strongly covalent bonded ^{35}Cl ion,[84] it follows that the lattice contribution to the quadrupole coupling constant is $\sim 5\%$ of the observed value. Assuming that the remaining 95% comes from the $ReCl_6^{2-}$ complex and that it is relatively insensitive to lattice distortions, one then finds that in the tetragonal phase the NQR line splitting of ~ 40 kHz corresponds to a rotation of the $ReCl_6^{2-}$ octahedra by $\sim 3°$. Effects of small tetragonal strains ($c/a - 1 < 10^{-2}$) typically associated with these transitions make far smaller contributions to the NQR line splitting. For a $3°$ rotation of the octahedra, the calculated asymmetry parameter is 0.003. One final interesting feature of this example calculation is that whereas the octahedra are rotated through $3°$, the principal axis of the electric field gradient tensor is rotated through $3.45°$, indicating a lead of $\sim 15\%$. This effect continued to angles of rotation of $8°$ where the lead was still ~ 15%. In this context, the experiment on K_2ReCl_6 described in Section III, Part B should be interpreted as yielding values of the order parameter associated with the 110.9 K phase transition which are ~ 15% too large.

IV. DYNAMIC INVESTIGATIONS

A. Resonant versus Relaxational Soft-Mode Behaviour

Nuclear quadrupolar relaxation is governed by the non-secular part of the Hamiltonian that describes the interaction of the quadrupolar moment of a nucleus with its local electric field gradient. Dynamic aspects associated with structural phase transitions often manifest themselves in the fluctuating components of the electric field gradient tensor and thereby in the spin–lattice relaxation processes. The increasing amplitudes of atomic motions as the transition temperature is approached can cause an anomalous behaviour of the spin–lattice relaxation time. Such anomalies in T_1 have been observed for various nuclei in the R_2MX_6 and RMX_3 families. However, the interpretation of these observations has proceeded along different lines for the two families. Although from the static point of view the structural phase transitions in the two families are similar, from the dynamic point of view they can be quite different.

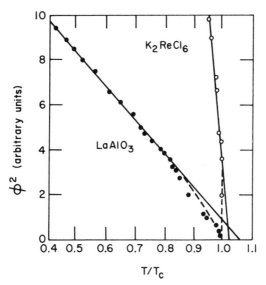

Fig. 31.   Temperature variation of the square of the order parameters in the perovskite $LaAlO_3$ and in the antifluorite K_2ReCl_6.

The temperature variation of the order parameters in the perovskite $LaAlO_3$ and in the antifluorite K_2ReCl_6 are shown in Fig. 31. One can note that a significant deviation of the order parameter from the mean field (Landau) behaviour begins further from T_c in the perovskite than in the antifluorite. This observation is predicted by the Ginzburg formula[17]

$$\varepsilon_c = \frac{T - T_c}{T_c} = \frac{\left(\dfrac{k_B}{\rho \Delta C_p}\right)^2}{l^6}$$

which relates the size ε_c of the critical region to the jump ΔC_p of the specific heat at constant pressure at the phase transition. The quantity ρ is the mass density of the crystal and l the zero temperature coherence length. In terms of the models proposed for the tilting phase transitions in RMX_3 and R_2MX_6 structure, l should be of the order of the separation between adjacent octahedra. This relation gives $\varepsilon_c = 0.1$ for $LaAlO_3$ and $\varepsilon_c \leq 0.01$ for K_2ReCl_6.

The observed anomaly in T_1 in the perovskite $NaNbO_3$ occurs completely within the critical region, whereas in the antifluorite K_2PtBr_6 the region of the anomaly extends far beyond the critical region. Although the term 'soft-mode behaviour' is used in connection with phase transitions, in both families the dynamic mechanisms involved may differ significantly.

Neutron scattering experiments in perovskites have shown that some soft modes, for example, that in $LaAlO_3$, experience significant damping[85]—they are classified as relaxational soft modes. On the other hand, some soft modes, for example, that in $SrTiO_3$, experience negligible damping[86]—they are classified as resonant soft modes. Although comprehensive neutron scattering results are not available for any antifluorite crystal, the fact that the infrared spectrum of the soft mode in K_2ReCl_6 does not exhibit anomalous broadening,[11] except very near to the tilting type phase transition at 103 K, suggests that the soft mode is resonant in character.

The dynamics of the phase transitions in R_2MX_6 salts has been analysed by treating the associated anharmonicity as a small perturbation on an otherwise harmonic lattice. Spin relaxation is treated in terms of the phonon states of the harmonic lattice with small spin–lattice coupling. Such an approach is incorrect for the strongly anharmonic RMX_3 lattices which exhibit relaxational soft-mode behaviour. In these substances the relaxation time data is analysed directly in terms of spectral densities appropriate to anharmonic lattices.

B. Relaxation Theory for Nearly Harmonic Crystalline Solids

A theory for quadrupolar nuclear spin–lattice relaxation in harmonic crystalline solids was first discussed by Van Kranendonk[87] in 1954. The electric field gradient is calculated using a point-charge model and its variation due to the influence of thermal vibrations, in the Debye approximation. Van Kranendonk evaluated the probability of relaxation transitions for the spin system and concluded that the dominant relaxation mechanism was the first-order phonon Raman process. The theory provided a successful explanation for the temperature dependence of T_1 in alkali halides, but was unable to provide even order of magnitude agreement with the observed values of T_1. This discrepancy was resolved by Van Kranendonk and Walker[88,89] in 1967 by the inclusion of the effects of lattice anharmonicity. They showed that the anharmonic Raman process provided a relaxation mechanism that was two orders of magnitude stronger than that provided by the ordinary Raman process, but that exhibited the same temperature dependence. The theory has also been extended to take into account covalent effects,[90,91] more realistic phonon spectra[92-95] and to provide an explanation of relaxation in R_2MX_6 compounds which undergo a tilting phase transition.[29,96]

The Van Kranendonk theory of quadrupolar nuclear spin–lattice relaxation may be summarized as follows. The Hamiltonian for the spin system in a crystal lattice may be written

$$\mathcal{H} = \mathcal{H}_s + \mathcal{H}_l + \mathcal{H}_{sl}$$

$\mathcal{H}_s$ describes the unperturbed spin system and for a pure NQR experiment is given by

$$\mathcal{H}_s = \mathcal{H}_Q + \mathcal{H}_{dd}$$

where $\mathcal{H}_Q$ is the static quadrupolar interaction and $\mathcal{H}_{dd}$ the magnetic dipole–dipole interaction. The pure spin eigenstates satisfy the relation $\mathcal{H}_0|m\rangle = E_m|m\rangle$. $\mathcal{H}_l$ represents the lattice vibrations and is given by

$$\mathcal{H}_l = \mathcal{H}_o + V_3 + V_4 + \ldots$$

where $\mathcal{H}_o$ includes the kinetic energy of all the nuclei in the lattice and the harmonic portion of the potential in which the nuclei move, and the terms V_3, $V_4 \ldots$ correspond to the cubic, quartic $\ldots$ terms of the lattice potential. The pure lattice eigenstates satisfy the relation $\mathcal{H}_l|n\rangle = E_n|n\rangle$. $\mathcal{H}_{sl}$ describes the spin–lattice coupling which for electric quadrupole relaxation corresponds to the non-secular part of the interaction between the nuclear electric quadrupole moment and the local electric field gradient. In particular

$$\mathcal{H}_Q + \mathcal{H}_{sl} = -\sum_i \sum_{m=-2}^{2} Q_i^{-m}(\nabla E_i)^m$$

where the Q_i^{-m} and the $(\nabla E_i)^m$ are the components for nucleus i of the nuclear quadrupole moment tensor and electric field gradient tensor respectively. The components $(\nabla E_i)^m$ are expanded as a power series in the displacements U_i of the nuclei such that

$$(\nabla E_i)^m = (\nabla E_i)_0^m + \sum_{j\alpha} f_\alpha^m(\mathbf{R}_{ij})U_{j\alpha} + \sum_{jk\alpha\beta} f_{\alpha\beta}^m(\mathbf{R}_{ij}, \mathbf{R}_{ik})U_{j\alpha}U_{k\beta} + \ldots$$

225

The $\mathbf{R}_{ij}$ are relative position vectors. The spin–lattice coupling is treated as a perturbation which induces transitions of the type $|mn\rangle \rightleftharpoons |m'n'\rangle$. In first order, the transition probability per unit time is given by

$$W_{mn,m'n'} = \frac{2\pi}{\hbar} |\langle m'n'|\mathcal{H}_{sl}|mn\rangle|^2 \delta(E_{m'} + E_{n'} - E_m - E_n)$$

If the lattice is in thermal equilibrium at a temperature $T = (k\beta)^{-1}$, the average transition probability per unit time for the spin transitions $|m\rangle \rightleftharpoons |m'\rangle$ is

$$W_{mm'} = \sum_{nn'} \frac{\exp(-\beta\mathcal{H}_l)}{Z_l} W_{mn,m'n'}$$

Assuming the existence of a spin temperature, it follows that the spin–lattice relaxation rate T_1^{-1} is given by the Hebel and Slichter formula[97]

$$T_1^{-1} = \frac{\frac{1}{2}\sum_{mm'} W_{mm'}(E_m - E_{m'})^2}{\sum_m E_m^2}$$

For a harmonic lattice the terms V_3, V_4, ... in the lattice Hamiltonian are zero. This was the case considered in Van Kranendonk's initial paper. He considered, in turn, the various terms in $\mathcal{H}_{sl}$ resulting from the series expansion of the components of the electric field gradient tensor. The term linear in the $U_{j\alpha}$ results in the direct process (absorption or emission of a phonon by the spin system) for relaxation. If it alone were operative the relaxation rate would be

$$(T_{1d})^{-1} \propto \frac{T}{\omega_Q} \sum_{\mathbf{k}j} |f(\mathbf{k}j)|^2 \delta(\varepsilon_{\mathbf{k}j} - \hbar\omega_Q)$$

where ω_Q is the NQR frequency as determined by the static part of the electric field gradient tensor. Because of the small number of phonons for which $\varepsilon_{\mathbf{k}j} \sim \hbar\omega_Q$, this process can be regarded as unimportant. The term quadratic in the $U_{j\alpha}$ results in the first-order phonon Raman process (the inelastic scattering of a phonon by the spin system) and the corresponding relaxation rate is given by

$$(T_{1,1R})^{-1} \alpha \sum_{\mathbf{k}j\mathbf{k}'j'} |f(\mathbf{k}j\mathbf{k}'j')|^2 n(\mathbf{k}j)(n(\mathbf{k}'j') + 1)$$

$$\times \delta(\varepsilon_{\mathbf{k}'j'} - \varepsilon_{\mathbf{k}j} - \hbar\omega_Q)$$

The factor $n(\mathbf{k}j)$ represents the thermal averaged population of the lattice mode $(\mathbf{k}j)$. This process leads to a T^2 temperature dependence at high temperatures $(T > \theta_D/2)$ and a T^7 temperature dependence at low temperatures, in agreement with the experimental observations in the alkali halides. Higher-order terms (multiple scattering processes) have been shown to give a negligible contribution to T_1^{-1} because of the smallness of the f coefficients.

For a nearly harmonic lattice, the displacement field operators $\mathbf{U}_{j\alpha}$ must be replaced by those appropriate to a lattice with cubic (or higher order) anharmonicities. Taking the part of $\mathcal{H}_{sl}$ linear in the displacements and considering only that portion which results from the anharmonicity of the lattice, one obtains the anharmonic Raman process. This process corresponds to the inelastic scattering of a phonon caused by a combination of the direct process and the three-phonon process associated with the V_3 term of the lattice potential (Fig. 32). The phonon associated with the direct process is one of the phonons of the three-phonon scattering process. Since this phonon is a virtual phonon, it follows that the temperature dependence of the anharmonic Raman process is identical to that for the ordinary Raman process. However, since the anharmonic Raman process occurs for a multitude of intermediate virtual phonon states, it represents an even stronger relaxation mechanism than the ordinary Raman process. The expression for $T_{1,AR}$ can be related to that for

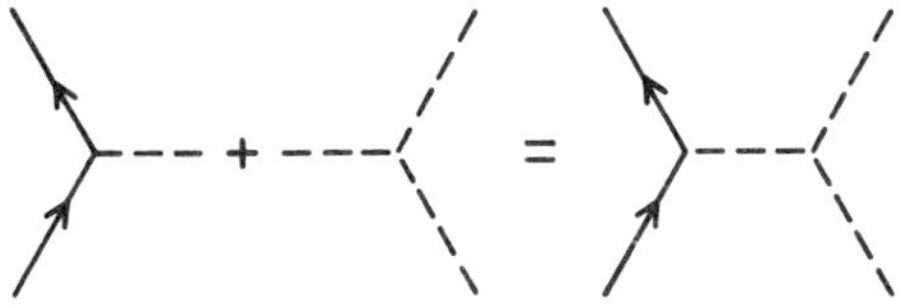

d V_3 aR

Fig. 32. Feynman diagrams for the direct process (d), the three phonon process due to the cubic anharmonic interaction (V_3) and the anharmonic Raman process (aR). The solid lines represent spin states and the dashed lines phonons. The 'equation' indicated is symbolic.

$T_{1,1R}$ to yield

$$T_{1,AR}^{-1} = C\gamma_G^2 \frac{c_1^2 + c_2^2}{d_1^2 + d_2^2} T_{1,1R}^{-1}$$

C is a numerical constant determined by the structure of the crystal, γ_G is taken to be the Grüneisen parameter for the crystal and corresponds to an average of the microscopic mode $\gamma(\mathbf{k}, j)$'s for a stable lattice, and the constants c_i, d_i are related to the magnitudes of the field gradient fluctuations. High-order terms resulting from the inclusion of the lattice anharmonicity have been shown to be completely negligible. Effects of the anharmonicity in producing finite lifetimes for the 'phonons' and altering the 'phonon eigenfunctions' of the Hamiltonian $\mathcal{H}_{sl}$ can be eliminated by appropriately renormalizing the phonon states to represent true eigenfunctions of the anharmonic Hamiltonian.

To this point in the discussion the various phonon processes have been assumed to act independently. Such of course is not the case; in general, interference effects arise between the different processes. However, for most crystals it is thought that the anharmonic Raman process by itself should

227

dominate the experimentally observed relaxation. This is difficult to confirm in a dynamically stable compound since both the $1R$ and AR processes lead to the T^2 dependence of T_1 on temperature for $T > \theta_D/2$. But the change in anharmonicity occurring within a dynamically unstable R_2MX_6 lattice as a phase transition is approached has provided convincing evidence in support of this conjecture.

C. Nuclear Quadrupole Relaxation Detection of Soft Modes

The relaxation theory of Van Kranendonk and Walker was extended to R_2MX_6 lattices by Armstrong and Jeffrey[96] and Van Driel *et al.*[29] For this family of compounds the relaxation rate for the X nuclei due to the anharmonic Raman process can be written as

$$T_{1AR}^{-1} = \frac{29}{N^2} \sum_{\mu=1,2} \frac{4\pi(eQ)^2}{3\hbar} \sum_{\mathbf{kk'}jj'} n(\varepsilon_j(\mathbf{k}))n(\varepsilon_{j'}(\mathbf{k'})+1)$$

$$\times \gamma_j(\mathbf{k})\gamma_{j'}(\mathbf{k'})|f^\mu(\mathbf{kk'}jj')|^2 \delta(\varepsilon_{j'}(\mathbf{k'}) - \varepsilon_j(\mathbf{k}) - \hbar\omega_Q)$$

In this expression $\gamma_j(\mathbf{k})$ is the microscopic Grüneisen parameter associated with mode j and wave vector $\mathbf{k}$, eQ is the scalar quadrupolar moment of nucleus X, $\varepsilon_j(\mathbf{k})$ is the energy of the $(\mathbf{k}j)$ phonon, $n(\varepsilon_j(\mathbf{k}))$ is the Bose–Einstein occupation number and N is the total number of unit cells in the crystal. The quantity $f^\mu(\mathbf{kk'}jj')$ is defined as

$$f^\mu(\mathbf{kk'}jj') = \frac{\hbar}{2[\omega_j(\mathbf{k})\omega_{j'}(\mathbf{k'})]^{1/2}} \sum_{\alpha,\beta} f_{\alpha\beta}^\mu \left[\frac{e_\alpha(X|\mathbf{k}j)}{m_X^{1/2}} - \frac{e_\alpha(M|\mathbf{k}j)}{m_M^{1/2}} \right]$$

$$\times \left[\frac{e_\beta(X|\mathbf{k'}j')}{m_X^{1/2}} - \frac{e_\beta(M|\mathbf{k'}j')}{m_M^{1/2}} \right]$$

where $f_{\alpha\beta}^\mu$ is the derivative with respect to U_α and U_β of the μth component of the electric field gradient tensor evaluated at the equilibrium position of the X nucleus. The double summation $\sum_{jj'}$ is taken over the fifteen internal modes of the MX_6 octahedron and the three rotary lattice modes according to the model previously adopted for the field gradient (Section III, Part C). The introduction of the high-temperature approximation, the replacement of the double summation $\sum_{\mathbf{kk'}}$ by the double integration $(NV)^2\int_{\mathbf{kk'}} d\mathbf{k}\, d\mathbf{k'}$, the assumption that the functions $\omega(\mathbf{k})$, $\gamma(\mathbf{k})$ are isotropic, and the neglect of the $\mathbf{k}$ dependence of the e_α's, permit the expression for $T_{1,AR}^{-1}$ to be reduced to

$$T_{1,AR}^{-1} = \frac{29\pi(eQ)^2}{3\hbar^2} \sum_{\mu=1,2} \sum_{jj'} \left| \sum_{\alpha\beta} f_{\alpha\beta}^\mu \left[\frac{e_\alpha(X|j)}{m_X^{1/2}} - \frac{e_\alpha(M|j)}{m_M^{1/2}} \right] \right.$$

$$\times \left. \left[\frac{e_\beta(X|j')}{m_X^{1/2}} - \frac{e_\beta(M|j')}{m_M^{1/2}} \right] \right|^2 4\pi^2 k_B^2 T^2 V^2$$

$$\times \int_k \gamma_j(k)\gamma_{j'}(k)\frac{k^4 dk}{\omega_j^2(k)\omega_{j'}^2(k)|d\omega_{j'}/dk|}$$

The prime over $\sum_{jj'}$ denotes that the summations over j, j' contain non-zero contributions only if j and j' are degenerate branches of the same mode. This follows from the conservation of energy of the phonon scattering processes because of the negligible amount of overlap of the initial modes. As a result of their relatively high frequencies, the internal modes make an insignificant contribution to the relaxation rate. Hence $\sum_{jj'}$ need only include the rotary lattice modes. Therefore

$$T_{1,AR}^{-1} = \frac{29\pi(eQ)^2}{3\hbar^2}\frac{k_B^2 T^2}{m_x^2}\sum_{\mu=1,2}^{\text{rot modes}}\sum_{j,j'}\left|\sum_{\alpha\beta}f_{\alpha\beta}^\mu e_\alpha(X|_j)e_\beta(X|_{j'})\right|^2$$

$$\times (4\pi)^2 V^2 \int_k \gamma_j(k)\gamma_{j'}(k)\frac{k^2 dk}{\omega_j^2(k)\omega_{j'}^2(k)|\partial\omega_{j'}/\partial k|}$$

The coefficients $f_{\alpha\beta}^\mu$ can be readily evaluated on the basis of a point charge model. For an M nucleus located at $(0, 0, 0)$ and an X nucleus at $(0, 0, R)$ these coefficients are given by

$$f_{\alpha\beta}^1 = \frac{3se}{R^5}\begin{vmatrix} 0 & 0 & -2 \\ 0 & 0 & -2i \\ -2 & -2i & 0 \end{vmatrix}; \qquad f_{\alpha\beta}^2 = \frac{3se}{2R^5}\begin{vmatrix} 1 & i & 0 \\ i & -1 & 0 \\ 0 & 0 & 0 \end{vmatrix}$$

where s is a factor introduced to take account of covalency and antishielding effects. An estimate of s may be obtained from the NQR frequency data through the point-charge expression

$$\nu_Q = \frac{eQ}{2h}\left(\frac{2se}{R^3}\right)$$

The expression for the relaxation rate becomes

$$T_1^{-1} = T_{1,AR}^{-1} = \frac{87\pi}{4}\left(\frac{\gamma Qe^2 sk_B}{\hbar m_x R^5}\right)^2\frac{T^2}{\bar\omega_{T_1}^5}$$

where the average rotary-mode frequency $\bar\omega_{T_1}$ is defined as

$$\bar\omega_{T_1} = \left(\frac{(4\pi)^2 V^2}{\gamma^2}\int_k \frac{\gamma_{rot}^2(k)k^4\,dk}{\omega_{rot}^4(k)|\partial\omega_{rot}/\partial k|}\right)^{-1/5}$$

with γ the average Grüneisen parameter for the rotary mode far from the phase transition.

For a dynamically stable compound such as Cs_2PtCl_6, $\bar\omega_{T_1}$ is expected to be a temperature independent constant and $T_{1,AR}^{-1}$ is predicted to have a T^2 temperature dependence. Figure 33 shows the ^{35}Cl T_1 data as measured in

229

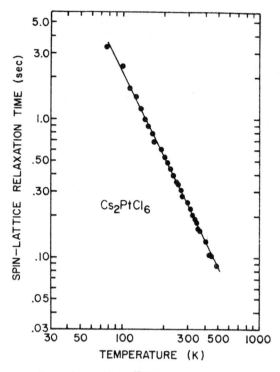

Fig. 33. Temperature dependence of the ^{35}Cl spin–lattice relaxation time in Cs_2PtCl_6. The straight line has slope -2.

Cs_2PtCl_6 from 77 to 500 K. The straight line has slope -2 in agreement with the theoretical prediction.

For a dynamically unstable compound such as K_2PtBr_6, $\bar{\omega}_{T_1}$ is expected to decrease as the temperature approaches the transition temperature from above. Therefore, $T_{1,AR}^{-1}$ is predicted to deviate from the T^2 law. Figure 34 shows the ^{79}Br T_1 data as measured in K_2PtBr_6 from 30 to 500 K. The temperatures at which phase transitions have been detected are indicated. The straight line has a slope of -2. The high-temperature phase transition at 169 K is clearly reflected in the T_1 data; the other phase transitions have no noticeable influence on the behaviour of T_1. Below 169 K the two T_1's observed correspond to the two NQR lines observed in this temperature regime (see Fig. 9). Above 320 K the relaxation is dominated by a non-resonant hindered-rotation mechanism.[98,99] This process was identified by its characteristic isotope independence. The T_1 data for temperatures in the range $169 < T < 300$ K were analysed to yield the temperature dependence of $\bar{\omega}_{T_1}$ shown in Fig. 35. As the temperature decreases, $\bar{\omega}_{T_1}$ decreases from 27 cm^{-1} at 300 K to 16 cm^{-1} at 170 K. That is, a 40% reduction of this average rotary-mode frequency is indicated.

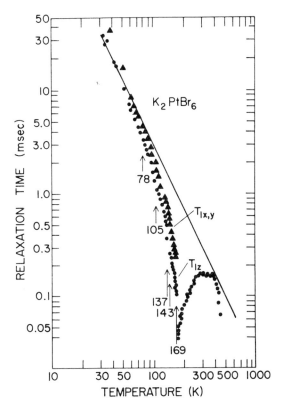

Fig. 34. Temperature variation of the ^{79}Br spin–lattice relaxation time in K_2PtBr_6. The straight line has slope -2.

In Section III, Part C, it was seen that $\bar{\omega}_\nu$, a different average rotary-mode frequency, experienced a 12% reduction over the same temperature range. This difference can be understood in terms of the model proposed for this type of phase transition in which the Γ point rotary-mode phonons soften to zero at the transition. In order to calculate the two averages, $\bar{\omega}_\nu$ and $\bar{\omega}_{T_1}$, a specific form for the rotary lattice mode dispersion curve must be adopted. The form of dispersion curve was chosen on the basis of the rigid-ion model calculation presented in Section III, Part E (see Fig. 27). An evaluation of $\bar{\omega}_\nu$ between 300 and 170 K indicates that a reduction of $\sim 12\%$ is easily realizable. To calculate $\bar{\omega}_{T_1}$, we take

$$\gamma_{rot}(k) = \frac{V}{\omega_{rot}(k)}\frac{\partial\omega_{rot}(k)}{\partial V} = \frac{V}{2\omega_{rot}^2(k)}\frac{\partial\omega_{rot}^2(k)}{\partial V}$$

231

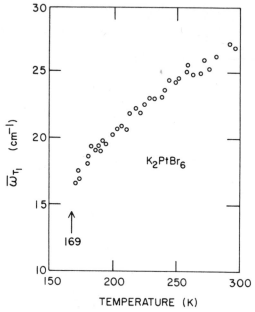

Fig. 35. Temperature dependence of the average rotary mode frequency $\bar{\omega}_{T_1}$ as deduced from the spin–lattice relaxation time data in the high-temperature cubic phase.

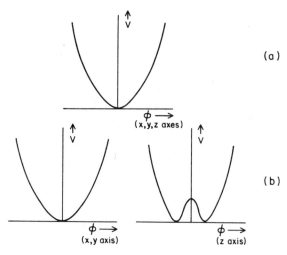

Fig. 36. (a) Potential governing the librational motions of $PtBr_6^{2-}$ octahedra about the x-, y- and z-axes for $T > T_c$. (b) Potentials governing the librational motions of $PtBr_6^{2-}$ octahedra about x- (or y-) and z-axes for $T < T_c$.

232

The frequency ω_{rot} reflects the delicate balance between short-range and long-range forces. Symbolically we may write

$$\omega_{rot}^2(k) = \frac{A(k)}{V^3} - \frac{B(k)}{V}$$

where each term on the right-hand side is much larger than their difference. At the stability limit, the two terms cancel. It follows that

$$\gamma_{rot}(k) \propto \frac{1}{\omega_{rot}^2(k)}$$

A softening of $\sim 40\%$ in $\bar{\omega}_{T_1}$ is easily realizable.

It is instructive to note that had it been assumed that the 1R process provided the dominant relaxation mechanism, the corresponding average rotary-mode frequency would have been

$$\bar{\omega}_{T_1}^0 = \left((4\pi)^2 V^2 \int_k \frac{k^4\, dk}{\omega_{rot}^4(k)|\partial\omega_{rot}(k)/\partial k|} \right)^{-1/5}$$

Using this expression and the dispersion curves indicated in Fig. 27, a 40% reduction of $\bar{\omega}_{T_1}^0$ cannot be explained. In fact, the predicted reduction for $\bar{\omega}_{T_1}^0$ is less than that for $\bar{\omega}_y$. This result provides substantial evidence in support of the generally held view that it is the anharmonic Raman process, and not the ordinary Raman process, which dominates quadrupolar controlled nuclear spin–lattice relaxation.

The dependence of T_1 on temperature below 169 K provides additional insight into the phase transitions in K_2PtBr_6. For example, the fact that T_1 does not reflect the presence of any of the other phase transitions constitutes evidence that these transitions are not induced by the rotary lattice mode.

Finally, let us consider the fact that for some range just below 169 K, $T_{1x,y} \sim 2T_{1z}$. As noted in the discussion above concerning the cubic phase, the only modes that could participate in the Raman process were those that,

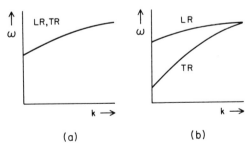

(a)   (b)

Fig. 37. (a) The dispersion curves for the longitudinal and transverse branches of the rotary mode in the cubic phase. (b) The dispersion curves for the longitudinal and transverse branches of the rotary mode in the tetragonal phase.

together with the change in nuclear spin energy $\hbar\omega_Q$, could satisfy energy conservation. In the cubic phase, the two branches of the rotary mode which are responsible for the displacement of a particular Br nucleus can yield four different types of Raman processes, namely $(j \to j)$, $(j \to j')$, $(j' \to j)$ and $(j' \to j')$. However, in the tetragonal phase, the potential governing the librational motions of the octahedra about the x- (or y-) and z-axes are altered as shown in Fig. 36 for $\mathbf{k} = 0$. Therefore, one expects that for librations about the z-axis of frequency $\omega_z \neq \omega_{x,y}$ the number of types of Raman processes that can contribute to the relaxation of an x (or y) nucleus is reduced from four to two, namely $(j \to j)$ and $(j' \to j')$, because of the change in the nature of the dispersion curves for the various branches of the rotary mode (see Fig. 37). It follows that

$$T_{1,x,y}^{-1} = \frac{87\pi}{16} \left(\frac{\gamma Qse^2 k_B}{\hbar m_{Br} R^5} \right)^2 \left[\frac{1}{(\bar{\omega}_z)_{T_1}^5} + \frac{1}{(\bar{\omega}_{x,y})_{T_1}^5} \right] T^2$$

Similarly, a consideration of librations about the x (or y) axis leads to the equation

$$T_{1,z}^{-1} = \frac{87\pi}{4} \left(\frac{\gamma Qse^2 k_B}{\hbar m_{Br} R^5} \right)^2 \frac{T^2}{(\bar{\omega}_{x,y})_{T_1}^5}$$

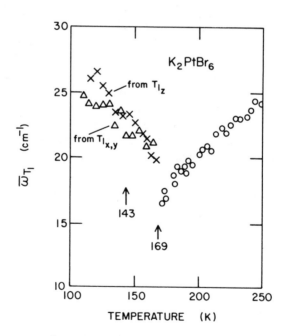

Fig. 38. Temperature dependence of the average rotary mode frequency $\bar{\omega}_{T_1}$ as deduced from the spin–lattice relaxation time data. For $143 < T < 169$ K, the $(\bar{\omega}_z)_{T_1}$ and $(\bar{\omega}_{x,y})_{T_1}$ values are, within experimental uncertainty, indistinguishable.

234

However, $(\bar{\omega}_z)_{T_1}$ and $(\bar{\omega}_{x,y})_{T_1}$ are nearly equal, from which it follows that

$$T_{1,x,y} \sim 2T_{1,z}$$

The temperature dependences of the average rotary mode frequencies $(\bar{\omega}_{x,y})_{T_1}$ and $(\bar{\omega}_z)_{T_1}$ as deduced from the data below 169 K are shown in Fig. 38 along with the temperature dependence of $\bar{\omega}_{T_1}$ as deduced from the data above 169 K, where the rotary mode again hardens.

The T_1 data measured for the ^{35}Cl nuclei in K_2PtCl_6 are shown in Fig. 39. It was initially considered to be anomalous since, at the higher temperatures, the temperature dependence is much weaker than the expected T^2 law behaviour. Although such behaviour is inconsistent with the Van Kranendonk theory for a dynamically stable lattice, it is just what is predicted for a substance on the verge of becoming unstable. In Fig. 40, the temperature dependence of $\bar{\omega}_{T_1}$ as deduced from the ^{35}Cl T_1 data in K_2PtCl_6 is compared with the temperature dependence of $\bar{\omega}_{T_1}$ as deduced from the ^{35}Cl T_1 data in Cs_2PtCl_6, which is dynamically stable, and from the ^{79}Br T_1 data in K_2PtBr_6, which experiences a structural phase transition at 169 K. The similarity of behaviour of $\bar{\omega}_{T_1}$ in

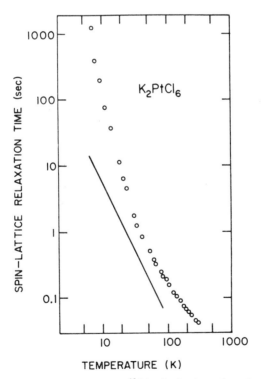

Fig. 39. Temperature dependence of the ^{35}Cl spin–lattice relaxation time in K_2PtCl_6. The straight line depicts a T^2 law.

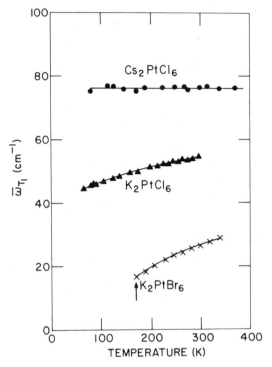

Fig. 40. Temperature dependence of the average rotary mode frequency $\bar{\omega}_{T_1}$ in Cs_2PtCl_6, K_2PtCl_6 and K_2PtBr_6.

K_2PtCl_6 provides convincing evidence of an incipient phase transition in K_2PtCl_6. A similar conclusion was reached from a consideration of the ^{35}Cl NQR frequency data taken in the same substance (Section III, Part C).

D. Effect of Volume Changes and Electronic Paramagnetism on Nuclear Quadrupole Spin–Lattice Relaxation Times

The Van Kranendonk theory for T_1 is a constant volume theory and therefore the experimental observations of a T^2 temperature dependence in a constant pressure experiment strongly suggests that the effect of volume changes is negligible. If it is assumed that the temperature dependences of $\bar{\omega}_{T_1}$ and R are of the form

$$\bar{\omega}_{T_1} = \bar{\omega}_{T_1}^0(1 - \alpha T); \qquad R = R^0(1 + \alpha T)$$

then from the T_1 data for Cs_2PtCl_6 it can be concluded that $\alpha < 2 \times 10^{-4}\ K^{-1}$. That is, since the T_1 measurements have a relatively large associated probable error of $\pm 5\%$, any value of α less than this amount could not be detected.

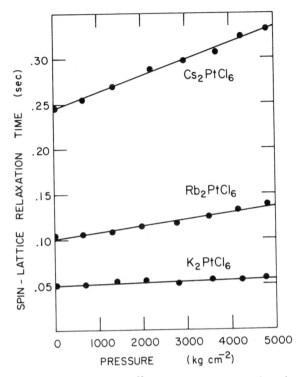

Fig. 41. Pressure dependence of the ^{35}Cl spin–lattice relaxation time in K_2PtCl_6, Rb_2PtCl_6 and Cs_2PtCl_6 at 295 K.

Figure 41 shows the pressure dependence[54,100] of T_1 for K_2PtCl_6, Rb_2PtCl_6 and Cs_2PtCl_6 at 295 K. In each case the data are consistent with a linear dependence of the form

$$T_1 = T_1^0(1 + \beta P)$$

with $\beta \sim 10^{-5}$ kg^{-1} cm^2. For temperatures in the vicinity of 300 K, the coefficient, β, as obtained from this type of experiment, is independent of the temperature. That is, even though volume changes have little effect on T_1 in an experiment in which the temperature is changed at constant pressure, they are easily detected in an experiment in which the pressure is changed at constant temperature.

K_2ReCl_6 and K_2IrCl_6 are paramagnetic compounds belonging to the R_2MX_6 family. These substances exhibit a magnetic hyperfine interaction[101] between the unpaired electron spin of the metal ion and the ligand nuclear spins. Relaxation caused by the presence of this interaction has been investigated theoretically by Moriya.[102] Figure 42 shows the ^{35}Cl T_1 data[103,104] for K_2ReCl_6 which exhibits several phase transitions over the temperature range

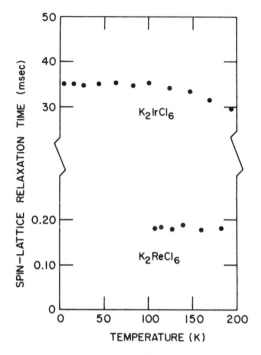

Fig. 42. Temperature dependence of the ^{35}Cl spin–lattice relaxation time in the paramagnetic compounds K_2IrCl_6 and K_2ReCl_6.

investigated, and for K_2IrCl_6, which exhibits no phase transition. In each case the data are essentially independent of temperature. No evidence of the onset of a phase transition can be seen from the K_2ReCl_6 data. Measurements of the ratio of relaxation times for the ^{35}Cl and ^{37}Cl nuclei are in excellent agreement with the theoretical prediction for a relaxation mechanism dominated by the magnetic hyperfine interaction. That is, spin–lattice relaxation time measurements in paramagnetic R_2MX_6 compounds provide no information concerning the dynamics of structural phase transitions.

E. Critical Dynamics and Quadrupolar Relaxation

The investigation of quadrupolar relaxation in the RMO_3 perovskites has in general been confined to the critical region about the phase transitions where large classical fluctuations of the order parameter can give rise to an anomalous behaviour in the spin–lattice relaxation time. Because of the large amplitudes of the fluctuations of the critical coordinates, and because of the anharmonicity associated with their motion, it is necessary to invoke a more fundamental approach for the calculation of relaxation times than that introduced in Section

IV, Part B. The quantity related to the experimental results is the power spectrum $S(\omega)$ of the fluctuations in the electric field gradient tensor. The manner in which the spin–lattice relaxation rate depends upon the temperature depends upon the details of the critical dynamics. In this Section we discuss the form of $S(\omega)$ in terms of the present understanding of the structural transitions. The work reported is mainly that of Borsa, Bonera and Rigamonti[105-107] following the general approach taken by Blinc and co-workers[108,109] in KDP and related systems.

The onset of a structural phase transition is associated with the dynamical behaviour of a particular local coordinate $\phi_l(t)$, the microscopic angle of rotation of an MO_6 octahedron. The label l refers to the unit cell. The Hamiltonian is constructed in terms of the critical coordinates and the remainder of the lattice is regarded as a thermal bath. That is

$$\mathcal{H} = \sum_l \mathcal{H}_l^{(s)} + \tfrac{1}{2} \sum_{ll'} V_{ll'} \phi_l \phi_{l'}$$

where the single particle Hamiltonian $\mathcal{H}_l^{(s)}$ is assumed to be of the form

$$\mathcal{H}_l^{(s)} = \frac{1}{2M} P_l^2 + V(\phi_l)$$

with M an effective mass and $V(\phi_l)$ the total potential for the coordinate ϕ_l in an otherwise rigid lattice. The interaction Hamiltonian is approximated by a sum of bilinear 'two-body' interactions. In the linear-response regime in the absence of any interactions, the application of an external stimulus $F \, e^{-i\omega t}$ would result in a time dependent $\phi_l(t)$ of the form

$$\phi_l(t) = \phi_l \, e^{-i\omega t} + \delta\phi_l(t)$$

where $\delta\phi_l(t)$ describes the spontaneous fluctuations. The effect of the interactions may be included in the mean field approximation by writing

$$\phi_l = \chi^s(\omega) \left[\sum_{l'} V_{ll'} \phi_{l'} + F_l \right]$$

$$\delta\phi_l(t) = \chi^s(\omega) \sum_{l'} V_{ll'} \delta\phi_{l'}(t) + \delta\phi_l^s(t)$$

where $\chi^s(\omega)$ is the 'single-particle' generalized susceptibility and $\delta\phi_l^s(t)$ is the spontaneous fluctuation in the absence of interactions. The quantities ϕ_l, F_l and $\chi^s(\omega)$ may be expressed in terms of the collective normal coordinates ϕ_k, the corresponding forces F_k and the collective susceptibility $\chi(\mathbf{k}, \omega) = \phi_k / F_k$ viz

$$\phi_l = \frac{1}{(N)^{-1/2}} \sum_k \phi_k \, e^{i\mathbf{k}.\mathbf{R}_l}$$

$$F_l = \frac{1}{(N)^{-1/2}} \sum_k \phi_k \, e^{i\mathbf{k}.\mathbf{R}_l}$$

239

$$\chi(\mathbf{k}, \omega) = \frac{\chi^s(\omega)}{1 - V_\mathbf{k} \chi^s(\omega)}$$

where

$$V_\mathbf{k} = \sum_{l'} V_{ll'} \, e^{i\mathbf{k}\cdot(\mathbf{R}_l - \mathbf{R}_{l'})}$$

For a more general formalism involving all of the coordinates of the lattice the poles of the generalized susceptibility $\chi(\mathbf{k}, \omega)$ correspond to values of ω which represent normal mode frequencies of the lattice. An instability of the system corresponds to a divergence of the static susceptibility for a given $\mathbf{k}$ with $\omega_\mathbf{k} \to 0$. For $T > T_c$, $V_\mathbf{k}\chi^s(0) < 1$ for all $\mathbf{k}$ whereas for $T = T_c$, there exists a $\mathbf{k} = \mathbf{k}_c$ such that $V_{\mathbf{k}_c}\chi^s(0) = 1$.

For a nuclear magnetic relaxation experiment, the relevant quantity is the value of the power spectrum $S(\mathbf{k}, \omega)$ of the fluctuations of the critical coordinate in the radiofrequency region. The power spectrum is of the form

$$S(\mathbf{k}, \omega) = \int_{-\infty}^{\infty} \langle \delta\phi_\mathbf{k}(0)\delta\phi_\mathbf{k}^*(t)\rangle \, e^{-i\omega t} \, dt$$

The amplitude at frequency ω of the collective fluctuation at $\mathbf{k}$ is

$$(\delta\phi_\mathbf{k})_\omega = \frac{(\delta\phi^s)_\omega}{1 - V_\mathbf{k}\chi^s(\omega)}$$

Therefore, as $\omega \to 0$ and $T \to T_c$, the Fourier amplitude at the critical wave vector $\mathbf{k}_c$ increases dramatically. An anomalous behaviour in the spin–lattice relaxation time is expected.

The dynamical properties of the system are related to this power spectrum through the fluctuation-dissipation theorem which in the high-temperature limit takes the form

$$S(\mathbf{k}, \omega) = -\frac{2k_B T}{\omega} \chi''(\mathbf{k}, \omega)$$

where $\chi''(\mathbf{k}, \omega)$ is the dissipative part of the generalized collective susceptibility. In order to proceed further, one requires a specific form for the single-particle response $\chi^s(\omega)$. For a harmonic oscillator of frequency $\omega(\mathbf{k})$, damping constant Γ, the collective susceptibility is

$$\chi(\mathbf{k}, \omega) = \frac{\chi(\mathbf{k}, 0)\omega^2(\mathbf{k})}{\omega^2(\mathbf{k}) - \omega^2 - i\Gamma\omega}$$

and the power spectrum

$$S(\mathbf{k}, \omega) = \frac{2k_B T\chi(\mathbf{k}, 0)\omega^2(\mathbf{k})\Gamma}{[\omega^2(\mathbf{k}) - \omega^2]^2 + \Gamma^2\omega^2}$$

which is that of a lightly-damped resonant mode. For a heavily-damped (diffusive) mode, the collective susceptibility in the radiofrequency regime is

$$\chi(\mathbf{k}, \omega) = \frac{\chi(\mathbf{k}, 0)}{1 + i\omega\tau(\mathbf{k})}$$

where

$$\tau(\mathbf{k}) = \frac{2\Gamma}{\omega^2(\mathbf{k})}$$

and the power spectrum

$$S(\mathbf{k}, \omega) = \frac{2k_B T \chi(\mathbf{k}, 0)\tau(\mathbf{k})}{1 + \omega^2\tau^2(\mathbf{k})}$$

For phonon soft modes,[110] defined by $\omega^2(\mathbf{k}_c) \propto |T - T_0|$ for $T_0 \simeq T_c$, the fluctuations in the critical variable, and hence, the power spectrum $S(\mathbf{k}_c, \omega)$, increase as $T \to T_0$. The static susceptibility $\chi(\mathbf{k}_c, 0)$ diverges as $(T - T_0)^{-1}$. For diffusive or relaxational soft modes the slowing down of the fluctuations at $\mathbf{k} = \mathbf{k}_c$ is described by the divergence of $\tau(\mathbf{k}_c)$. Again $\chi(\mathbf{k}_c, 0)$ diverges as $(T - T_0)^{-1}$.

Neutron scattering experiments to study structural phase transitions[111] in perovskites have indicated a scattering peak at $\omega = 0$, called a 'central mode', in addition to the soft-mode behaviour. The intensity of the central mode increases as $T \to T_c$. It has been concluded that a description of these phase transitions in terms of linear response theory, and in the framework of the mean field approximation, is incomplete. On approaching the transition temperature, clusters of short-range order are formed characterized by correlation lengths and lifetimes which increase as $T \to T_c$. The presence of these clusters results in zero-frequency fluctuations of macroscopic size.

Schneider and Stoll[112] have proposed a two-dimensional model system which is expected to exhibit structural phase transitions of the type that we are discussing. Their model indicates the presence of a central peak in the dynamic form factor as the transition point is approached. Their results suggest that this peak is associated with the formation of clusters and their dynamics.

Schwabl[113] has given an equation for the collective variable $\delta\phi_\mathbf{k}$ which accounts for both the soft mode and the central mode near T_c. From the Schwabl equation it follows that

$$S(\mathbf{k}, \omega) = \frac{k_B T \, \text{Re}\,[\Gamma(\omega)]}{\{\omega^2 - \omega^2(\mathbf{k}) - \omega \, \text{Im}\,[\Gamma(\omega)]\}^2 + \{\omega \, \text{Re}\,[\Gamma(\omega)]\}^2}$$

where $\text{Re} \equiv$ real, $\text{Im} \equiv$ imaginary. A phenomenalogical Ansatz was proposed for the damping term, namely

$$\Gamma(\omega) = \frac{ib^2}{\omega + i\gamma} + \sigma$$

with non-critical coefficients b, γ, σ. For $\omega \ll \omega(\mathbf{k})$, $S(\mathbf{k}, \omega)$ reduces to a central peak of width

$$\Gamma_c = \frac{\omega^2(\mathbf{k})}{\omega^2(\mathbf{k}) + b^2}$$

For larger frequencies a soft-mode resonance appears at

$$\omega_\pm = \pm(\omega_0^2 + b^2)^{1/2} - \frac{i}{2}[\gamma b^2(\omega_0^2 + b^2)^{-1} + \sigma]$$

The strength of the central peak rises much faster than that of the soft modes on approaching T_c for $\omega(\mathbf{k}) \gg b$. Near T_c, the soft mode moves towards b and its strength is transferred to the central peak. Therefore, sufficiently near to T_c, the soft mode can be neglected and the central mode becomes diffusive in form. In this case

$$S(\mathbf{k}, \omega) = \frac{2 \langle |\delta\phi_\mathbf{k}|^2 \rangle \, \tau(\mathbf{k})}{1 + \omega^2\tau^2(\mathbf{k})}$$

$$= \frac{2k_B T\chi(\mathbf{k}, 0)\tau(\mathbf{k})}{1 + \omega^2\tau^2(\mathbf{k})}$$

The $\mathbf{k}$ dependence of the spatial correlation near T_c can be described as a modification[113,114] of the Ornstein–Zernike expression. In particular,

$$\langle |\delta\phi_{\mathbf{k}_c}|^2 \rangle = \frac{\langle |\delta\phi\mathbf{k}_c|^2 \rangle}{[1 - (1 - \Delta)\mathbf{k}_c^2 + \kappa^2]^{1-\eta/2}}$$

where $\kappa \propto \varepsilon^\nu = [(T - T_c)/T_c]^\nu$ is the inverse of the correlation length, Δ is a parameter which takes into account the possible anisotropy near $\mathbf{k}_c$ and the z-direction is that of the axis of rotation of the octahedra. Bonera et $al.$[114] have taken

$$\tau(\mathbf{k}) \propto \langle |\delta\phi_\mathbf{k}|^2 \rangle$$

In the most general case, the correlation time can be related to the correlation length by means of the dynamic scaling hypothesis.

F. Quadrupolar Relaxation Studies of Structural Phase Transitions in Perovskites

For the RMO_3 perovskites only the R and M nuclei possess quadrupole moments. This limits the usefulness of the quadrupolar relaxation technique as a means to study the structural phase transition, since according to the model adopted to account for them, the critical dynamics is mainly associated with the O atoms. In the cubic phase, the static quadrupolar energy of the R and M nuclei vanishes because of symmetry. Therefore, a nuclear magnetic resonance experiment is employed to measure the quadrupolar relaxation rates. The transition probability between the Zeeman levels m and $m + \mu$, induced by the time dependent electric quadrupole interaction $\mathcal{H}_Q(t)$ at the nuclear sites, can be written

$$W_{m,m+\mu} = \frac{1}{\hbar^2} \int_{-\infty}^{\infty} e^{-i\omega\mu t} \langle (m|\mathcal{H}_Q(0)|m+\mu)(m+\mu|\mathcal{H}_Q(t)|m) \rangle \, dt$$

$$= \frac{1}{\hbar^2} |Q_{\mu m}|^2 \int_{-\infty}^{\infty} e^{-i\omega_L\mu t} \langle V_\mu(0) V_\mu^*(t) \rangle \, dt$$

where ω_L is the nuclear Larmor frequency, $Q_{\mu m}$ the appropriate matrix element of the nuclear quadrupole moment operator and $V_\mu(t)$ is the μth component of the electric field gradient tensor. To establish the relation between the transition probability and the power spectrum of the spontaneous fluctuations, the $V_\mu(t)$ are expanded with respect to the critical coordinates to yield

$$V_\mu(t) = V_\mu(0) + \sum_l A_l^\mu \phi_l(t) + \tfrac{1}{2} \sum_{ll'} B_{ll'}^\mu \phi_l(t) \phi_{l'}(t) + \ldots$$

with

$$A_l^\mu = \frac{\partial V^\mu}{\partial \phi_l}, \qquad B_{ll'}^\mu = \frac{\partial^2 V^\mu}{\partial \phi_l \, \partial \phi_{l'}}$$

The second term in this expansion gives rise to relaxation by means of the direct phonon process and the third term by means of the phonon Raman process. Because of the strong damping and the critical slowing down that occurs near the phase transition, the direct process is expected to dominate. The transition probability may be written[105]

$$W_{m,m+\mu} = \frac{1}{\hbar^2} |Q_{\mu m}|^2 \sum_{ll'} A_l^\mu A_{l'}^{\mu*} J_{ll'}(\mu\omega_L)$$

where

$$J_{ll'}(\omega) = \int_{-\infty}^{\infty} e^{-i\omega t} \langle \delta\phi_l(0) \, \delta\phi_{l'}^*(t) \rangle \, dt$$

In the random-phase approximation,

$$W_{m,m+\mu} = \frac{1}{\hbar^2} |Q_{\mu m}|^2 \sum_k \sum_{ll'} A_l^\mu A_{l'}^{\mu*} e^{i\mathbf{k}\cdot\mathbf{R}_{ll'}} S(\mathbf{k}, \mu\omega_L)$$

$$= \frac{1}{\hbar^2} |Q_{\mu m}|^2 \sum_k A_\mathbf{k} A_{-\mathbf{k}} S(\mathbf{k}, \mu\omega_L)$$

Since near to the phase transition the largest component of the spectral density is the one for which $\mathbf{k} = \mathbf{k}_c$, and since the factor $A_\mathbf{k} A_{-\mathbf{k}}$ depends only slightly on

k, it follows that to a good approximation

$$W_{m,m+\mu} = \frac{1}{\hbar^2}|Q_{\mu m}|^2 A^\mu_{k_c}A^{\mu*}_{-k_c}\sum_k S(k, \mu\omega_L)$$

$$= \frac{1}{\hbar^2}|Q_{\mu m}|^2 A^\mu_{k_c}A^{\mu*}_{-k_c}\sum_k \frac{2\langle|\delta\phi_k|^2\rangle\tau(k)}{1+(\mu\omega_L)^2\tau^2(k)}$$

Using the Ornstein–Zernike expression with $\tau(k)\propto\langle|\delta\phi_k|^2\rangle$, and performing the summation assuming $\mu\omega_L \ll \tau^{-1}(k)$, yields

$$W_{m,m+\mu} \propto \frac{\kappa^{-(1-2\eta)}}{\Delta^{1/2}} \arctan\left(\frac{\pi\Delta^{1/2}}{\kappa a}\right)$$

where a is the lattice constant. For three-dimensional correlations $\Delta = 1$ and inside the critical region where $\kappa a \ll 1$

$$W_{m,m+\mu} \propto \kappa^{-(1-2\eta)} \propto \varepsilon^{-\nu(1-2\eta)}$$

For two-dimensional correlations $\Delta = 0$ and

$$W_{m,m+\mu} \propto \kappa^{-2(1-2\eta)} \propto \varepsilon^{-2\nu(1-2\eta)}$$

For $\mu\omega_L \simeq \tau^{-1}(k)$ the relaxation rate cannot be expressed as a simple power law; for $\mu\omega_L \gg \tau^{-1}(k)$ the relaxation rate is predicted to be a constant.

The temperature dependence of the ^{23}Na relaxation rate in the region of the cubic to tetragonal phase transition[105,115] in NaNbO$_3$ is shown in Fig. 43. An anomalous increase in T_1^{-1} by a factor of ~ 30 is indicated. In addition, the T_1^{-1} data show a double peak. The $A_{k_c}A_{-k_c}$ factor is zero for the ^{23}Na nuclei by symmetry at the R point of the Brillouin zone, and a maximum for the M point. The relaxation data therefore indicate that the phase transition is brought about by the condensation of the M point pseudo-rotary mode. It is interesting to note that for a softening of all of the modes between the R and M points the $A_{k_c}A_{-k_c}$ factor would be only $\frac{1}{2}$ of the value of the M point mode alone.

The presence of the double peak is more difficult to interpret. Bonera *et al.*[105] have made several suggestions in terms of the model for the phase transition discussed in the previous section. X-ray scattering observations[116] in NaNbO$_3$ have shown that for $\varepsilon \geq 10^{-2}$ the fluctuations are strongly two dimensional. For a two-dimensional Ising model $\nu = 1$, $\eta = \frac{1}{4}$ and $n = 1.5$; for a three-dimensional Ising model $\nu = 2/3$, $\eta \approx 0$ and $n = 1.33$. For a mean field theory $\nu = \frac{1}{2}$, $\eta = 0$ and $n = 1.0$. The relaxation rate data give $n = 1.2$. The minimum of the relaxation rate observed before the transition temperature is reached could result from a decrease of the factor $A_{k_c}A_{-k_c}$ associated with an increase in the correlation of the fluctuations between adjacent planes following the softening of the R^{25} mode. This could be possible even though recent X-ray measurements[117] show that the tetragonal phase corresponds to the 'freezing in' of the M mode. Another possibility arises if the slowing down assumption is removed. It is possible to obtain a maximum in the relaxation

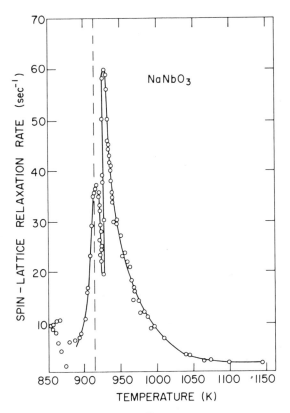

Fig. 43. Temperature dependence of the ^{23}Na spin–lattice relaxation rate in NaNbO$_3$
in the region of the cubic to tetragonal phase transition.

rate if $\tau(\mathbf{k}_c)^{-1} \sim \omega_L$ through an application of the dynamic scaling hypothesis. More recently Avogadro *et al.*[115] have reanalysed the data in a manner similar to that used by Muller *et al.*[121] to analyse EPR data in SrTiO$_3$ (see Section IV, Part G).

The temperature dependence of the ^{27}Al relaxation rate[105] in LaAlO$_3$ shows no anomalous increase near the phase transition. This result is expected since the ^{27}Al nuclei reside at the centre of the oxygen octahedra and the $A_\mathbf{k} A_{-\mathbf{k}}$ factor is zero for both the R^{25} and M modes.

The temperature dependence of the ^{87}Sr relaxation rate[105,118] in SrTiO$_3$ is shown in Fig. 44. The data show no anomalous behaviour near the phase transition. This observation indicates that the soft mode that drives the cubic-tetragonal phase transition is a quasi-pure R^{25} mode so that the rotational fluctuations are practically three-dimensionally correlated.

Because of the many assumptions made in the development of the theory it is difficult to make numerical comparisons with the data. This difficulty is

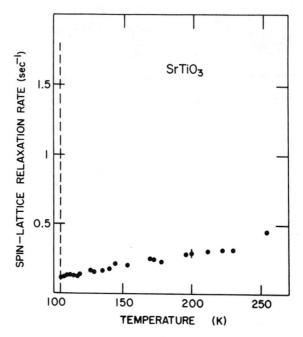

Fig. 44.   Temperature dependence of the ^{87}Sr spin–lattice relaxation rate in SrTiO$_3$
near the 105.6 K cubic to tetragonal phase transition.

enhanced by the lack of knowledge as to when the central mode behaviour may begin to dominate the spin–lattice relaxation rate in the compounds studied.

G. Fluctuations and Correlations in SrTiO₃ as Deduced from Electron Paramagnetic Resonance Experiments

A discussion of the use of magnetic resonance techniques to probe local fluctuations near structural phase transitions must include the results of electron paramagnetic resonance studies. Müller et al.[119–121] have interpreted the temperature dependence of the EPR line-width of the $Fe^{3+} - V_O$ centre (see Fig. 45) in SrTiO$_3$ to obtain detailed dynamical information about the 105.6 K structural phase transition. This transition is driven by the softening of one branch of the R^{25} mode. An anomalous behaviour of the line-width occurs in the same temperature interval for $T \geq T_c$ in which Riste et al.[111,112] observed a central mode using neutron diffraction.

The EPR results reflect the local orientation of the TiO$_6$ octahedra. The high sensitivity attainable permitted a study of the stochastic variation of $\delta\phi_i(t) = \phi_i(t) - \langle\phi_i\rangle$. The time-dependent departure $\Delta H(t)$ of the resonance field from its mean value $H_0 + C\langle\phi_i\rangle$ is proportional to $\delta\phi_i(t)$. For **H** near [110]

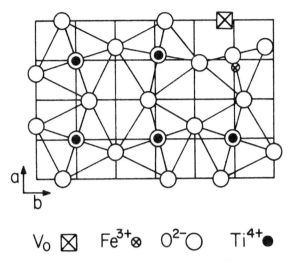

Fig. 45. Schematic representation of rotated oxygen octahedra and oxygen pyramid of an $Fe^{3+} - V_0$ centre in the tetragonal phase of $SrTiO_3$.

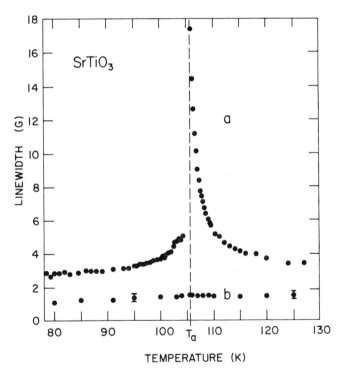

Fig. 46. Temperature dependence of the EPR linewidth of an $Fe^{3+} - V_0$ centre in $SrTiO_3$: (a) for the high-field line; (b) for the low-field line.

the resonance field of the centre with its axis parallel to [110] (high field line) is linearly sensitive to the $\phi^{[010]}$ component and quadratically sensitive to the $\phi^{[010]}$ component. This latter, and much smaller, effect, has been measured separately by the broadening of the low-field line which contains no $\phi^{[001]}$ contribution. The experimental line-width broadening for both lines is depicted in Fig. 46, The background contribution due to fluctuations in the stable lattice modes must first be subtracted. The remaining critical line-width for the high-field line can be related to the autocorrelation function of the fluctuations in $\phi^{[001]}$ so that

$$\Delta H(\varepsilon) \propto \int_0^\infty \langle \phi^{[001]}(0)\phi^{[001]}(t)\rangle_\varepsilon \, dt$$

$$= \int_0^{\pi/a} S(\mathbf{k}, \, \varepsilon, \, \omega = 0) \, d^3\mathbf{k}$$

for the fast motion regime ($\omega \simeq 0$, $T > T_c$). If Schwabl's temperature dependent dynamical structure factor is used to describe the rotational fluctuations, it follows that

$$\Delta H(\varepsilon) \propto \frac{\kappa(\varepsilon)^{-(1-2\eta)}}{\Delta^{1/2}} \arctan \frac{[\pi\Delta^{1/2}]}{[\kappa(\varepsilon)a]}$$

For the temperature range $106.6 < T < 152$ K an excellent fit to the data is achieved for $T_c = 105.6$ K, $\nu = 0.63 \pm 0.07$, $\Delta = 0.017 \pm 0.010$ and $\eta = 0$. the value of κ was taken from neutron diffraction data.[122] The deduced value of the anisotropy parameter Δ implies that about 60 octahedral units in a (001) plane

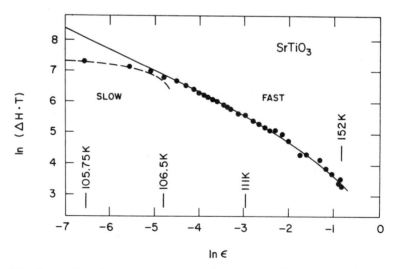

Fig. 47. Comparison between experimental and theoretical broadened EPR line-widths of an $Fe^{3+} - V_0$ centre for the fast- and slow-motion regimes in $SrTiO_3$.

are correlated when two such units in adjacent planes are correlated. This behaviour approximates to the planar Heisenberg model as suggested by Stanley[123] which predicts $\eta = 0$, $\beta = 1/3$, $\nu = 2/3$, and $\gamma = 1.29 \pm 0.010$ as calculated by Schneider and Stoll.[124]

For temperatures in the range $105.75 < T < 106.6$ K, the line-width was observed to be non-divergent. The assumption of the fast fluctuation regime limit breaks down for $T < 106.6$ K. This is to be expected since ΔH is a local variable and the direct coupling of the octahedra limits the local value of ϕ_i^a to $2°$. As T_c is approached the fluctuations are critically slowed down by the central mode. (Slow means relative to the characteristic spin precession frequency of about 100 MHz.) In this regime the spins see the fluctuations as static and the resonance line becomes inhomogeneously broadened. Indeed a change from a Lorentzian line shape to a Gaussian form is observed experimentally at $T = 106.6$ K (see Fig. 47). Schwabl's expression yields for $\kappa \to 0$

$$\langle \Delta H_{\mathrm{crit}}^2(\varepsilon) \rangle \propto \left\{ C - \frac{\kappa(\varepsilon)}{\Delta^{1/2}} \arctan \frac{[\pi \Delta^{1/2}]}{[\kappa(\varepsilon)a]} \right\}$$

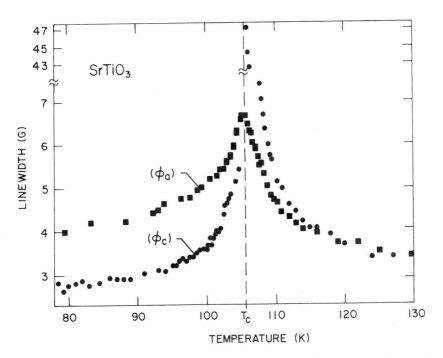

Fig. 48.   Anisotropy of the temperature dependence of the EPR anisotropic line-width of an $Fe^{3+} - V_0$ centre in $SrTiO_3$.

with C a constant. Since the second term vanishes for $\kappa \to 0$, a finite maximum line-width is predicted at the transition. Therefore, one can write

$$\Delta H_{crit}^2(\varepsilon) = \Delta H_{max}^2(1 - C\varepsilon^\nu)$$

An analysis of the data in this regime gives $\Delta H_{max} = 20.1$ G, $C = 17.7$ and $\nu = 0.63$.

Figure 48 shows the line-width measurements for **H** along [110] and [101]. For the former case, for $T < T_c$, one measures fluctuations ϕ^a around the tetragonal axis; for the latter, fluctuations ϕ^c around a perpendicular axis. For $T > T_c$, where the crystal possesses cubic symmetry, the line-width behaviour is anomalous and cannot be explained in terms of the static symmetry properties of the crystal. The anisotropy must be dynamic in origin and is believed to be related to the formation of clusters. That is, for $T > T_c$, regions of correlated rotational fluctuations, larger than a few unit cells in size, exist in which the local fluctuations of a particular component of the equivalent $\phi^{[100]}$ rotations are correlated as the crystal 'senses' the symmetry it will assume in the low-temperature phase. Dynamically, the three [100] directions are not equivalent and the static susceptibility $\chi(\mathbf{k}, \varepsilon) \propto (\kappa^2 + \mathbf{k}^2)^{-1}$ is larger for the correlated component since κ is smallest in this case. According to Schwabl's expression the ϕ^c fluctuation is therefore expected to be larger as $T \to T_c$ from above.

Finally, it is noteworthy that the ϕ^a broadened line-width is remarkably symmetric with respect to T_c in contrast to the behaviour of the ϕ_c broadened line-width. This result is interpreted to mean that $\langle \phi^a \rangle = 0$ in both phases and therefore that the potential well for the ϕ^a motion is not split into two minima $\pm \phi_0^a(T)$ below T_c as is the case for the ϕ^c motion. Rather, the curvature of the well is a minimum and the fluctuations maximum at T_c.

V. CONCLUDING REMARKS

Nuclei possessing quadrupole moments provide useful local probes for the study of both the static and dynamic aspects associated with structural phase transitions. The purely tilting type of displacive phase transition in the perovskite and antifluorite structures has provided the subject material for this article. Although considerable progress has been made towards achieving an understanding of the basic nature of these transitions, many important experiments remain for future investigations. For example, no comprehensive NMR or NQR studies have as yet been reported for an X nucleus in a perovskite nor for an R or M nucleus in an antifluorite. However, a complete understanding of these phase transitions will require not only additional magnetic resonance experiments, but also more neutron scattering, light scattering, thermal expansion, infrared and Raman scattering and other types of measurements. Because of the obvious similarity of the two types of structures, future work might well focus on an attempt to provide a common theoretical understanding.

REFERENCES

1 W. A. Hines and W. P. Knight, *Phys. Rev. B*, **4,** 893 (1971).
2 M. Rotter and B. Sedlak, *Czech. J. Phys. B*, **20,** 1285 (1970).
3 P. Heller and G. B. Benedek, *Phys. Rev. Lett.*, **8,** 428 (1962).
4 G. Bonera, F. Borsa, and A. Rigamonti, *Phys. Rev. B*, **2,** 2784 (1970).
5 R. Blinc, *J. Phys. Chem. Sol.*, **13,** 204 (1960).
6 A. Zussman and S. Alexander, *J. Chem. Phys.*, **49,** 3792 (1968).
7 R. G. Wyckoff, *Crystal Structures*, Vol. III, Second Edition, Wiley–Interscience, New York–London, 1964, p. 339.
8 R. G. Wyckoff, *Crystal Structures*, Vol. II, Second Edition, Wiley–Interscience, New York–London, 1964, p. 390.
9 G. Shirane, *Structural Phase Transitions and Soft Modes*, Eds.: E. J. Samuelsen, E. Andersen, and J. Feder, Universitetsforlaget, Oslo, Norway, 1971, p. 217.
10 J. K. Kjems, G. Shirane, K. A. Müller, and H. J. Scheel, *Phys. Rev. B*, **8,**1119 (1973).
11 G. P. O'Leary and R. G. Wheeler, *Phys. Rev. B*, **1,** 4409 (1970).
12 G. P. O'Leary, *Phys. Rev. Lett.*, **23,** 782 (1969).
13 I. D. Brown, *Can. J. Chem.*, **42,** 2758 (1964).
14 M. J. Buerger, *Sov. Phys.-Crystallogr.*, **16,** 959 (1972).
15 L. D. Landau and E. M. Lifshitz, *Statistical Physics*, Addison-Wesley, Reading, Massachusetts and London, 1958, p.430.
16 K. S. Aleksandrov, V. I. Zinenko, L. M. Mikhel'son, and Yu. I. Sirotin, *Sov. Phys.-Cryst.*, **14,** 256 (1969).
17 V. L. Ginzburg, *Sov. Phys.-Uspekhi*, **5,** 649 (1960).
18 R. A. Cowley, *Phil. Mag.*, **11,** 673 (1965).
19 P. B. Miller and P. C. Kwok, *Sol. St. Commun.*, **5,** 57 (1967).
20 K. S. Aleksandrov and V. I. Zinenko, *Sov. Physics-Sol. St.*, **12,** 1662 (1971).
21 F. E. Goldrich and J. L. Birman, *Phys. Rev.*, **167,** 528 (1968).
22 H. E. Stanley, *Introduction to Phase Transitions and Critical Phenomena*, Clarendon Press, Oxford, 1971, p. 170.
23 K. A. Müller and W. Berlinger, *Phys. Rev. Lett.*, **26,** 13 (1971).
24 J. D. Axe, B. Dorner and G. Shirane, *Phys. Rev. Lett.*, **26,** 519 (1971).
25 E. Pytte and J. Feder, *Phys. Rev.*, **187,** 1077 (1969).
26 R. A. Cowley and A. D. Bruce, *J. Phys. C: Sol. St. Phys.*, **6,** L191 (1973).
27 D. Nakamura and M. Kubo, *J. Phys. Chem.*, **68,** 2986 (1964).
28 R. Ikeda, A. Sasane, D. Nakamura, and M. Kubo, *J. Phys. Chem.*, **70,** 2926 (1966).
29 H. M. van Driel, M. Wiszniewska, B. M. Moores, and R. L. Armstrong, *Phys. Rev. B.*, **6,** 1596 (1972).
30 M. Wiszniewska and R. L. Armstrong, *Can. J. Phys.*, **51,** 781 (1973).
31 A. G. Brown, R. L. Armstrong, and K. R. Jeffrey, *Phys. Rev. B.*, **8,** 121 (1973).
32 N. Tovborg-Jensen, *J. Chem. Phys.*, **50,** 559 (1969).
33 C. K. Møller, *Kyl. Danske Videnskab. Selskab Mat. Fys. Medd.*, **32,** 1 (1959).
34 C. K. Møller, *Nature*, **182,** 981 (1957).
35 S. Hirotsu, *J. Phys. Soc. Jap.*, **31,** 552 (1970).
36 E. H. Carlson, *J. Chem. Phys.*, **55,** 4662 (1971).
37 Y. Fujii, S. Hoshino, Y. Yamada, and G. Shirane, *Phys. Rev. B* (to be published).
38 H. Ohta, H. Harada, and S. Hirotsu (to be published).
39 K. Uchino, K. Yagi, and G. Honjo (to be published).
40 F. Borsa, *Phys. Rev. B*, **7,** 913 (1973).
41 F. F. Borsa, M. L. Crippa, and B. Derighetti, *Phys. Lett. A.*, **34,** 5 (1971).
42 H. Bayer, *Z. Physik*, **130,** 227 (1951).
43 T.-C. Wang, *Phys. Rev.*, **99,** 566 (1955).

44 T. Kushida, *J. Sci. Hiroshima Univ.*, **A19,** 327 (1955).
45 R. L. Armstrong, G. L. Baker, and K. R. Jeffrey, *Phys. Rev. B.*, **1,** 2847 (1970).
46 A. A. Maradudin, E. W. Montroll, and G. H. Weiss, *Theory of Lattice Dynamics in the Harmonic Approximation*, 1st Edn., Academic Press, New York–London, 1963, p. 236.
47 M. Debeau and H. Poulet, *Spectrochim. Acta*, **A25,** 1553 (1969).
48 D. F. Cooke and R. L. Armstrong, *Can. J. Phys.*, **49,** 2381 (1971).
49 R. L. Armstrong and H. M. van Driel, *Can. J. Phys.*, **50,** 2048 (1972).
50 G. A. Samara and B. Morosin, *Phys. Rev. B*, **8,** 1256 (1973).
51 S. H. Wemple, *Phys. Rev.*, **137,** A1575 (1964).
52 T. Kushida, G. B. Benedek, and N. Bloembergen, *Phys. Rev.*, **104,** 1364 (1956).
53 W. R. Abel, *Phys. Rev. B.*, **4,** 2696 (1971).
54 R. L. Armstrong and K. R. Jeffrey, *Can. J. Phys.*, **47,** 1095 (1969).
55 R. L. Armstrong and G. L. Baker, *Can. J. Phys.*, **48,** 2411 (1970).
56 G. L. Baker and R. L. Armstrong, *Can. J. Phys.*, **48,** 1649 (1970).
57 R. L. Armstrong, G. L. Baker, and H. M. van Driel, *Phys. Rev. B.*, **3,** 3072 (1971).
58 G. P. O'Leary, *Phys. Rev. B*, **3,** 3075 (1971).
59 H. W. Willemsen, C. A. Martin, R. L. Armstrong, and P. P. M. Meincke (to be published).
60 R. Ikeda and D. Nakamura, *J. Phys. Chem.*, **69,** 2101 (1965).
61 T. E. Haas and E. P. Marram, *J. Chem. Phys.*, **43,** 3985 (1965).
62 M. Kubo and D. Nakamura, *Advan. Inorg. Chem. Radiochem.*, **8,** 257 (1966).
63 R. Bersohn and R. G. Shulman, *J. Chem. Phys.*, **45,** 2298 (1966).
64 J. H. M. Thornley, *J. Phys. C.: Sol. St. Phys.*, **1,** 1024 (1968).
65 A. G. Brown, R. L. Armstrong, and K. R. Jeffrey, *J. Phys. C.: Sol. St. Phys.*, **6,** 532 (1973).
66 R. H. Busey, H. H. Dearman, and R. B. Bevan Jr., *J. Phys. Chem.*, **66,** 82 (1962).
67 L. A. Woodward and M. J. Ware, *Spectrochim. Acta*, **20,** 711 (1964).
68 A. A. Maradudin, E. W. Montroll, G. H. Weiss, and I. P. Ipatova, *Theory of Lattice Dynamics in the Harmonic Approximation*, 2nd Edn., Academic Press, New York–London, 1971, p. 221.
69 H. M. van Driel, M. M. McEnnan, and R. L. Armstrong (to be published).
70 A. A. Maradudin, E. W. Montroll, G. H. Weiss, and I. P. Ipatova, *Theory of Lattice Dynamics in the Harmonic Approximation*, 2nd Edn., Academic Press, New York–London, 1971, p. 230.
71 D. Gribier, B. Farraux, and B. Jacrot, *Inelastic Scattering of Neutrons in Solids and Liquids,* Vol. II, International Atomic Energy Agency, Vienna, 1963, p. 225.
72 G. Dolling, R. A. Cowley, and A. D. B. Woods, *Can. J. Phys.*, **43,** 1397 (1964).
73 P. Dorain, *J. Chem. Phys.* (to be published).
74 A. D. B. Woods, W. Cochran, and B. N. Brockhouse, *Phys. Rev.*, **119,** 980 (1960).
75 E. W. Kellerman, *Phil. Trans. Roy. Soc. London*, **238,** 513 (1940).
76 A. Baldereschi, *Phys. Rev. B*, **7,** 5212 (1973).
77 D. J. Chadi and M. L. Cohen, *Phys. Rev. B*, **8,** 5747 (1973).
78 D. J. Chadi and M. L. Cohen, *Phys. Rev. B*, **7,** 1572 (1973).
79 C. K. Coogan, *Aust. J. Chem.*, **17,** 1 (1964).
80 R. Bersohn, *J. Chem. Phys.*, **29,** 326 (1958).
81 H. C. Bolton, W. Fawcett, and I. D. C. Gurney, *Proc. Phys. Soc.* (*London*), **80,** 199 (1962).
82 G. G. Belford, R. A. Bernheim, and H. S. Gutowsky, *J. Chem. Phys.*, **35,** 1032 (1961).
83 F. W. de Wette, *Phys. Rev.*, **123,** 103 (1961).
84 E. A. C. Lucken, *Nuclear Quadrupole Coupling Constants*, Academic Press, New York–London, 1969, p. 94.

85 J. D. Axe, G. Shirane, and K. A. Müller, *Phys. Rev.*, **183**, 820 (1969).

86 G. Shirane and Y. Yamada, *Phys. Rev.*, **177**, 858 (1969).

87 J. Van Kranendonk, *Physica*, **20**, 781 (1954).

88 J. Van Kranendonk and M. B. Walker, *Phys. Rev. Lett.*, **18**, 701 (1967).

89 J. Van Kranendonk and M. B. Walker, *Can. J. Phys.*, **46**, 2441 (1968).

90 K. Yoshida and T. Moriya, *J. Phys. Soc. Jap.*, **11**, 33 (1956).

91 J. Kondo and J. Yamashita, *Phys. Chem. Sol.*, **10**, 245 (1959).

92 E. G. Wikner, W. E. Blumberg, and E. L. Hahn, *Phys. Rev.*, **118**, 631 (1960).

93 B. I. Kochelaev, *Sov. Phys.–JETP*, **37**, 171 (1960).

94 S. K. Joshi, R. Gupta, and T. P. Das, *Phys. Rev.*, **134**, 693 (1964).

95 F. Bridges and W. G. Clark, *Phys. Rev.*, **164**, 288 (1967).

96 R. L. Armstrong and K. R. Jeffrey, *Can. J. Phys.*, **49**, 49 (1971).

97 L. C. Hebel and C. P. Slichter, *Phys. Rev.*, **113**, 1504 (1959).

98 K. R. Jeffrey and R. L. Armstrong, *Phys. Rev.*, **174**, 359 (1968).

99 S. Alexander and A. Tzalmona, *Phys. Rev.*, **138**, A845 (1965).

100 R. L. Armstrong and D. F. Cooke, *Can. J. Phys.*, **49**, 2389 (1971).

101 J. Owen and J. H. M. Thornley, *Rept. Progr. Phys.*, **29**, 675 (1966).

102 T. Moriya, *Progr. Theo. Phys.* (*Kyoto*), **16**, 641 (1956).

103 H. M. van Driel and R. L. Armstrong (to be published).

104 K. R. Jeffrey, R. L. Armstrong, and K. E. Kisman, *Phys. Rev. B*, **1**, 3770 (1970).

105 G. Bonera, F. Borsa, and A. Rigamonti, *Revista del Nuovo Cimento, Serie 2*, **2**, 325 (1972).

106 F. Borsa, *Structural Phase Transitions and Soft Modes*, Eds.: E. J. Samuelsen, E. Andersen, and J. Feder, Universitetsforlaget, Oslo, Norway, 1971, p. 371.

107 G. Bonera, F. Borsa, M. L. Crippa, and A. Rigamonti, *Phys. Rev. B*, **4**, 52 (1971).

108 R. Blinc and B. Zeks, *Adv. Phys.*, **21**, 693 (1972).

109 R. Blinc, *Stuctural Phase Transitions and Soft Modes*, Eds.: E. J. Samuelsen, E. Andersen, and J. Feder, Universitetsforlaget, Oslo, Norway, 1971, p. 97.

110 R. A. Cowley, *Adv. Phys.*, **12**, 421 (1963).

111 T. Riste, E. J. Samuelsen, and K. Otnes, *Structural Phase Transitions and Soft Modes*, Eds.: E. J. Samuelsen, E. Andersen, and J. Feder, Universitetsforlaget, Oslo, Norway, 1971, p. 395.

112 T. Schneider and E. Stoll, *Phys. Rev. Lett.*, **31**, 1254 (1973).

113 F. Schwabl, *Phys. Rev. Lett.*, **28**, 500 (1972).

114 G. Bonera, F. Borsa, and A. Rigamonti, *Magnetic Resonance and Related Phenomena*, Ed.: V. Hovi, North Holland, London–Amsterdam, 1973, p. 50.

115 A. Avogadro, G. Bonera, F. Borsa, and A. Rigamonti (to be published).

116 F. Denoyer, R. Comes, and M. Lambert, *Solid State Comm.*, **8**, 1979 (1970).

117 A. M. Glazer and H. D. Megaw, *Phil. Mag.*, **25**, 1119 (1972).

118 G. Angeleni, G. Bonera, and A. Rigamonti, *Magnetic Resonance and Related Phenomena*, Ed.: V. Hovi, North Holland, London–Amsterdam, 1973, p. 346.

119 Th. von Waldkirch, K. A. Müller, W. Berlinger, and H. Thomas, *Phys. Rev. Lett.*, **28**, 503 (1972).

120 Th. von Waldkirch, K. A. Müller, and W. Berlinger, *Phys. Rev. B*, **7**, 1052 (1973).

121 K. A. Müller, *Structural Phase Transitions and Soft Modes*, Eds.: E. J. Samuelsen, E. Andersen, and J. Feder, Universitetsforlaget, Oslo, Norway, 1971, p. 85.

122 T. Riste, E. J. Samuelsen, K. Otnes, and J. Feder, *Solid State Comm.*, **9**, 1455 (1971).

123 H. E. Stanley, *Structural Phase Transitions and Soft Modes*, Eds.: E. J. Samuelsen, E. Andersen, and J. Feder, Universitetsforlaget, Oslo, Norway, 1971, p. 271.

124 T. Schneider and E. Stoll, *Structural Phase Transitions and Soft Modes*, Eds.: E. J. Samuelsen, E. Andersen, and J. Feder, Universitetsforlaget, Oslo, Norway, 1971, p. 383.

AUTHOR INDEX

A

Abe, Y., 118, 120
Abel, W. R., 206
Abragam, A., 2, 11, 12, 96
Ainbinder, N. E., 10, 11, 12
Aleksandrov, K. S., 184, 189
Alexander, S., 10, 11, 12, 179, 230
Allen, L. C., 152, 165, 172
Alrich, R., 148
Alymov, I. M., 12
Amirkhanov, B. F., 10, 11, 12
Anderson, R. E., 1, 4, 6
Andrew, E. R., 12
Anferov, V. P., 12
Angeleni, G., 245
Antheunis, D. A., 21, 22
Armstrong, R. L., 194, 199, 201, 203, 206, 208, 209, 211, 218, 219, 221, 225, 228, 230, 237
Arrighini, G. P., 160, 165, 167
Attia, A. I., 21
Avogadro, A., 244, 245
Axe, J. D., 190, 224
Ayant, Y., 1, 10
Azumi, T., 15

B

Bagus, P. S., 143, 162, 168
Bak, B., 168
Baker, G. L., 201, 206, 209, 211
Baldereschi, A., 220
Ballester, M., 56
Barnes, R. G., 58, 66, 67, 120

Bartell, L. S., 168
Bayer, H., 1, 2, 4, 66, 130, 199
Bazhulin, P. A., 8
Bazley, N. W., 140
Becker, R. S., 15
Belford, G. G., 222
Bendazzoli, G. L., 160, 161, 164, 166, 167
Bender, C. F., 163
Benedek, G. B., 179, 205
Berlinger, W., 190, 246
Bernheim, R. A., 222
Bernstein, T., 118
Bersohn, R., 2, 7, 66, 67, 208, 222
Bethe, H. A., 148
Bevan, Jr., R. B., 211
Birman, J. L., 189
Bitter, F., 19
Bjorkstam, J. L., 9
Blake, F. C., 121
Blinc, R., 8, 80, 84, 87, 95, 107, 179, 239
Bloembergen, N., 205
Blumberg, W. E., 225
Bobkov, Yu. N., 11
Boguslavskii, A. A., 11,
Bolton, H. C., 222
Bonaccorsi, R., 165
Bonera, G., 179, 239, 242, 243, 244, 245
Borsa, F., 179, 197, 198, 237, 242, 243, 244, 245
Boswijk, K. H., 118, 120, 121, 124
Bowmaker, G. A., 118, 119, 120, 122, 123, 124, 125
Boys, S. F., 143, 149
Brandon, R. W., 20

255

Bray, P. J., 56, 58, 66, 67
Breiland, W. G., 21
Breneman, G. L., 118
Brenner, H. C., 21
Bridges, F., 225
Brill, T. B., 127
Brock, J. C., 21
Brockhouse, B. N., 212
Broekema, J., 118
Brossel, J., 19
Brown, A. G., 194, 199, 209, 211
Brown, I. D., 183, 184
Brown, T. L., 127, 131
Brown, W. B., 151
Browne, J. C., 156, 160, 167
Bruce, A. D., 191
Brueckner, K. A., 147, 148
Buckley, M. J., 20, 21, 34, 41, 60, 65, 66, 67
Buerger, M. J., 184
Burbelo, V. M., 12
Busey, R. H., 211
Buyle-Bodin, M., 1

C

Calvin, M., 18
Carlson, E. H., 196
Caron, P., 1, 10
Carrington, A., 35, 39, 65
Cartmell, E., 117
Cederbaum, L. S., 168
Chadi, D. J., 220
Chan, I. Y., 20
Cheesman, G. H., 118, 119, 120, 121
Chiba, T., 9
Chinn, S. R., 19
Christensen, D., 168
Clark, W. G., 225
Clauser, M. J., 158
Clementi, E., 148
Cochran, W., 212
Cohen, M. H., 36, 55, 66
Cohen, M. L., 220
Cohen, R. L., 19
Comes, R., 244
Coogan, C. K., 222
Cook, Jr., E. C., 161
Cooke, D. F., 203, 237
Coolidge, A. S., 148
Cornwell, C. D., 118
Cowley, R. A., 188, 191, 211, 241

Crippa, M. L., 198, 239
Czimadia, I. G., 160, 161, 163, 164

D

Dailey, B. P., 128
Dalgarno, A., 151
Das, G., 148
Das, T. P., 1, 4, 9, 11, 36, 55, 66, 135, 139, 225
Dearman, H. H., 211
Debeau, M., 201, 203, 211, 215, 220
De Groot, M. S., 18, 20, 34, 64, 65
Dehmelt, H. G., 126
Denoyer, F., 244
Depireux, J., 171
Derighetti, B., 198
Devlin, G. E., 19
Dewar, M. J. S., 58
De Wette, F. W., 222
Diercksen, G. H. F., 168
Diner, S., 145
Dixon, M., 160, 161, 164, 166
Docken, K. K., 156, 160, 167
Dodgen, H. W., 1, 4, 6, 11
Dolling, G., 211
Dorain, P., 211
Dorner, B., 190
Dorp, W. G. Van (see Van Dorp, W. G.)
Driel, H. M. Van (see Van Driel, H. M.)
Dunning, Jr., T. H., 161, 164, 166

E

Ebbing, D., 161
Eberhardt, J. J., 160
Eckart, C., 140
Edmiston, C., 143
Edmonds, D. T., 87, 89, 95, 96, 106
Egorov, V. A., 12
Ehrenberg, L., 80, 87, 95, 107
El-Sayed, M. A., 15, 20, 21, 30, 54
Epstein, S. T., 151
Ermler, W. C., 156, 161, 162, 163, 166

F

Falle, H. R., 56
Fankhauser, H. R., 140
Farraux, B., 211

Fawcett, W., 222
Feder, J., 191, 248
Fedin, E. I., 9
Fink, W., 165
Finney, A. J. T., 118, 119, 120, 121
Foley, H. M., 152
Fowles, G. W. A., 117
Franchini, P. F., 160, 161, 166
Francis, A. H., 21, 66
French, D., 118
Fryer, C. W., 131
Fujii, Y., 196

G

Gabes, W., 130
Gearhart, R. C., 127
Gerkin, R. E., 20
Geschwind, S., 19, 135
Gibson, G., 34
Gilbert, M., 145
Gilbert, T. L., 143
Gilmore, E., 34
Ginzburg, V. L., 188, 224
Glasbeek, M., 21
Glazer, A. M., 244
Goldman, M., 96
Goldrich, F. E., 189
Goldstone, J., 145, 148
Gondo, Y., 55
Goodisman, J., 140
Gordeev, A. D., 6, 12
Gornostansky, S. D., 161, 166
Goshinski, O., 137
Gouterman, M., 34, 65
Grechishkin, V. S., 6, 12
Green, S., 156, 167
Gribier, D., 211
Grigolini, P., 156, 157, 158, 164, 169
Grimaldi, F., 143
Groot, M. S. De (see De Groot, M. S.)
Guidotti, C., 160, 165, 167
Gunther-Mohr, G. R., 135
Gupta, R., 225
Gurney, I. D. C., 222
Gutowsky, H. S., 1, 6, 10, 11, 12, 222
Gwaiz, A. A., 21

H

Haas, T. E., 208, 210

Hach, R. J., 118, 128, 130
Hacobian, S., 118, 119, 120, 122, 123, 124, 125
Hagiwara, S., 118, 120
Hahn, E. L., 1, 4, 9, 11, 36, 55, 66, 71, 83, 86, 95, 101, 109, 135, 139, 225
Hameka, H. F., 18, 32
Hand, R. W., 168
Handy, N. C., 149
Hansen-Nygaard, L., (and Hansen, L.), 168
Harada, H., 118, 120, 122, 123, 124, 125, 130, 196
Hartland, A., 80
Hartmann, S. R., 71
Harris, C. B., 20, 21, 22, 34, 41, 48, 60, 65, 66, 67
Harris, P. M., 121
Hashi, T., 80
Havinga, E. E., 118, 124
Hayakawa, N., 118
Hazell, A. C., 120
Hebel, L. C., 226
Heller, P., 179
Hensley, E. B., 21
Herbstein, F. H., 118
Hesselmann, I. A. M., 18, 20, 64, 65
Hines, W. A., 179
Hinze, J., 156, 160, 167
Hirota, N., 65
Hirotsu, S., 196
Hirschfelder, J. O., 151
Hoff, C. A. Van't (see Van't Hoff, C. A.)
Hollister, C., 159
Honjo, G., 196
Hoover, R. J., 21, 22
Horsfield, A., 56
Hoshino, S., 196
Hunt, M. J., 87, 89, 95, 96, 106
Hunt, W. J., 168
Hunter, S. J., 21
Hurley, A. C., 146
Hutchison, Jr., C. A., 18, 20, 55, 65
Hylleraas, E. A., 148, 151

I

Ikeda, R., 194, 208, 209
Ipatova, I. P., 211, 222
Ivanov, E. N., 10
Izmestiev, I. V., 10, 11, 12

J

Jacrot, B., 211
James, H. M., 148
James, W. J., 118
Jeffrey, K. R., 194, 199, 201, 206, 209, 211, 225, 228, 230, 237
Jennings, P., 140
Johnson, G. S., 117
Jordan, F., 145
Joshi, S. K., 225

K

Kalman, O. F., 30
Kari, K. E., 160, 161, 163, 164
Kasha, M., 48
Kasha, M. J., 18
Kastler, A., 19
Kellerman, E. W., 213, 222
Kelly, H. P., 145, 146
Kemble, E. C., 137
Kent, L. G., 127, 131
Kern, C. W., 156, 161, 162, 163, 166
Ketudat, S., 9
Khotsyanova, V., 121
Kimball, G. E., 117
Kinoshito, M., 15
Kisman, K. E., 237
Kitaigorodskii, I. I., 121
Kjems, J. K., 183
Klemperer, W., 140
Knight, W. P., 179
Kochelaev, B. I., 225
Kojima, S., 118, 120
Kondo, J., 225
Koo, J., 83, 95, 96
Korst, N. N., 10, 11
Kosulin, A. T., 12
Kothandaraman, G., 21, 48, 65
Kottis, Ph., 19
Kracht, D., 130
Kraemer, W., P., 168
Kranendonk, J. Van (see Van Kranendonk, J.)
Krohn, B. J., 156, 161, 166
Kuan, T. S., 20
Kubo, M., 118, 119, 120, 122, 123, 124, 125, 127, 130, 131, 194, 208, 209
Kukolich, S. G., 157
Kunitomo, M., 80
Kurita, Y., 118

Kushida, T., 130, 200, 205
Kutzelnigg, W., 148
Kvasnička, V., 145
Kwiram, A. L., 19, 21
Kwok, P. C., 188
Kyuntsel, I. A., 6, 12

L

Lambert, M., 244
Landau, L. D., 1, 11, 184, 186, 187
Landesman, A., 96
Lefebvre, R., 19
Lennard-Jones, J. E., 146
Leppelmeier, G. W., 86, 95
Lewis, G. N., 18
Lifschitz, E. M., 1, 11, 184, 186, 187
Lin, C. T., 21
Linderberg, L., 158
Liskov, D. H., 162, 163
Lister, D. G., 160, 167
Liu, B., 168
Loo, B. H., 21
Lotfullin, R. Sh., 2, 3, 8, 10, 11, 12
Löwdin, P. O., 137, 143, 148
Lower, S. K., 15
Lucken, E. A. C., 36, 55, 56, 58, 66, 135, 139, 222
Luckhurst, G. R., 56
Ludwig, G. W., 120
Lurie, F. M., 71, 107

M

McClure, D., 34
McDugle, Jr., W. G., 131
McEnnan, M. M., 211, 218, 219, 221
McGlynn, S. P., 15
McGrath, J. W., 9
Mack, Jr., E., 121
Mackay, A. L., 87, 89, 95, 96, 106
McLachlan, A. D., 35, 39, 65
McWeeny, R., 136, 140, 146, 148, 152, 172
Magera, R. V., 11
Maki, A. H., 20, 21, 40, 41, 55
Mali, M., 80, 84, 87, 95, 107
Malrieu, J. P., 145
Mangum, B. W., 18, 65
Maradudin, A. A., 201, 211, 222
March, N. H., 145, 146

Marram, E. P., 1, 10, 208, 210
Martin, C. A., 208, 211
Matcha, R. L., 156, 161, 166
Matsen, F. A., 156, 160, 167
Matuo, T., 131
Mazziotti, A., 140
Megaw, H. D., 244
Meincke, P. P. M., 208, 211
Mellor, J. W., 117
Meyer, W., 146, 148
Migchelsen, T., 118, 123, 124, 127, 128
Mikhel'son, L. M., 184
Miller, P. B., 188
Moccia, R., 149, 152, 153, 155, 156, 157, 158, 160, 161, 164, 166, 169
Möeller, C., 141
Moffit, W., 34, 65
Møller, C. K., 196
Mokarram, M., 160, 162, 171
Montroll, E. W., 201, 211, 222
Mooney, R. C. L., (and Mooney-Slater, R. C. L.), 117, 118, 119, 120, 127, 128
Moores, B. M., 194, 225, 228
Moriya, T., 225, 237
Morosin, B., 205
Moskowitz, J. W., 159, 161, 163, 166
Mössbauer, R. L., 158
Mukherjee, N. G., 148
Müller, K. A., 183, 190, 224, 245, 246

N

Nakamura, D., 118, 119, 120, 122, 123, 124, 125, 127, 130, 131, 194, 208, 209
Nesbet, R. K., 147
Neumann, D. B., 161, 163, 166
Nicholas, J. V., 55
Niessen, W. Von (see Von Niessen, W.)
Nijman-Meester, M. A. M., 130
Nishimura, A. M., 21
Nozières, P., 145, 146

O

Ogawa, S., 118
Ohta, H., 196
O'Konski, C. T., 161, 162, 163, 164, 165
O'Leary, G. P., 183, 189, 194, 201, 207, 211, 224
Osipenko, A. N., 10, 11, 12
Osredkar, R., 80, 84, 87, 95, 107

Ossman, G. W., 9
Otnes, K., 241, 246, 248
Owen, J., 237
Owens, D., 54
Owens, O. V., 30

P

Pakhomov, V. I., 6
Palmer, P., 65
Palmieri, P., 160, 161, 164, 166, 167
Panos, R. M., 21, 65, 66, 67
Parker, H., 21
Pauling, L., 117
Peierls, R. F., 152
Pekeris, C. L., 148
Peng, H., 151, 172
Pimentel, G. C., 117, 118, 130
Pines, A., 21
Plesset, M. C., 141
Pooley, D., 56
Pople, J. A., 146
Popov, A. I., 117
Poulet, H., 201, 203, 211, 215, 220
Pound, R. V., 1, 9, 126
Prelesnik, A., 80, 87, 95
Primas, H., 145
Proctor, W. G., 96
Pytte, E., 191
Pyykkö, P., 158

R

Ragle, J. L., 1, 4, 5, 6, 8, 10, 11, 160, 162, 171
Rakhimov, A. A., 8
Rakov, A. V., 8
Rastrup-Andersen, J., 168
Rehn, V., 9
Reif, F., 36, 55, 66
Rez, I. S., 6
Rigamonti, A., 179, 239, 242, 243, 244, 245
Rinné, M., 171
Riste, T., 241, 246, 248
Rose, M. E., 138
Rosenberg, Yu. I., 6, 12
Rothemberg, S., 163, 165, 166, 167
Rotter, M., 179
Ruedenberg, K., 143
Rundle, R. E., 118, 128, 130

S

Salem, L., 169,170
Samara, G. A., 205
Sampanthar, S., 145, 146
Samuelsen, E. J., 241, 246, 248
Sarasvati, V., 9
Sasane, A., 118, 119, 120, 130, 131, 194
Schaefer III, H. F., 143, 162, 163, 165, 166, 167
Scheel, H. J., 183
Schempp, E., 56
Schlupp, R. L., 21
Schmidt, J., 20, 21, 22, 56
Schneider, T., 241, 246, 249
Schuch, H., 21
Schulman, J. M., 159
Schwabl, F., 241, 242
Scott, G. W., 55
Scrocco, E., 135, 139, 165
Sedlak, B., 179
Segel, S. L., 58, 120
Seliger, J., 80, 84, 87, 95, 107
Semin, G. K., 2, 3, 6, 9, 10, 11, 12
Sharnoff, M., 19, 21
Shimauchi, A., 118, 120
Shirane, G., 183, 190, 196, 224
Shulman, R. G., 208
Sinanoğlu, O., 146, 147, 148
Sirotin, Yu. I., 184
Slater, J. C., 128
Slichter, C. P., 11, 71, 107, 226
Slusher, R. E., 83, 101
Smith, J. A. S., 6, 11, 131
Soda, G., 9
Soifer, G. B., 10, 11, 12
Speight, P. A., 87
Stanley, H. E., 190, 249
Stehlik, D., 109
Sternheimer, R., 152
Stewart, A. L., 151
Stoicheff, B. P., 168
Stoll, E., 241, 246, 249
Strandberg, M. W. P., 34, 65
Struchkov, M., 121
Summers, C. P., 87, 89, 95, 106
Sutcliffe, B., 140, 148

T

Tae-Kyu Ha, 161, 162, 163, 164, 165
Tasman, H. A., 118, 120, 121

Terao, T., 80
Thomas, H., 246
Thornley, J. H. M., 208, 237
Tinkham, M., 34, 65
Tinti, D. S., 20, 21, 30, 48, 65
Tokuhiro, T., 1, 6, 10
Tomasi, J., 160, 165, 167
Tovberg-Jensen, N., 196
Townes, C. H., 128, 135
Tsukada, K., 118
Tsutsumi, Y., 80
Tsyganov, V. V., 10, 11
Tycko, D., 152
Tzalmona, A., 10, 11, 12, 230

U

Uchino, K., 196

V

Valiev, K. A., 10
Van der Waals, J. H., 18, 20, 21, 22, 34, 56, 64, 65
Van Dorp, W. G., 21
Van Driel, H. M., 194, 203, 206, 211, 218, 219, 221, 225, 228, 237
Van Kranendonk, J., 224, 225
Van't Hoff, C. A., 21
Veeman, W. S., 21, 56
Veillard, A., 148
Verbeck, P. J. F., 21
Versilov, V. S., 12
Vijayaraghava, R., 9
Vincent, J. S., 40, 55
Von Niessen, W., 168
Von Waldkirch, Th., 246
Vos, A., 118, 123, 124, 127, 128

W

Waals, J. H. Van der (see Van der Waals, J. H.)
Wahl, A. C., 148
Waldkirch, Th. Von (see Von Waldkirch, Th.)
Walker, M. B., 225
Wang, T.-C., 168, 199
Ware, M. J., 211, 215
Watkins, G. D., 1

Webb, R. H., 48
Weinberg, H. F., 140
Weinhold, F., 140
Weiss, G. H., 201, 211, 222
Welsh, W. A., 127
Wemple, S. H., 205
Wette, F. W. de (see de Wette, F. W.)
Wheeler, R. G., 183, 189, 194, 211, 224
Whiffen, D. H., 56
Wiebenga, E. H., 118, 124, 130
Wikner, E. G., 225
Willemsen, H. W., 208, 211
Willett, R. D., 118
Willimorth, J. H., 9
Wilson, Jr., E. B., 140
Winscom, C. J., 21
Wiszniewska, M., 194, 225, 228
Woessner, D. E., 1, 6, 10, 11, 12
Wofsy, S. C., 157
Woods, A. D. B., 211, 212
Woodward, L. A., 211, 215
Wyckoff, R. G., 180, 182

Y

Yagi, K., 196
Yamada, Y., 196, 224
Yamasaki, R. S., 118
Yamashita, J., 225
Yarkony, D. R., 168
Yee, E. M., 30
Yoshida, K., 225
Young, W. H., 145, 146
Yusupov, M. Z., 12

Z

Zandomeneghi, M., 160, 161, 166
Zeks, B., 239
Zewail, A. H., 21
Zhukov, A. P., 6
Ziegler, S., 54
Zinenko, V. I., 184, 189
Zupančič, I., 80, 87, 95
Zussman, A., 10, 11, 179

SUBJECT INDEX

A

Acoustic waves, induction of electron-spin transitions by, 21

Adiabatic demagnetization, 21, 87 *et seq.*, 112 *et seq.*

measurement of short $T_{1\rho}$ by, 114

Aluminium hydride, AlH, theoretical calculation of electric field gradient in, 162

Ammonia, theoretical calculation of electric field gradient in, 160, 164, 175

Ammonium dihydrogen phosphate, $^1H/^{14}N$ double resonance in, 108

Ammonium hexabromoplatinate, temperature dependence of ^{79}Br quadrupole resonance frequency in, 193 *et seq.*

Ammonium hexachloroiridate, π-bonding effects in, 208

Ammonium triiodide

anomalous temperature coefficient of ^{127}I quadrupole resonance frequency in, 130 *et seq.*

crystal structure of, 117, 119 *et seq.*

^{127}I quadrupole resonance in, 118 *et seq.*, 128

Anharmonic effects in displacive phase transitions, 191, 238 *et seq.*

Antifluorite compounds, nuclear quadrupole resonance in, 179 *et seq.*

Antishielding factor, 172 *et seq.*

A priori calculations of nuclear quadrupole coupling constants, 135 *et seq.*

Atomic Units, 159

B

Barium titanate, phase transition in, 190

Bayer theory of effect of torsional modes on nuclear quadrupole resonance frequencies, 199

Benzene, theoretical calculation of electric field gradient in, 162, 163

Beryllium difluoride, theoretical calculation of electric field gradient in, 167

Beryllium hydride, BeH, theoretical calculation of electric field gradient in, 160

Born–Oppenheimer approximation, 135, 155

Boron difluoride, BF_2, theoretical calculation of electric field gradient in, 167

Boron hydride, BH, theoretical calculation of electric field gradient in, 160

Brillouin's theorem, 141, 143, 155

Brillouin zone

in cubic antifluorite structure, 180, 181

in perovskite structure, 182

C

Caesium hexachloroplatinate

^{35}Cl spin–lattice relaxation in, 229, 230

pressure dependence of ^{35}Cl spin–lattice relaxation time in, 237

rotary mode frequency in, 236

Caesium lead trichloride, $CsPbCl_3$

phase transition in, 185

Caesium lead trichloride—*Contd.*
temperature dependence of ^{35}Cl quadrupole resonance frequency in, 195 *et seq.*
Caesium octaiodide
crystal structure of, 124
^{127}I quadrupole resonance in 125 *et seq.*
Caesium triiodide
crystal structure of, 121
^{127}I quadrupole resonance in, 120 *et seq.*
Carbon difluoride, CF_2, theoretical calculation of electric field gradient in, 163, 168
Carbon hydride, CH, theoretical calculation of electric field gradient in, 160
Carbon monoxide, theoretical calculation of electric field gradient in, 163, 166
Chloramine, NH_2Cl, theoretical calculations of electric field gradient in, 167
Chlorine hyperfine interactions in free radicals, 55
Chlorobenzene, $^3\pi\pi^*$ state of, 44 *et seq.*
8-Chloroquinoline, $^3\pi\pi^*$ state of
^{14}N quadrupole interactions in, 52 *et seq.*
^{35}Cl quadrupole interactions in, 52 *et seq.*
Chloroiodides, nuclear quadrupole resonance in, 118
Cluster expansion, 146
Configuration interaction (CI) method in calculation of nuclear quadrupole coupling constants, 144 *et seq.*
Correlation time for molecular motion, effect on nuclear quadrople resonance frequencies, 1
Critical dynamics
and electron spin resonance linewidths, 246 *et seq.*
and spin–lattice relaxation, 238 *et seq.*
Cross-relaxation time, 99, 104, 106, 108
Cyanate ion, theoretical calculations of electric field gradient in, 165
Cyanide ion, theoretical calculation of electric field gradient in, 165
Cyanoacetylene, NCCCH, theoretical calculations of electric field gradient in, 165
Cyanogen, theoretical calculation of electric field gradient in, 165
Cyanogen chloride, theoretical calculation of electric field gradient in, 165

Cyanogen fluoride, theoretical calculation of electric field gradient in, 165
Cyclopentadiene, theoretical calculations of electric field gradient in, 168
Cytosine, ^{1}H/^{14}N double resonance in, 94

D

Density matrix, 78, 140 *et seq.*, 226
Dibromoethane, molecular reorientation in, 4 *et seq.*
Dichloroethane, molecular reorientation in, 4 *et seq.*
p-Dichlorobenzene, molecular reorientation in, 8
quadrupole interaction in $^3\pi\pi^*$ state, 60 *et seq.*
bent C—Cl bond in excited state, 67
'solid effect' in, 96
Dichloroiodide ion, structure of, 117
Difluorinemonoxide, F_2O, theoretical calculation of electric field gradient in, 165, 168
Diglycine, ^{14}N quadrupole coupling constant in, 107
Dimethylammonium triiodide, ^{127}I quadrupole resonance in, 123 *et seq.*
Di-*n*-propyl ammonium triiodide, ^{127}I quadrupole resonance in, 123 *et seq.*
Dipolar spin-lattice relaxation time, see T_{1D}
Displacive phase transititions, 179 *et seq.*
Double resonance detection of nuclear quadrupole resonance, 71 *et seq.*
in the rotating frame, 71 *et seq.*, 87 *et seq.*, 109 *et seq.*
in the laboratory frame, 71 *et seq.*, 95 *et seq.*
between the laboratory and rotating frame, 71 *et seq.*, 82 *et seq.*
Double rotating frame representation, 78

E

Electric field gradient
dependence on local point symmetry, 139
point-charge calculation of, 222
theoretical calculation of, 139 *et seq.*

Electron spin resonance (e.s.r.) of molecules in triplet states, 19, 30 *et seq.*

Electron–electron double resonance (EEDOR), 20

ENDOR
experimental methods, 50 *et seq.*
optically detected, 15 *et seq.*, 20
sensitivity of, 30 *et seq.*, 43 *et seq.*

Electronic quadrupole polarization, 157 *et seq.*

Electrostatic Hamiltonian, 136 *et seq.*

Ewald theta transformation, 222

Excitons, study by optically detected magnetic resonance, 21

Hydrocyanic acid, theoretical calculation of electric field gradient in, 161, 165

Hydrogen-bonding
effect on nuclear quadrupole resonance frequencies, 131
proton exchange, in triglycine sulphate, 82
proton exchange, in KH_2PO_4, 84 *et seq.*

Hydrogen chloride, theoretical calculation of electric field gradient in, 162

Hydroxylamine, theoretical calculation of electric field gradient in, 165

Hyperfine interaction, electron-nuclear, 39 *et seq.*

Hypochlorous acid, theoretical calculation of electric field gradient in, 167

F

Ferric fluoride, FeF_3, theoretical calculation of electric field gradient in, 168

Fermi contact term, 40

First-order observables, calculation of
by perturbation theory, 149 *et seq.*
by variation theorem, 152 *et seq.*

Fluorescence, 17

Fluorine monochloride, theoretical calculation of electric field gradient in, 167

Force constant, relationship to nuclear quadrupole coupling constant, 169 *et seq.*

Formaldehyde, theoretical calculation of electric field gradient in 161, 163, 166

Furan, theoretical calculation of electric field gradient in, 168

I

Independent electron-pairs method, 147 *et seq.*

Inner core polarization, SCF calculation at ^{14}N in NH_3, 175

Internal conversion, 18

Internal vibrational modes, effect on nuclear quadrupole resonance frequencies, 201 *et seq.*

Intersystem crossing rates, 18, 22 *et seq.*

Intramolecular energy transfer, 17, 21

K

Krypton difluoride, theoretical calculation of electric field gradient in, 168

K-shells, quadrupole polarization of, 152, 173, 174, 175

Kushida theory of effect of torsional modes on nuclear quadrupole resonance frequencies, 200

Kushida, Benedek, Bloembergen theory of pressure effects on nuclear quadrupole resonance frequencies, 205 *et seq.*

G

Geminal product wave functions, 146

Grüneisen parameter, 227, 228

Guanine, $^1H/^{14}N$ double resonance in, 94

H

Heat pulses, induction of electron-spin transitions by, 21

Helmann–Feynman theorem, 153, 169

Hydrazine, theoretical calculations of electric field gradient in, 165

L

Landau–Lifshitz theory of second-order transitions, 186 *et seq.*

Lanthanum aluminate, $LaAlO_3$
^{27}Al spin–lattice relaxation in, 245
phase transition in, 183, 198, 223, 224
neutron scattering in, 224
Level-crossing
multiple cycle, sensitivity of, 92 *et seq.*
single cycle, sensitivity of, 91 *et seq.*
use in double resonance, 86, 87 *et seq.*
use in T_1 measurements, 112 *et seq.*
Line-width of nuclear quadrupole resonance lines, relation to relaxation times, 9
Lithium fluoride, 'solid effect' in, 96
Lithium hydride, theoretical calculation of electric field gradient in, 160, 167
Localized orbitals, use in interpretation of nuclear quadrupole coupling constants, 143 *et seq.*

M

Magnesium hydride, MgH, theoretical calculation of electric field gradient in, 162
Manganese hydride, MnH, theoretical calculation of electric field gradient in, 162
Many-body theory, in calculation of nuclear quadrupole coupling constants, 145 *et seq.*
Mean field theory for second-order transitions, 186 *et seq.*
Methane, theoretical calculation of electric field gradient in, 160
Methylamine, theoretical calculation of electric field gradient in, 165
Methyl chloride, theoretical calculation of electric field gradient in, 167
Methyl cyanide, theoretical calculation of electric field gradient in, 163
Methyl isocyanide, theoretical calculation of electric field gradient in, 163
Methyl negative ion, CH_3^-, theoretical calculation of electric field gradient in, 160, 163
Methylsilane, theoretical calculation of electric field gradient in, 162, 163
Molecular reorientation (see reorientation, molecular)
Multi-configuration SCF method, 148

N

Naphthalene, e.s.r. spectrum in triplet state, 18
Nitrogen, theoretical calculations of electric field gradient in, 164
Nitrogen hyperfine interactions in free radicals, 55
^{14}N, double resonance with ^1H, sensitivity of, 89 *et seq.*
^{14}N spin–lattice relaxation time, effect on sensitivity of double resonance experiments, 91 *et seq.*
Nitrogen difluoride, NF_2, theoretical calculations of electric field gradient in, 165, 168
Nonradiative energy transfer, 17, 22, 29
Nuclear polarization, 135

O

Octaiodide ion, charge distribution in, 129
Optical detection of e.s.r. in
$\bar{E}$ (2E) state of Cr^{3+} in Al_2O_3, 19
lowest triplet state of naphthalene, 19
triplet state of phenanthrene, 19
triplet state of quinoxaline-d_6, 20
Optically detected magnetic resonance (ODMR), 15 *et seq.*
experimental methods, 48 *et seq.*
sensitivity of, 22 *et seq.*
spectrum of 8-chloroquinoline, 52 *et seq.*
spectrum of p-dichlorobenzene, 60 *et seq.*
Orbach processes, 28
Overhauser effect, 30

P

Perovskites, nuclear quadrupole resonance in, 179 *et seq.*
Perturbation theory
in calculation of nuclear quadrupole coupling constants, 137 *et seq.*, 158
Perturbed coupled Hartree-Fock method (PCHF), 152
Phosphole, theoretical calculation of electric field gradient in, 168

Phosphorescence, 17, 18, 54

Phosphorescence microwave double resonance (PMDR), 20, 21, 22 *et seq.*
  modulation by microwave radiation, 21

Phosphoridine, theoretical calculation of electric field gradient in, 168

Phosphorus hydride, PH, theoretical calculation of electric field gradient in, 162

Polyglycine, ^{1}H/^{14}N double resonance in, 106 *et seq.*

Polyiodide ions, charge distribution in, 13) *et seq.*

Potassium chlorate, reorientation of ClO$_3^-$ ion in, 11

Potassium dihydrogen phosphate, ^{1}H/^{17}O double resonance in, 84 *et seq.*

Potassium hexabromoplatinate
  dispersion curves for longitudinal and transverse rotary mode frequencies, 233
  phase transitions in, 185, 193
  rotary mode frequency in, temperature dependence of, 203, 205, 232, 234, 236
  temperature dependence of ^{79}Br quadrupole resonance frequency in, 193, 203
  temperature dependence of ^{79}Br spin–lattice relaxation time in, 230

Potassium hexabromotellurate, phase transition in, 183

Potassium hexachloroiridate
  hindered rotation of IrCl$_6^{2-}$ ion in, 207
  π-bonding in, 208 *et seq.*, 209 *et seq.*
  pressure effect on ^{35}Cl quadrupole resonance frequency, 206 *et seq.*
  temperature effect on ^{35}Cl quadrupole resonance frequency, 209

Potassium hexachloroosmate
  π-bonding in, 208
  thermal expansion coefficient in, 208

Potassium hexachloroplatinate
  angular displacement of PtCl$_6^{2-}$ ion in, 221
  hindered rotation of PtCl$_6^{2-}$ ion in, 207
  π-bonding in, 208, 209 *et seq.*
  pressure dependence of ^{35}Cl quadrupole resonance frequency in, 206 *et seq.*
  pressure dependence of ^{35}Cl spin-lattice relaxation time in, 237

Potassium hexachloroplatinate—*Contd.*
  rotary mode frequency in, 204 *et seq.*, 236
  temperature dependence of ^{35}Cl quadrupole resonance frequency in, 203 *et seq.*, 221
  temperature dependence of ^{35}Cl spin-lattice relaxation time in, 235

Potassium hexachlororhenate
  angular displacement of ReCl$_6^{2-}$ ion in, 221
  crystal structure of, 212
  dispersion curves for internal modes, 214
  dispersion curves for lattice modes, 216
  dispersion curves for rotary modes, 217
  internal vibrational mode frequencies in, 202
  lattice dynamical model for, 211 *et seq.*
  lattice force constants, effect on rotary and normal mode frequencies, 218, 219
  lattice mode frequencies for, 214
  π-bonding in, 208
  phase transitions in, 183, 192
  point-charge calculations for, 222
  rotation angle of ReCl$_6^{2-}$ octahedra in, 198 *et seq.*, 222, 223
  temperature dependence of ^{35}Cl quadrupole resonance frequency in, 192 *et seq.*, 203, 207

Potassium hexaiodoplatinate, phase transition in, 185

Potassium manganese trifluoride, KMnF$_3$, ^{39}K quadrupole coupling constant in, 196 *et seq.*

Potassium tantalate, KTaO$_3$, ferroelectric transition in, 205

Pressure
  effect on nuclear quadrupole resonance frequencies, 205 *et seq.*
  effect on nuclear quadrupole spin–lattice relaxation times, 237

Proton migration, effect on nuclear quadrupole resonance frequencies, 6, 82, 85

Pseudo-harmonic approximation, 189, 219

Pyrazine, theoretical calculation of electric field gradient in, 162, 163, 165

Pyridine, theoretical calculation of electric field gradient in, 162, 163, 165, 168

Pyrrote, theoretical calculations of electric field gradient in, 168

Rubidium triiodide, ^{127}I quadrupole resonance in, 120 *et seq.*

Q

Quadrupole Hamiltonian, 2, 36 *et seq.*, 191
effect of molecular motion, 2 *et seq.*
Quadrupole double resonance in high magnetic fields
experimental methods, 73 *et seq.*
sensitivity of, 79
Quinoline, $^3\pi\pi^*$ state of, 41
^{14}N quadrupole interaction in $^3\pi\pi^*$ state of, 53

R

Radiative energy transfer, 22, 25 *et seq.*
Radiofrequency induced coupling between spin systems, 96 *et seq.*
single coupling process, 103 *et seq.*
multiple coupling process, 105 *et seq.*
Radius ratio, use in predicting distortion of antifluorite lattice, 184
Reorientation, molecular
effect on nuclear quadrupole resonance frequencies, 1 *et seq.*, 4
effect on spin–lattice relaxation times, 10 *et seq.*, 207
Rotary mode frequencies, temperature dependence of, in Rb$_2$PtCl$_6$, K$_2$PtCl$_6$, K$_2$PtBr$_6$, 205, 232
Rotating-frame representation, 77
Rotating-frame spin–lattice relaxation time, see $T_{1\rho}$
Rotation, hindered
effect on nuclear quadrupole resonance frequencies, 7 *et seq.*, 201 *et seq.*
effect on spin–lattice relaxation times, 10 *et seq.*
Rotational state, dependence of nuclear quadrupole coupling constant on, 157
Rubidium dihydrogen phosphate, ^{1}H/^{87}Rb double resonance in, 109
Rubidium hexachloroplatinate
pressure dependence of ^{35}Cl spin–lattice relaxation time in, 237
rotary mode frequency in, 205
temperature dependence of ^{35}Cl quadrupole resonance frequency in, 203 *et seq.*

Second-order phase transitions, 184
Landau–Lifshitz theory of, 186 *et seq.*
Silicon hydride, SiH, theoretical calculation of electric field gradient in, 162
Sodium fluoride, ^{19}F/^{23}Na double resonance in, 108
Sodium hydride, NaH, theoretical calculation of electric field gradient in, 162
Sodium niobate
^{23}Na spin–lattice relaxation in, 224, 244 *et seq.*
X-ray scattering in, 244
Soft mode phonons, 180, 188 *et seq.*, 196, 203, 241
effect on nuclear quadrupole resonance frequencies, 199 *et seq.*
Solid effect
use in double resonance, 86, 95 *et seq.*
use in T_1 measurements by double resonance, 113 *et seq.*
Spin diffusion, 79
Spin-echo from excited triplet state, 21
multiple spin echoes, 21
Spin–lattice relaxation time (see also T_1)
critical dynamics, effect on, 238 *et seq.*
effect on sensitivity of optically detected magnetic resonance, 27 *et seq.*
in Cs$_2$PtCl$_6$, 230, 237
in K$_2$PtBr$_6$, 231
in K$_2$PtCl$_6$, 235, 237
in Rb$_2$PtCl$_6$, 237
in paramagnetic compounds, 237
lattice modes, effect on, 224 *et seq.*, 228 *et seq.*
measurement by double resonance in the rotating frame, 109 *et seq.*
measurement in the laboratory frame via the 'solid effect', 113 *et seq.*
phase transitions, effect on, 244 *et seq.*
pressure dependence, in K$_2$PtCl$_6$, Rb$_2$PtCl$_6$ and Cs$_2$PtCl$_6$, 237
quadrupole relaxation, 10, 11
temperature dependence, in K$_2$IrCl$_6$ and K$_2$ReCl$_6$, 238
Spin–lattice relaxation time, dipolar, see T_{1D}

Spin–lattice relaxation time for quadrupole relaxation, see T_1

Spin–lattice relaxation time in the rotating frame, see $T_{1\rho}$

Spin–locking, 21

Spin–orbit coupling
effect on energy transfer rates, 18
effect on spin Hamiltonian, 17, 32

Spin quenching, 95

Spin temperature, 75, 76, 87

Spin-vibronic coupling
effect on energy transfer rates, 18

Spontaneous polarization, relationship of quadrupole coupling constants to, 86

Sternheimer polarization correction, 172 *et seq.*, 222

Strontium titanate, $SrTiO_3$
crystal structure of, 247
electron spin resonance of V_0 centre in, 245, 246 *et seq.*
neutron scattering in, 224
phase transition in, 183, 190
^{87}Sr spin–lattice relaxation in, 245

Sulphur hydride, SH, theoretical calculation of electric field gradient in, 162

T

T_1 for quadruple relaxation
^{79}Br, in K_2PtBr_6, 231
calculation by density-matrix theory, 11
calculation by perturbation theory, 10
^{35}Cl, in $KClO_3$, 11
^{35}Cl, in Cs_2PtCl_6, 230
^{35}Cl in K_2IrCl_6, 238
^{35}Cl, in K_2PtCl_6, 235
^{35}Cl, in K_2ReCl_6, 238
effect of molecular motion on, 10

T_{1D}, dipolar spin–lattice relaxation time, 83, 91
effect on sensitivity of double-resonance experiments, 91 *et seq.*

$T_{1\rho}$, spin–lattice relaxation time in the rotating frame, 74, 76, 78, 79, 109 *et seq.*, 114

Temperature dependence of nuclear quadrupole resonance frequencies
anomalous ^{127}I temperature coefficient in ammonium, thallium(I), and tetramethylammonium triiodides, 130 *et seq.*
Bayer theory, 199

Temperature dependence—*Contd.*
soft-mode frequencies and, 199 *et seq.*
volume contribution to, 205 *et seq.*

Tetra-*n*-butylammonium triiodide, ^{127}I
quadrupole resonance in, 123 *et seq.*

Tetraethylammonium triiodide
crystal structure of, 122 *et seq.*, 128
^{127}I quadrupole resonance in, 122 *et seq.*

Tetramethylammonium triiodide
anomalous temperature coefficient of ^{127}I frequency in, 130 *et seq.*
crystal structure of, 124
^{127}I quadrupole resonance in, 124 *et seq.*

Tetraphenylarsonium iodide, crystal structure of, 127 *et seq.*

Tetra-*n*-propylammonium triiodide
crystal structure of, 123
^{127}I quadrupole resonance in, 123 *et seq.*

Thallium triiodide
anomalous temperature coefficient of, ^{127}I frequency in, 130 *et seq.*
^{127}I quadrupole resonance in, 120 *et seq.*, 128

Theoretical calculations of electric field gradients at
beryllium, 167
boron, 167
carbon, 163
chlorine, 167
fluorine, 167, 168
hydrogen, 160 *et seq.*
iron, 168
krypton, 168
lithium, 167
nitrogen, 164 *et seq.*
oxygen, 166
zinc, 168

Theoretical calculations of nuclear quadrupole coupling constants, 135 *et seq.*

Thiocyanate ion, theoretical calculations of electric field gradient in, 165

Thiophene, theoretical calculations of electric field gradient in, 168

Thymine, $^1H/^{14}N$ double resonance in, 94, 106 *et seq.*

Torsional oscillations
effect on nuclear quadrupole resonance frequencies, 199 *et seq.*
effect of pressure on, 210
effect on spin–lattice relaxation time, T_1, 11

Transcorrelated wave function method, 148

Triglycine, ^{14}N quadrupole coupling constant in, 107

Triglycine sulphate, ^{1}H/^{14}N double resonance in, 80 *et seq.*

Triiodide ion, electronic structure of, 117 *et seq.*, 128 *et seq.*

Triplet state
  measurement of nuclear quadrupole interactions in, 15 *et seq.*
  detection of e.s.r. in, 19 *et seq.*
  interaction with resonant microwave fields, 21

U

Uncoupled perturbed Hartree–Fock method (PUHF), 152

Uracil, ^{1}H/^{14}N double resonance in, 94

V

Van Kranendonk theory of spin–lattice relaxation, 224 *et seq.*
  anharmonic Raman process, 225, 227
  Raman process, 225, 226

Variation theorem, in calculation of nuclear quadrupole coupling constants, 144 *et seq.*

Vibrational corrections to quadrupole coupling constants, 155 *et seq.*, 160 *et seq.*

Vibrational relaxation, 17

Volume changes, effect on nuclear quadrupole resonance frequencies, 205 *et seq.*

W

Water molecule
  reorientation of, in hydrates, 9
  theoretical calculation of electric field gradient in, 161, 166

Z

Zero-field Hamiltonian, 32 *et seq.*

Zero-field quadrupole double resonance, 82

Zero-field splitting, 17, 34

Zero-field splitting parameters
  benzene-d$_2$, 63, 65
  8-chloroquinoline, 53
  f-dichlorobenzene, 63, 65
  quinoline, 53
  tetrachlorobenzene, 65

Zero-field transitions of molecules in triplet states, 20
  in 2,3-dichloroquinoxaline, 20

Zinc difluoride, theoretical calculation of electric field gradient in, 168

CHEMICAL COMPOUNDS INDEX

Nuclei are arranged in alphabetical order; for any one nucleus the compounds are arranged in alphabetical order of the element to which the quadrupolar nucleus is attached.

B

^{79,81}Br, quadrupole coupling constants of
C—^{79}Br
  trans-1,2-dibromoethane, 4
Pt—^{79}Br
  K$_2$PtBr$_6$, 193 et seq., 231
  (NH$_4$)$_2$PtBr$_6$, 193 et seq.

C

^{35,37}Cl, quadrupole coupling constants of
C—^{35}Cl
  trans-1,2-dichloroethane, 4, 5
  p-dichlorobenzene, 8, 49, 66
  8-chloroquinoline, $^3\pi\pi^*$ state of, 53 et seq., 66
  p-dichlorobenzene, $^3\pi\pi^*$ state of, 60 et seq.
  6-chloroquinoline, 58
  7-chloroquinoline, 58
  1,2,4,5-tetrachlorobenzene, $^3\pi\pi^*$ state of, 66
Ir—^{35}Cl
  K$_2$IrCl$_6$, 206, 209, 238
Pb—^{35}Cl
  CsPbCl$_3$, 195 et seq.
Pt—^{35}Cl
  K$_2$PtCl$_6$, 204, 206, 209, 235, 237
  Cs$_2$PtCl$_6$, 230, 237
  Rb$_2$PtCl$_6$, 204, 237

Re—^{35}Cl
  K$_2$ReCl$_6$, 192 et seq., 207, 238

H

^{2}H, quadrupole coupling constants of
  N—^{2}H, ND$_3$ groups in triglycylphosphate, 8
  O—^{2}H, hydrates, 9

I

^{127}I, quadrupole coupling constants of
I—^{127}I
  ammonium triiodide, 118 et seq., 130 et seq.
  rubidium triiodide, 118 et seq.
  caesium triiodide, 118 et seq.
  tetraethylammonium triiodide, 122 et seq.
  tetra-n-propylammonium triiodide, 123 et seq.
  dimethylammonium triiodide, 123 et seq.
  di-n-propylammonium triiodide, 123 et seq.
  tetra-n-butylammonium triiodide, 123 et seq.
  tetramethylammonium triiodide, 124 et seq., 130 et seq.

caesium octaiodide, 124 *et seq.*
thallium triiodide, 128 *et seq.*, 130 *et seq.*

K

^{39}K, quadrupole coupling constants of
KMnF$_3$, 196 *et seq.*

N

^{14}N, quadrupole coupling constants of
C—^{14}N $^3\pi\pi^*$ state   of   8-chloro-
quinoline, 53 *et seq.*
$^3\pi\pi^*$ state of quinoline, 53 *et seq.*
triglycine sulphate, by double resonance, 80
thymine, by double resonance, 94,
106 *et seq.*
cytosine, by double resonance, 94
uracil, by double resonance, 94

^{14}N, quadrupole coupling constants of—
Contd.
guanine, by double resonance, 94
amino acids, by double resonance, 95
polyglycine, by double resonance,
106 *et seq.*
diglycine, 107
triglycine, 107
H—^{14}N
NH$_4$H$_2$PO$_4$, by double resonance,
108

O

^{17}O, quadrupole coupling constant of
P—^{17}O
KH$_2$PO$_4$, 84 *et seq.*

R

^{87}Rb, quadrupole coupling constants of
RbH$_2$PO$_4$, 109